Springer Series in
CHEMICAL PHYSICS 84

Springer Series in
CHEMICAL PHYSICS

The purpose of this series is to provide comprehensive up-to-date monographs in both well established disciplines and emerging research areas within the broad fields of chemical physics and physical chemistry. The books deal with both fundamental science and applications, and may have either a theoretical or an experimental emphasis. They are aimed primarily at researchers and graduate students in chemical physics and related fields.

70 **Chemistry of Nanomolecular Systems**
Towards the Realization of Molecular Devices
Editors: T. Nakamura, T. Matsumoto, H. Tada, K.-I. Sugiura

71 **Ultrafast Phenomena XIII**
Editors: D. Miller, M.M. Murnane, N.R. Scherer, and A.M. Weiner

72 **Physical Chemistry of Polymer Rheology**
By J. Furukawa

73 **Organometallic Conjugation**
Structures, Reactions and Functions of d–d and d–π Conjugated Systems
Editors: A. Nakamura, N. Ueyama, and K. Yamaguchi

74 **Surface and Interface Analysis**
An Electrochmists Toolbox
By R. Holze

75 **Basic Principles in Applied Catalysis**
By M. Baerns

76 **The Chemical Bond**
A Fundamental Quantum-Mechanical Picture
By T. Shida

77 **Heterogeneous Kinetics**
Theory of Ziegler–Natta–Kaminsky Polymerization
By T. Keii

78 **Nuclear Fusion Research**
Understanding Plasma–Surface Interactions
Editors: R.E.H. Clark and D.H. Reiter

79 **Ultrafast Phenomena XIV**
Editors: T. Kobayashi, T. Okada, T. Kobayashi, K.A. Nelson, S. De Silvestri

80 **X-Ray Diffraction by Macromolecules**
By N. Kasai and M. Kakudo

81 **Advanced Time-Correlated Single Photon Counting Techniques**
By W. Becker

82 **Transport Coefficients of Fluids**
By Byung Chan Eu,

83 **Quantum Dynamics of Complex Molecular Systems**
Editors: D.A. Micha, I. Burghardt

84 **Progress in Ultrafast Intense Laser Science I**
Editors: K. Yamanouchi, S.L. Chin, P. Agostini, P.G. Ferrante

Kaoru Yamanouchi · See Leang Chin
Pierre Agostini · Gaetano Ferrante

Progress in Ultrafast Intense Laser Science

Volume I

With 162 Figures

Springer

Professor Kaoru Yamanouchi

University of Tokyo, Department of Chemistry
Hongo 7-3-1, 113-0033 Tokyo, Japan
e-mail: kaoru@chem.s.u-tokyo.ac.jp

Professor See Leang Chin

Laval University Quebec
Center for Optics, Photonics and Laser (COPL)
& Department of Physics, Engineering Physics and Optics
Quebec, Qc G1K 7P4 , Canada
e-mail: slchin@phy.ulaval.ca

Professor Pierre Agostini

Ohio State University
Department of Physics
191 W Wodruff Ave Columbus OH 43210
e-mail: agostini@mps.ohio-state.edu

Professor Gaetano Ferrante

Università di Palermo
Dipto. di Fisica e Tecnologie Relative
Viale delle Scienze, 90128 Palermo, Italy
e-mail: ferrante3@unipa.it

Series Editors:

Professor A.W. Castleman, Jr.

Department of Chemistry, The Pennsylvania State University
152 Davey Laboratory, University Park, PA 16802, USA

Professor J. P. Toennies

Max-Planck-Institut für Strömungsforschung
Bunsenstraße 10, 37073 Göttingen, Germany

Professor W. Zinth

Universität München, Institut für Medizinischen Optik
Öttingerstraße 67, 80538 München, Germany

ISSN 0172-6218

ISBN 978-3-642-07075-4   e-ISBN 978-3-540-34422-3

Springer is a part of Springer Science+Business Media
springer.com

Softcover reprint of the hardcover 1st edition 2006

Cover design: eStudio Calamar Steinen
Cover production: *design & production*, Heidelberg, Germany

Preface

Light and Optical Science have undoubtedly deepened our understanding of nature. The history of science shows that each invention of a new optical technique and research instrument has spurred new discoveries in various disciplines. Of particular importance is laser radiation, discovered in 1960, which opened up new research fields. Later, key technologies such as modelocking and chirped-pulse-amplification have accelerated the development and initiated the era of optical science in which pulses as short as a few femtoseconds are commonly used while subfemtosecond pulses are rapidly coming of age.

Such ultrashort pulses have enabled the generation of extremely intense light fields whose magnitude can far exceed that of the atomic Coulomb field. Thus, light, which had long been used only as a probe for matter, has now achieved such huge intensity that it can strongly distort atomic and molecular quantum states. That is, the character of light-matter interaction becomes qualitatively different from that in the weak light field regime. New tools for controlling and fine-designing ultrashort laser pulses have introduced entirely new possibilities for controlling the dynamics of atoms and molecules. Indeed, the response of matter to an ultrashort intense laser field has been a very exciting and challenging subject of research from the beginning, and exceedingly so during the past 15 years.

During this period, tremendous advances have been achieved both in fundamental understanding and practical applications. For instance, the irradiation of liquid and solid materials by intense laser light generates short pulses of hard-X rays as well as high-energy electrons, ions and neutrons, which are now regarded as promising sources for time-resolved measurements. The interaction between gaseous media and ultrashort intense laser pulses induces quite remarkable phenomena such as filamentation, plasma formation, and high-order harmonic generation. The latter forms the basis of tabletop soft X-ray sources and of the Fourier synthesis of "attosecond" light pulses, the shortest light pulses ever made.

The research field of ultrafast intense laser science, clustering both the fundamental and applied aspects of ultrashort (femtosecond to attosecond), intense (10^{14} to 10^{18} $\mathrm{Wcm^{-2}}$) pulses, is now rapidly growing worldwide, not only in physics (nonlinear optics, time-resolved ultrafast X-ray spectroscopy, metrology, non-linear physics) and chemistry (physical chemistry, molecular science) but also in biology. It is, therefore, much desired that researchers and students in an even wider range of research fields share the enthusiasm for this provocative interdisciplinary research field.

It is with this hope that a new series of symposia called "International Symposium on Ultrafast Intense Laser Science (ISUILS)" was launched in 2002 (http://www.isuils.jp). These very successful meetings have encouraged us to pursue the diffusion of ultrafast intense laser science under another form. This is the origin of the present series. Considering the interdisciplinary nature of Ultrafast Intense Laser Science, we thought it appropriate to complement the ISUILS symposia with a new review book series, in which concise review-style articles written by researchers at the forefront of their sub-fields are compiled, so that researchers with different research backgrounds and graduate students could grasp easily the essential aspects of the corresponding field. Although conference proceedings may be useful for researchers with specific interests, and long and complete review articles are necessary for researchers interested in a specific theme, there is room for a new style of review articles more fitted to the interdisciplinary interaction in UILS and the rapidly growing character of the field.

The research areas treated in this series will be: (i) atoms, molecules, and clusters in intense laser fields, (ii) control of molecules and clusters in intense laser fields, (iii) attosecond pulse generation, metrology, and applications, (iv) wavepacket control for high-order harmonics, (v) generation, metrology and interaction of intense few-cycle pulses, (vi) non-linear dynamics in chaotic tunneling for understanding ionization in intense laser fields, (vii) non-linear propagation and fs-ablation, (viii) short-pulsed laser plasma interaction, (ix) non-linear optics in nano plasmas, (x) X-ray imaging, (xi) short-pulsed electron diffraction, (xii) nuclear transitions in laser fields, (xiii) relativistic quantum dynamics, (xiv) laser pulse interaction with materials having nano structure, (xv) femtosecond biology.

Each book of the new series addresses such needs through a compilation of around fifteen 15–25 page chapters. All chapters are written with the goal of providing researchers with a different expertise a clear assessment of the motivation and significance of the research theme as well as a description of the authors' most recent results. The editors hope that the reader will find, in this first volume, not only answers to specific questions but also insights into other subfields and new perspectives for his/her own future research.

Invitations to contribute have been extended to the invited participants of the first three ISUILS symposia. Hence, the first three or four volumes of PUILS will be by invitation only. The PUILS series has been edited in cooperation with the activities of MEXT Priority Area Program on Control of Molecules in Intense Laser Fields (FY2002–2005), JSPS Core-to-Core Program on Ultrafast Intense Laser Science (FY2004–) and JILS (Japan Intense Light Field Science Society).

We take this opportunity to thank all the authors who have kindly contributed to the new PUILS series by describing frontiers of ultrafast intense laser science. We also thank reviewers who have served for this book project by reading carefully the submitted manuscripts. One of the co-editors (KY) thanks Ms. Miyuki Kusunoki and Ms. Chie Sakuta for their help with the editing processes. Dr. Claus Ascheron, Physics Editor of Springer Verlag at Heidelberg kindly agreed to our idea and helped us co-edit the first volume of PUILS. We very much appreciate his kind cooperation and support. We hope this book series of PUILS will convey the excitement of Ultrafast Intense Laser Science to the readers, and stimulate interdisciplinary interactions among researchers, thus paving the way for explorations of new frontiers.

Series editor of PUILS and Co-editor of PUILS I
University of Tokyo *Kaoru Yamanouchi*

Co-editors of PUILS I
Laval University *See Leang Chin*
Ohio State University *Pierre Agostini*
University of Palermo *Gaetano Ferrante*

March, 2006

Contents

List of Contributors

Mikahail V. Fedorov
General Physics Institute
Russian Academy of Science
38 Vavlov st.
Moscow 119991, Russia
fedorov@ran.gpi.ru

Kiyohiko Someda
Department of Basic Science
Graduate School
of Arts and Sciences
The University of Tokyo
Komaba, Meguro-ku
Tokyo 153-8902, Japan
kiyon@isolde.c.u-tokyo.ac.jp

Tomokazu Yasuike
Department of Basic Science
Graduate School
of Arts and Sciences
The University of Tokyo
Komaba, Meguro-ku
Tokyo 153-8902, Japan

Chunlei Guo
The Institute of Optics
University of Rochester,
Rochester, NY 14627 USA
guo@optics.rochester.edu

Wendell T. Hill, III
Institute for Physical Science and
Technology, University of Maryland,
College Park, MD 20742
Department of Physics,
University of Maryland,
College Park, MD 20742

Kun Zhao
Institute for Physical Science and
Technology, University of Maryland,
College Park, MD 20742

Lee N. Elberson
Institute for Physical Science and
Technology, University of Maryland,
College Park, MD 20742

Getahun M. Menkir
Institute for Physical Science and
Technology, University of Maryland,
College Park, MD 20742

Kazuhiro Yabana
Institute of Physics and
Center for Computational
Sciences,
University of Tsukuba,
Tsukuba 305-8571, Japan
yabana@nucl.ph.tsukuba.ac.jp

Tomohito Otobe
Advanced Photon Research Center,
Japan Atomic Energy Research
Institute, Kizu, Kyoto 619-0215,
Japan
otobe@apr.jaeri.co.jp

Jun-Ichi Iwata
Center for Computational
Sciences,
University of Tsukuba,
Tsukuba 305-8571, Japan
iwata@comas.frsc.tsukuba.ac.jp

Lora Ramunno
Center for Photonics Research,
University of Ottawa,
150 Louis Pasteur, Ottawa,
K1N 6N5 ON, Canada

Christian Jungreuthmayer
Center for Photonics Research,
University of Ottawa,
150 Louis Pasteur, Ottawa,
K1N 6N5 ON, Canada

Carlos F. Destefani
Center for Photonics Research,
University of Ottawa,
150 Louis Pasteur, Ottawa,
K1N 6N5 ON, Canada

Thomas Brabec
Center for Photonics Research,
University of Ottawa,
150 Louis Pasteur, Ottawa,
K1N 6N5 ON, Canada

Peter Schlagheck
Institut für Theoretische Physik,
Universität Regensburg,
93040 Regensburg, Germany

Christopher Eltschka
Institut für Theoretische Physik,
Universität Regensburg,
93040 Regensburg, Germany

Denis Ullmo
CNRS, Univ. Paris Sud,
UMR8626, LPTMS,
91405 Orsay Cedex, France
denis.ullmo@u-psud.fr

Mauro Nisoli
National Laboratory for Ultrafast
and Ultraintense Optical Science –
CNR-INFM
Department of Physics, Politecnico,
Milan, Italy
mauro.nisoli@fisi.polimi.it

Sandro De Silvestri
National Laboratory for Ultrafast
and Ultraintense Optical Science –
CNR-INFM
Department of Physics,
Politecnico,
Milan, Italy

Giuseppe Sansone
National Laboratory for Ultrafast
and Ultraintense Optical Science –
CNR-INFM
Department of Physics,
Politecnico,
Milan, Italy

Luca Poletto
Laboratory for Ultraviolet
and X-Ray Optical Research –
CNR-INFM
D.E.I. – Università di Padova,
Padova, Italy

Paolo Villoresi
Laboratory for Ultraviolet
and X-Ray Optical Research –
CNR-INFM
D.E.I. – Università di Padova,
Padova, Italy

Salvatore Stagira
National Laboratory for Ultrafast
and Ultraintense Optical Science –
CNR-INFM
Department of Physics,
Politecnico,
Milan, Italy

Caterina Vozzi
National Laboratory for Ultrafast
and Ultraintense Optical Science –
CNR-INFM
Department of Physics,
Politecnico,
Milan, Italy

Orazio Svelto
National Laboratory for Ultrafast and Ultraintense Optical Science – CNR-INFM
Department of Physics, Politecnico, Milan, Italy

Jean-Claude Gauthier
Centre Lasers Intenses et Applications (CELIA)
UMR 5107 CNRS, CEA, Université Bordeaux 1,
33405 Talence (France)
gauthier@celia.u-bordeaux1.fr

Maurizio Lontano
Plasma Physics Institute "P. Caldirola", C.N.R.,
Milan, Italy
lontano@ifp.cnr.it

Matteo Passoni
Plasma Physics Institute "P. Caldirola", C.N.R.,
Milan, Italy
Nuclear Engineering Dept., Polytechnic of Milan,
Milan, Italy
matteo.passoni@polimi.it

Gaetano Ferrante
Dipartimento di Fisica e Tecnologie Relative dell'Universitá di Palermo,
Viale delle Scienze, Edificio 18,
90128 Palermo, Italy

Michelangelo Zarcone
Dipartimento di Fisica e Tecnologie Relative dell'Universitá di Palermo,
Viale delle Scienze, edificio 18,
90128 Palermo, Italy
zarcone@unipa.it

Sergei A. Uryupin
P.N. Lebedev Physics Institute,
Leninsky pr. 53,
119991, Moscow, Russia
uryupin@sci.lebedev.ru

Saverio Bivona
Dipartimento di Fisica e Tecnologie Relative,
Universita di Palermo,
Viale delle Scienze, Edificio 18,
90128 Palermo, Italy

Riccardo Burlon
Dipartimento di Fisica e Tecnologie Relative,
Universita di Palermo,
Viale delle Scienze, Edificio 18,
90128 Palermo, Italy

Claudio Leone
Dipartimento di Fisica e Tecnologie Relative,
Universita di Palermo,
Viale delle Scienze, Edificio 18,
90128 Palermo, Italy
leone@unipa.it

Arnaud Couairon
Centre de Physique Théorique,
École Polytechnique,
CNRS UMR 7644,
F-91128, Palaiseau Cedex, France
couairon@cpht.polytechnique.fr

Andre Mysyrowicz
Laboratoire d'Optique Appliquée,
École Nationale Supérieure des Techniques Avancées – École Polytechnique,
CNRS UMR 7639,
F-91761 Palaiseau Cedex, France
mysyr@ensta.fr

Ruxin Li
State Key Laboratory of High Field Laser Physics and Shanghai Institute of Optics and Fine Mechanics,
Chinese Academy of Sciences,
P.O. Box 800–211, Shanghai 201800,
P. R. China

Xiaowei Chen
State Key Laboratory of High Field Laser Physics and Shanghai Institute of Optics and Fine Mechanics, Chinese Academy of Sciences, P.O. Box 800–211, Shanghai 201800, P. R. China

Jun Liu
State Key Laboratory of High Field Laser Physics and Shanghai Institute of Optics and Fine Mechanics, Chinese Academy of Sciences, P.O. Box 800–211, Shanghai 201800, P. R. China

Yuxin Leng
State Key Laboratory of High Field Laser Physics and Shanghai Institute of Optics and Fine Mechanics, Chinese Academy of Sciences, P.O. Box 800–211, Shanghai 201800, P. R. China

Jiansheng Liu
State Key Laboratory of High Field Laser Physics and Shanghai Institute of Optics and Fine Mechanics, Chinese Academy of Sciences, P.O. Box 800–211, Shanghai 201800, P. R. China

Yi Zhu
State Key Laboratory of High Field Laser Physics and Shanghai Institute of Optics and Fine Mechanics, Chinese Academy of Sciences, P.O. Box 800–211, Shanghai 201800, P. R. China

Xiaochun Ge
State Key Laboratory of High Field Laser Physics and Shanghai Institute of Optics and Fine Mechanics, Chinese Academy of Sciences, P.O. Box 800–211, Shanghai 201800, P. R. China

Haihe Lu
State Key Laboratory of High Field Laser Physics and Shanghai Institute of Optics and Fine Mechanics, Chinese Academy of Sciences, P.O. Box 800–211, Shanghai 201800, P. R. China

Lihuang Lin
State Key Laboratory of High Field Laser Physics and Shanghai Institute of Optics and Fine Mechanics, Chinese Academy of Sciences, P.O. Box 800–211, Shanghai 201800, P. R. China

Zhizhan Xu
State Key Laboratory of High Field Laser Physics and Shanghai Institute of Optics and Fine Mechanics, Chinese Academy of Sciences, P.O. Box 800–211, Shanghai 201800, P. R. China

Sergio Martellucci
University of Rome "Tor Vergata", Department of "Ingegneria dell'Impresa", Via del Politecnico n. 1, 00133 Rome, Italy
`smart@uniroma2.it` or `francucci@ing.uniroma2.it`

Massimo Francucci
University of Rome "Tor Vergata", Department of "Ingegneria dell'Impresa", Via del Politecnico n. 1, 00133 Rome, Italy
`smart@uniroma2.it`
`francucci@ing.uniroma2.it`

Patrizio Ciuffa
Elettronica s.p.a.,
Systems Technology and
Processing Department,
00131 Rome, Italy
patrizio.ciuffa@elt.it

Yasuhiko Shimotsuma
Kyoto University International
Innovation Organization
Katsura Nishikyo-ku,
Kyoto, 615-8520 Japan
shimo@iic.kyoto-u.ac.jp

Jianrong Qiu
Department of Materials Science,
Zhejiang University,
Hangzhou 310027, China
jrqiu@zju.edu.cn

Kiyotaka Miura
Department of Material Chemistry,
Kyoto University,
Katsura Nishikyo-ku,
Kyoto, 615-8510 Japan
miura@
collon1.kuic.kyoto-u.ac.jp

Kazuyuki Hirao
Department of Material Chemistry,
Kyoto University,
Katsura Nishikyo-ku,
Kyoto, 615-8510 Japan
hirao@bisco1.kuic.kyoto-u.ac.jp

1 Stabilization of Atoms in a Strong Laser Field

M.V. Fedorov

General Physics Institute, Russian Academy of Science, 38 Vavlov st., Moscow 119991 Russia
fedorov@ran.gpi.ru

The main mechanisms of atomic stabilization in a strong light field are surveyed. The main features and physics of Kramers–Henneberger and interference stabilization are discussed. Some experiments confirming existence of stabilization effects are briefly overviewed. Two-color interference stabilization is described in its connection with the single frequency interference stabilization and effects like the laser-induced continuum structures.

1.1 Introduction

Stabilization of atoms in a strong light field is a counterintuitive and rather unusual effect. Normally, the rate of ionization of atoms or the ionization yield per pulse are some growing functions of the laser intensity, which is quite natural: a stronger field provides faster ionization. Stabilization means a violation of this general rule. According to the ideas of stabilization, under some conditions, the rate of ionization or the ionization yield per pulse can saturate at some level where atoms are not yet completely ionized or even start to fall with growing laser intensity. This is the definition of stabilization.

There are several mechanisms of stabilization, the main of which are interference stabilization of Rydberg atoms [1] and the so-called Kramers–Henneberger stabilization [2]. Many features of these phenomena are described in the book [3] and two review papers [4] and [5]. Here we will not go too deep into details of theoretical derivations. Instead we will outline briefly the main ideas of the phenomena. Then we will give a short description of the most important relevant experiments and then discuss some interesting generalization.

1.2 Kramers–Henneberger Stabilization

The idea of Kramers–Henneberger stabilization has emerged from the observation that the laser-atom interaction can be reduced to the form of an oscillating potential energy. Indeed, let us start from the Hamiltonian

$$H = \frac{1}{2m}\left(-\mathrm{i}\hbar\frac{\partial}{\partial \boldsymbol{r}} - \frac{e}{c}\boldsymbol{A}(t)\right)^2 + U(\boldsymbol{r}) \ , \tag{1.1}$$

where $U(\boldsymbol{r})$ is the atomic potential, $\boldsymbol{A}(t) = -c\int_{-\infty}^{t}\boldsymbol{\varepsilon}(t')\,\mathrm{d}t'$ and $\boldsymbol{\varepsilon}(t) = \boldsymbol{\varepsilon}_0\cos(\omega t)$ are the vector potential and field strength of a laser pulse, $\boldsymbol{\varepsilon}_0$ and ω are its amplitude and frequency. In reality, of course, the pulse envelope is not constant, $\boldsymbol{\varepsilon}_0 = \boldsymbol{\varepsilon}_0(t)$. But often a slow time dependence of $\boldsymbol{\varepsilon}_0(t)$ can be ignored.

The electron wave function $\Psi(\boldsymbol{r}, t)$ obeys the nonstationary Schrödinger equation $\mathrm{i}\hbar\partial\Psi/\partial t = H\Psi$. Let us make a wave function substitution $\Psi = S\Psi'$ with the unitary operator S given by

$$S(t) = \exp\left(\mathrm{i}\frac{e}{mc}\int_{\infty}^{t}\mathrm{d}t'\boldsymbol{A}(t')\cdot\frac{\partial}{\partial\boldsymbol{r}} - \mathrm{i}\frac{e^2}{2\hbar mc^2}\int_{\infty}^{t}\mathrm{d}t'\boldsymbol{A}^2(t')\right) . \tag{1.2}$$

Transformation of the Schrödinger equation corresponding to this transformation of the wave function is determined by the equation

$$S^{-1}\left(\mathrm{i}\hbar\frac{\partial}{\partial t} - H\right)S \equiv \mathrm{i}\hbar\frac{\partial}{\partial t} - H' , \tag{1.3}$$

where the transformed Hamiltonian is given by

$$H' = -\frac{\hbar^2}{2m}\frac{\partial^2}{\partial\boldsymbol{r}^2} + U\left(\boldsymbol{r} + \boldsymbol{\alpha}_0\cos(\omega t)\right) , \tag{1.4}$$

where $\boldsymbol{\alpha}_0 = |e|\boldsymbol{\varepsilon}_0/m\omega^2$ is the quiver motion amplitude of a free classical electron in the field $\boldsymbol{\varepsilon}(t)$.

So, indeed, after the unitary transformation described above all the laser-atom interaction appears to be reduced to a quiver motion in the argument of the potential energy. Physically, this transformation corresponds to a transition into the frame, the origin of which oscillates with respect to the laboratory frame as $-\boldsymbol{\alpha}_0\cos(\omega t)$. Such a frame can be referred to as the Kramers frame because the possibility of reducing the Hamiltonian to the form of (1.4) was demonstrated for the first time by Kramers [6].

The described transformation is exact, and the Schrödinger equation for Ψ' (with the Hamiltonian H' (1.4)) is absolutely equivalent to the original Schrödinger equation for Ψ (with the Hamiltonian H (1.1)). However, the Kramers Hamiltonian (1.4) appears to be much more appropriate than the original one (1.1) for approximate consideration of a strong-field limit. As suggested for the first time by Henneberger [7], in the strong-field limit the potential energy in the Hamiltonian (1.4), periodically depending on time t, can be approximated by its time-averaged value to give

$$H_{\mathrm{KH}} = -\frac{\hbar^2}{2m}\frac{\partial^2}{\partial\boldsymbol{r}^2} + U_{\mathrm{KH}}(\boldsymbol{r}) , \tag{1.5}$$

where $U_{\mathrm{KH}}(\boldsymbol{r})$ is the Kramers–Henneberger potential energy

$$U_{\mathrm{KH}}(\boldsymbol{r}) = \frac{\omega}{2\pi}\int_0^{2\pi/\omega}\mathrm{d}t\,U\left(\boldsymbol{r} + \boldsymbol{\alpha}_0\cos(\omega t)\right) . \tag{1.6}$$

Note that the exact potential energy in the Hamiltonian H' (1.4) can be expanded in a Fourier series

$$U\left(\boldsymbol{r} + \boldsymbol{\alpha}_0 \cos(\omega t)\right) = \sum_n U_n(\boldsymbol{r})\, \mathrm{e}^{\mathrm{i}n\omega t} . \tag{1.7}$$

The zeroth term of this expansion, $U_0(\boldsymbol{r})$, coincides with $U_{\mathrm{KH}}(\boldsymbol{r})$ (1.6). All other harmonics, $U_n(\boldsymbol{r})$ with $n \neq 0$, can be considered as a perturbation to the approximate averaged Kramers–Henneberger Hamiltonian (1.5). If this perturbation is relatively small, in the zero approximation the atomic Hamiltonian in a strong laser field H_{KH} (1.5) appears to be stationary. If this Hamiltonian has bound states, and if these states appear to be populated, this population appears to be stabilized against ionization, which corresponds just to strong-field Kramers–Henneberger stabilization of an atom. And the next important question to be discussed is: under what conditions the Kramers–Henneberger Hamiltonian (1.5) can be considered as a good zero-order approximation and perturbation determined by the sum of non-zero harmonics in the Fourier expansion (1.7) is really small?

To answer these questions it is useful to compare the forms of the KH and field-free atomic potentials. For the one-dimensional case and atomic potential having no singularity at the origin $x = 0$, the two potential curves $U(x)$ and $U_{\mathrm{KH}}(x)$ are shown in Fig. 1.1. This picture shows clearly that, in contrast to the single-well field-free potential energy, the Kramers–Henneberger potential $U_{\mathrm{KH}}(x)$ has two wells, and they are located at $x = -\alpha_0$ and $x = \alpha_0$. The wells are much shallower than of the field-free one. With a growing field strength, spacing between wells grows $\propto \varepsilon_0$ and they become increasingly shallow. For this reason at sufficiently high field strength, inevitably, the electron binding energy in the Kramers–Henneberger potential becomes significantly smaller than the photoelectron kinetic energy (equal

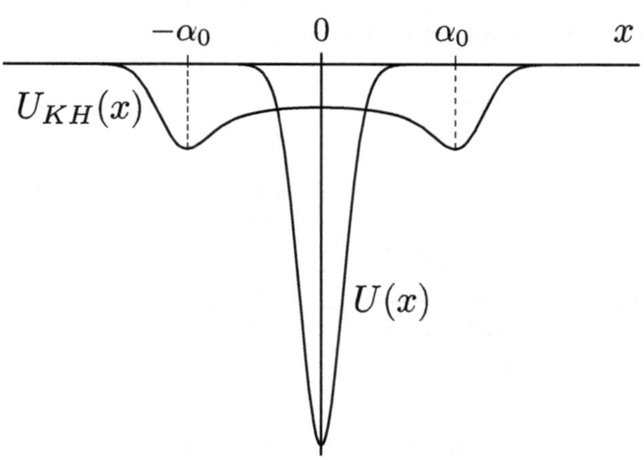

Fig. 1.1. Field-free and Kramers–Henneberger potential energies

to $\hbar\omega$ minus binding energy). This explains why the Kramers–Henneberger bound states are stable: the wave functions of a high-energy free electron in the KH atom are very fast oscillating compared to those of a bound KH electron and, hence, integrals determining matrix elements of bound-free transitions in the KH atom appear to be very small.

The condition under which the Kramers–Henneberger and field-free potentials differ strongly from each other is evident: $\alpha_0 \gg a_0$, where a_0 is a size of the localization region of the field-free atomic ground state wave function. But this is not the only condition of stabilization. Indeed, for stabilization the field must be strong enough whereas, e.g., at small values of frequency ω the condition $\alpha_0 \gg a_0$ can be fulfilled already at rather low fields. In reality, for the Kramers–Henneberger stabilization to occur, in addition to the condition $\alpha_0 \gg a_0$ one has to require the field strength ε_0 to be larger than the so called Barrier-Suppression-Ionization field, $\varepsilon_0 \gg \varepsilon_{\rm BSI}$ [5]. The concept of BSI arises from consideration of an atom in a static homogeneous electric field. The sum of the Coulomb potential and that of a static electric field has the form of a barrier-like curve (Fig. 1.2). The BSI mechanism works if the barrier's height falls below the field-free atomic ground state energy E_0. For a hydrogen atom $\varepsilon_{\rm BSI} = \frac{1}{16}\,\varepsilon_{\rm at}$, where $\varepsilon_{\rm at} = me^4/\hbar^2 \approx 5 \times 10^9$ V/cm is the atomic field. So, finally, the conditions under which the Kramers–Henneberger stabilization can occur are

$$\alpha_0 = \frac{|e|\varepsilon_0}{m\omega_0^2} \gg a_0 \quad \text{and} \quad \varepsilon_0 \gg \varepsilon_{\rm BSI}\,. \tag{1.8}$$

In accordance with the interpretation of [8], in the regime determined by the conditions (1.8) at the very beginning of the laser pulse the field-free atomic potential turns into the Karmers–Henneberger one. This happens so quickly that the atomic function does not change much an keeps a form of the ground-state field-free atomic wave function located at $x = 0$. As in this region the Kramers–Henneberger potential energy is small and has a zero gradient, at this stage the electron is almost free and its wave function

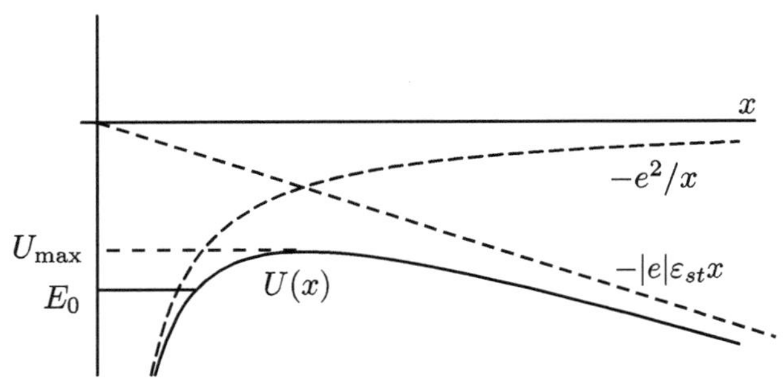

Fig. 1.2. Potential energy of an atom in a static electric field $U(x) = -e^2/|x| - |e|\varepsilon_{\rm st}x$; $\varepsilon_{\rm BSI}$ corresponds to $U_{\rm max} = E_0$; E_0 is the energy of the field-free ground-state atomic level

spreads like a free wave packet. This free spreading continues until the wave packet width $\Delta x(t)$ remains smaller than α_0. Then, at $\Delta x(t) \sim \alpha_0$, some part of the population appears to be captured by the Kramers–Henneberger potential wells, and this part of population remains there for a long time. This is a simple qualitative explanation of reasons and mechanism of the Kramers–Henneberger stabilization of atoms.

1.3 Interference Stabilization

Interference stabilization [1] is a different phenomenon with a different mechanism of ionization suppression. In this case it is assumed that an atom is excited initially to some Rydberg level E_n with $n \gg 1$, and that such an atom is ionized by a field with a frequency $\omega > |E_n|/\hbar$. If the field is strong enough, the direct photoionization process can be accompanied by Raman-type transitions to neighboring Rydberg levels (Fig. 1.3). These transitions can result in a coherent re-population of Rydberg levels and interference of transitions to the continuum from neighboring Rydberg levels. As it appears, the interference is destructive and suppresses photoionization. This is a very brief qualitative explanation of the phenomenon of interference stabilization. Though very simple, this explanation and a scheme of transitions in Fig. 1.3 are sufficient for qualitative determination of the main condition when this phenomenon can occur. The curve at the upper part of Fig. 1.3 describes the probability density $\mathrm{d}w^{(n)}/\mathrm{d}E$ of photoionization from an isolated level E_n in

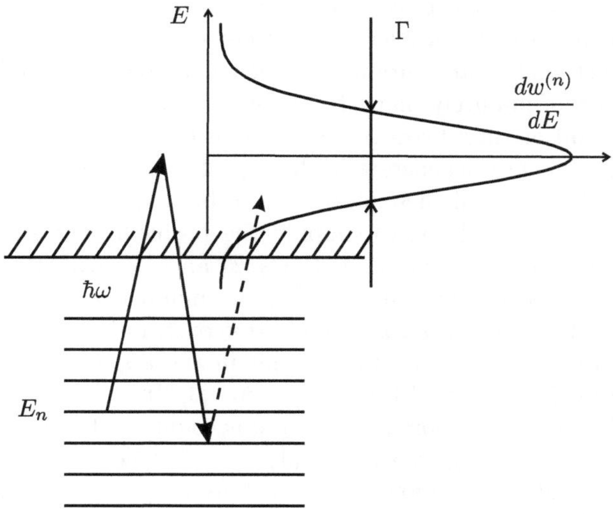

Fig. 1.3. Direct photoionization from a Rydberg level E_n, Λ-type Raman transitions to Rydberg levels and interfering transitions to the continuum (*solid and dashed lines with arrows at the top*)

its dependence on the photoelectron energy E. The width of the curve Γ is the ionization width determined by the Fermi Golden Rule ionization rate

$$\Gamma = \frac{\pi}{2} \left| \boldsymbol{d}_{E,n} \cdot \boldsymbol{\varepsilon}_0 \right|^2 , \tag{1.9}$$

where $\boldsymbol{d}_{E,n}$ is the bound-free matrix element of the dipole moment. If the width Γ is small compared to spacing between Rydberg levels $E_n - E_{n-1}$, the curves $\mathrm{d}w^{(n)}/\mathrm{d}E$ with different n do not overlap with each other and transitions to the continuum from various Rydberg levels do not interfere. Interference and interference stabilization can occur at just an opposite condition, when the width Γ exceeds spacing between neighboring levels

$$\Gamma \geq E_n - E_{n-1} \,. \tag{1.10}$$

This condition can be reduced to a much simpler and much more specific form with the help of so called quasiclassical, or WKB, expressions for the dipole matrix elements in atomic units given by [9]:

$$d_{E,n} \sim \frac{1}{n^{3/2}|E - E_n|^{5/3}} \approx \frac{1}{n^{3/2}\omega^{5/3}} \,. \tag{1.11}$$

Together with the definition of Γ (1.9) and with an approximate estimate of the Rydberg level spacing $E_n - E_{n-1} \approx 1/n^3$, (1.10) and (1.11) give

$$\varepsilon_0 \geq \omega^{5/3} \quad \text{or} \quad I = \varepsilon_0^2 \geq \omega^{10/3} \,, \tag{1.12}$$

where I is the laser intensity and all quantities are in atomic units.

In a general form the problem of interference stabilization is too complicated for exact analytical solution. In principle, one has to take into account the interaction of infinitely many Rydberg levels with all states of the continuum. In addition, in a hydrogen atom, each Rydberg level and each state of the continuum are degenerate with respect to the angular momentum, and the population of all these states has to be taken into account. Several approaches were worked out to simplify the problem. The simplest one is a model of two nondegenerate close levels E_1 and E_2 (with the indexes “1” and “2” not related to the principal quantum number of a real atom) interacting with a nondegenerate continuum. Moreover, in this model interaction with the continuum is taken into account in the approximation known as the “adiabatic elimination of the continuum” [3]. In this approximation the bound-continuum interaction is taken into account via the tensor of ionization widths $\Gamma_{\alpha,\beta} = \frac{\pi}{2} \boldsymbol{d}_{\alpha,E} \cdot \boldsymbol{\varepsilon}_0 \times \boldsymbol{d}_{E,\beta} \cdot \boldsymbol{\varepsilon}_0 \big|_{E=E_\alpha+\hbar\omega}$, where $\alpha = 1,\, 2$. At last, it is assumed that all the components of this tensor are approximately equal to each other and equal to the Fermi Golden Rule ionization width Γ (1.9). With all these approximations, assumptions and simplifications used, the problem is reduced to the following two equation for the probability amplitudes to find an atom at the levels E_1 and E_2:

$$\mathrm{i}\dot{C}_1(t) - E_1 C_1(t) = -\frac{\mathrm{i}}{2}\varGamma\,[C_1(t) + C_2(t)]$$
$$\mathrm{i}\dot{C}_1(t) - E_2 C_1(t) = -\frac{\mathrm{i}}{2}\varGamma\,[C_1(t) + C_2(t)]\ . \tag{1.13}$$

Equations (1.13) are stationary, and hence they have solutions of the form $C_{1,2} \propto \exp(-\mathrm{i}\gamma\, t)$, where γ is a complex quasienergy that can be easily found to be given by

$$\gamma_\pm = \frac{1}{2}\left\{E_1 + E_2 - \mathrm{i}\,\varGamma \pm \sqrt{(E_2 - E_1)^2 - \varGamma^2}\right\}\ . \tag{1.14}$$

This equation shows that drastic changes in the solutions occur when the interaction constant $\varGamma$ becomes larger than the level spacing $E_2 - E_1$. The point $\varGamma = E_2 - E_1$ is the branching point, below which (at $\varGamma < E_2 - E_1$) the root square is real in (1.14), whereas above the branching point (at $\varGamma > E_2 - E_1$) it becomes imaginary. The real parts of quasienergies $\gamma_\pm$ (1.14) are shown in Fig. 1.4a in their dependence on the parameter $\varGamma$ (which is proportional to ε_0^2). The curves of Fig. 1.4b describe the dependence on $\varGamma$ of the widths of quasienergy levels determined by imaginary parts of quasienergies, $\varGamma_\pm = 2\,|\mathrm{Im}\,(\gamma_\pm)|$. At last, the picture of Fig. 1.4c describes the broadened quasienergy levels in the form of quasienergy zones. "Gravity centers" of these zones are given by $\mathrm{Re}\,(\gamma_\pm)$ whereas widths of zones are equal to $\varGamma_\pm$. As it's seen from these pictures, in a strong field ($\varGamma > E_2 - E_1$) one of two quasienergy levels narrows with a growing field strength (the darker shadowed level in Fig. 1.4c), and its width ($\varGamma_+$) corresponds to a falling branch of the curves in Fig. 1.4b. The "gravity centers" of both quasienergy levels in the strong-field region appear to be localized exactly at the middle between the field-free energies E_1 and E_2.

Formation of a narrowing quasienergy level means that one of the two atomic eigenstates in a strong field has an increasing life-time and, hence, the population accumulated at this level appears resistant against ionization by a strong laser field, and the atom is stabilized.

Of course, the described two-level model is a rather imperfect imitation of a real Rydberg atoms. But, nevertheless, the main features of this model

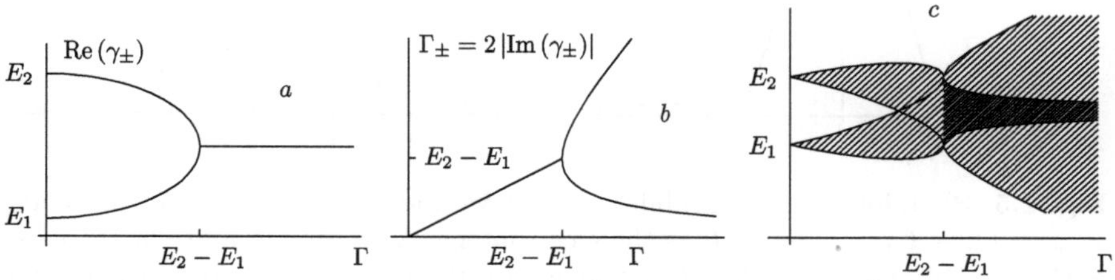

Fig. 1.4. **a** Real parts of quasienergies $\gamma_\pm$, **b** width of quasienergy levels $\varGamma_\pm$, and **c** quasienergy zones in dependence on the parameter $\varGamma \propto \varepsilon_0^2$

find their reflection in more complicated and realistic theoretical descriptions. In particular, it has been shown [1, 3] that in a system of infinitely many Rydberg levels (though, still, non-degenerate) in a strong-field limit all quasienergy levels are narrowing and they appear to be localized just in the middles between neighboring Rydberg levels, at $\frac{1}{2}(E_n - E_{n-1})$. Such a reconstruction of quasienergy levels manifests itself in the energy spectrum of photoelectrons shown in Fig. 1.5 for two cases, of weak (a) and strong (b) fields. A drastic difference between the weak- and strong-field cases is evident. If in a weak field the photoelectron spectrum has a Lorenz-like shape centered exactly at initial-state energy plus $\hbar\omega$, in a strong field the spectrum gets a multipeak structure and the peaks are located at energies equal to real parts of quasienergies $\mathrm{Re}(\gamma_\pm)$ plus $\hbar\omega$, i.e., $\frac{1}{2}(E_n + E_{n+1}) + \hbar\omega$. If in a weak field the photoelectron peak broadens, in a strong field peaks narrow with a growing field strength. These are effects accompanying and related to strong-field interference stabilization.

Degeneracy of Rydberg levels over angular momentum has been taken into account [10] to describe strong-field ionization and stabilization in realistic atoms. Some results are illustrated by two pictures of Fig. 1.6. At the first of these two pictures the probability of ionization is plotted in its dependence on fluence F at two given values of the pulse duration τ. By definition, fluence is given by the product of the peak pulse intensity by the pulse duration, $F = I_{\max} \times \tau$. In Fig. 1.6a fluence is evaluated in some relative units around $F \sim 7\,\mathrm{J/cm}^2$. The curves of Fig. 1.6a show that with growing fluence, the ionization probability saturates at a level smaller than one. Moreover, at a shorter pulse duration and at a given fluence the ionization probability is seen to be smaller than at a longer pulse duration. As at a given fluence a shorter pulse duration corresponds to a higher intensity, the

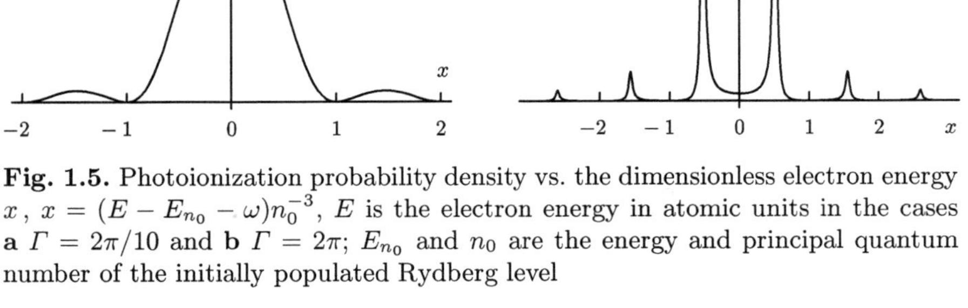

Fig. 1.5. Photoionization probability density vs. the dimensionless electron energy x, $x = (E - E_{n_0} - \omega)n_0^{-3}$, E is the electron energy in atomic units in the cases **a** $\Gamma = 2\pi/10$ and **b** $\Gamma = 2\pi$; E_{n_0} and n_0 are the energy and principal quantum number of the initially populated Rydberg level

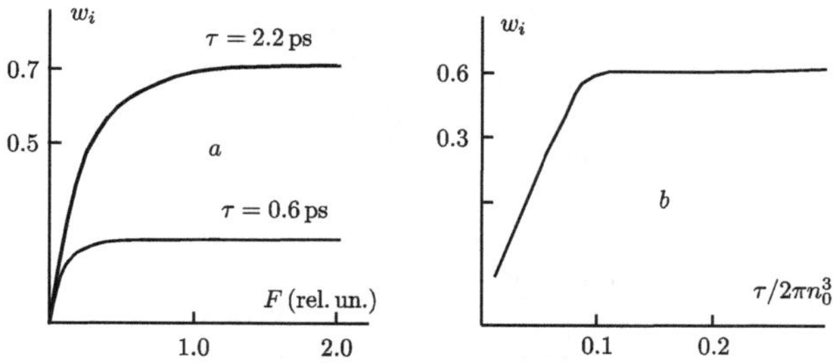

Fig. 1.6. a Ionization probability w_i vs. fluence F (in relative units) at two given values of the pulse duration τ and **b** w_i vs. τ at a given value of fluence F [10]

curves of Fig. 1.6a show that the ionization yield decreases with a growing pulse intensity. This is a clear indication of strong-field atomic stabilization.

In the second picture, the ionization probability is plotted in its dependence on the pulse duration at a given fluence. The pulse duration is evaluated in units of the Kepler period $T_{\rm K}$ for the initial Rydberg state ψ_{n_0}; in atomic units $t_K = 2\pi n_0^3$. The plateau of the curve at Fig. 1.6b corresponds to the first-order perturbation theory, in which the Fermi-Golden-Rule ionization rate is time independent and proportional to intensity. On the other hand, the fall of the curve $w_i(\tau)$ at small values of the pulse duration is a manifestation of strong-field stabilization, because at a given fluence a decreasing pulse duration corresponds to a growing pulse intensity and, in accordance with the curve of Fig. 1.6b, to a falling ionization yield.

To conclude this section, let us mention another approach to the theory of interference stabilization [11]. In this approach the WKB was used not for evaluation of dipole matrix elements as in earlier works but for a direct analytical solution of the 3D non-stationary Schrödinger equation. Such a solution was found in the case of relatively short pulses, $t < t_{\rm K}$. The result was found to be given by

$$w_i = \frac{1}{4\pi n_0^3} \int_{-\infty}^{\infty} \mathrm{d}t \left\{ 1 - \int_0^1 \mathrm{d}x \, J_0^2 \left(2^{2/3} 3^{1/6} \Gamma \left(\frac{2}{3} \right) \frac{\varepsilon_0(t)}{\omega^{5/3}} x \right) \right\} , \quad (1.15)$$

where J_0 is the zero-order Bessel function, $\Gamma(2/3)$ is the gamma-function, and $\varepsilon_0(t)$ is the laser pulse envelope; ε, ω, and t are in atomic units. The dependence of the ionization probability on the peak laser intensity is plotted in Fig. 1.7 (the solid line). For comparison the dashed line at this picture shows the ionization probability determined by the Fermi Golden Rule (the first order perturbation theory). Dots in the picture of Fig. 1.7 are the results of exact numerical solution of the 3D Schrödinger equation. As can be seen, the analytical solution given by (1.15) agrees very well with exact results. Both analytical and numerical solutions describe saturation of the ionization probability at a level smaller than one, which corresponds to stabilization of an atom in a strong light field.

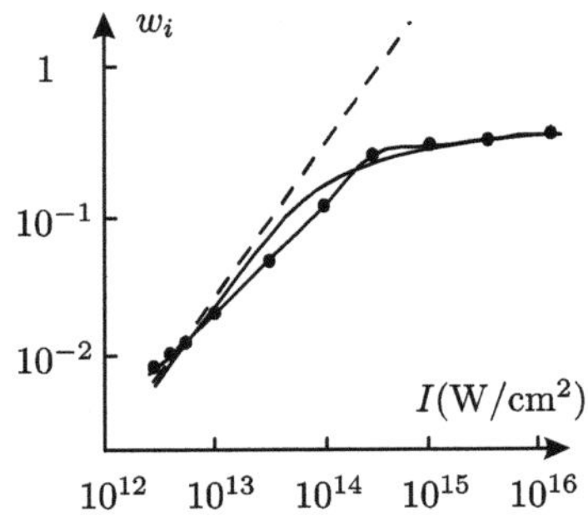

Fig. 1.7. Probability ionization of a hydrogen atom by a short laser pulse. The *solid curve*–the WKB solution (1.15), *dashed line*–the lowest order perturbation theory, *dots*–the exact numerical solution of [12]

1.4 Experiment

There were several experiments where the strong-field stabilization of atoms was observed or attempts to observe it were made. Not dwelling on all of them let us describe here only some of them where the existence of stabilization was proved in the most unambiguous way.

Suppression of photoionization in Ba atoms, excited preliminary to the Rydberg state $6s27d$ was observed experimentally in the work [13]. The electron yield due to one-photon ionization was measured with a pulse duration varied in the interval 2.7–0.25 ps. The electron yield was measured in its dependence on the pulse duration at a given value of fluence $F = 7.8\,\mathrm{J/cm^2}$. The experimental curve describing this dependence was found to have the form very similar to the theoretical one shown above in Fig. 1.6a. As said above, the fall of this curve at small values of τ (and, hence, high values of intensity I) can be considered as a direct evidence of strong-field ionization suppression or atomic stabilization. In the second experiment of the same work [13] the electron yield was measured in its dependence on fluence at two given values of the pulse duration, $\tau = 2.7$ and 0.6 ps. Note that exactly these values of the pulse duration were taken in the calculations of [10] to reproduce most closely the experimental conditions of the work [13]. The measured dependencies $w_i(F)$ are shown in Fig. 1.8.

The curves of this figure are similar qualitatively to the initial stages of the curves in Fig. 1.6b. Moreover, the dependence $w_i(F)$ calculated with the help of (1.15) under experimental conditions of the work [13] for $\tau = 0.6$ ps appears to be practically indistinguishable from the lower curve in Fig. 1.8. All this indicates clearly that the atomic stabilization was observed in the work in [13] and the observed phenomenon is nothing else but interference stabilization.

The second experimental work on the strong-field stabilization to be discussed here is that by De Boer et al. [14]. In this work Ne atoms were excited

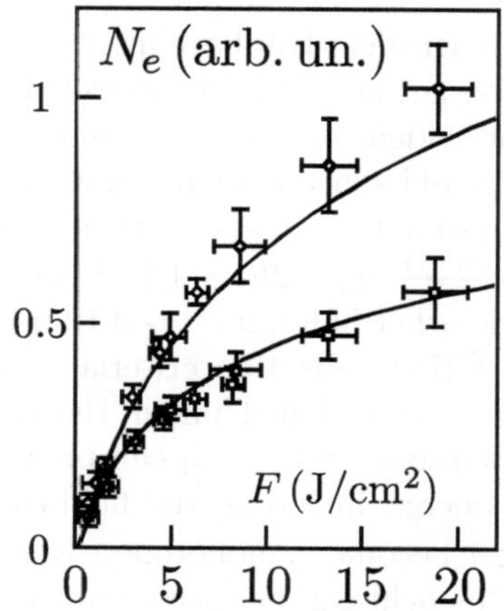

Fig. 1.8. Experimentally measured electron yield N_e vs. the fluence F at $\tau = 2.7$ ps (*the upper curve*) and $\tau = 0.6$ ps (*the lower curve*)

initially to a so called circular state. Such states are characterized by maximal possible values of the electron angular momentum l and its projection m_l, $l = m_l = n - 1$, where n is the principal quantum number. Specifically the quantum numbers of the excited state in the experiment [14] were $n = 5$ and $l = m_l = 4$. Than ionization of an atom in this state was provided by a strong picosecond laser pulse. In the same manner as in the experiment [13], the electron yield was measured in its dependence on fluence F at two given values of the pulse duration τ. The results of measurements are shown in Fig. 1.9. As well as in the case of Ba in the experiment [13], the Ne ionization probability observed at shorter pulse duration (and, hence, higher intensity) appears to be smaller than that observed in the case of longer pulse duration (and lower intensity). This is the same unambiguous proof of the stabilization effect as in the case of Ba atoms [13].

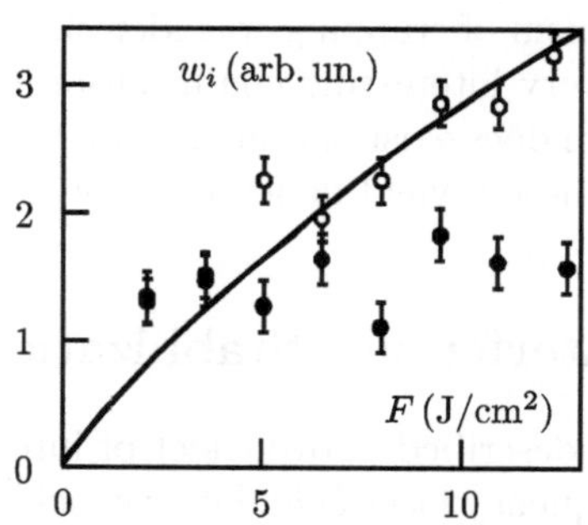

Fig. 1.9. Experimentally measured electron yield $N_e \propto w_i$ vs. fluence F at $\tau = 1$ ps (*blank circles*) and $\tau = 0.1$ ps (*black circles*)

The authors of the experiment [14] have interpreted their results in terms of the Kramers–Henneberger stabilization model. However, validity of this model to this case is not evident because the circular bound-state wave function has a rather large size of its localization region. This size can be estimated as that of the oscillator ground-state wave function near the minimum of the potential $U(r) = -\frac{1}{r} + \frac{l(l+1)}{r^2}$, $a = [2l(l+1)]^{3/4}$ (in atomic units). For $l = 4$ this gives $a \approx 16$. On the other hand, for $\varepsilon \sim 0.1$ and $\omega \sim 0.1$, $\alpha_0 \sim 10 < \alpha_0$, contrary to one of the KH-stabilization criteria (1.8). But the other criterion, $\varepsilon_0 \gg \varepsilon_{\mathrm{BSI}}$, is perfectly satisfied under the conditions of the experiment [14]. Actually, for circular states $\varepsilon_{\mathrm{BSI}}$ can be determined as such a field in which the potential energy including the field-electron interaction term, $U_{\mathrm{tot}}(r) = -\frac{1}{r}+\frac{l(l+1)}{r^2}-\boldsymbol{\varepsilon}\cdot\boldsymbol{r}$ becomes a monotonously falling (well-less) function in the direction of $\boldsymbol{\varepsilon}$. This field can be estimated as $\varepsilon_{\mathrm{BSI}} = 1/(27\,l^2(l+1)^2)$, which is very small and, hence, $\varepsilon_0 \gg \varepsilon_{\mathrm{BSI}}$.

So, the ionization suppression represents a very peculiar case when the Kramers–Henneberger interpretation appears to be only half applicable. Interpretation in terms of interference stabilization is completely inapplicable in this case because bound-free dipole matrix elements for transitions from circular states ar very small. In the work [16] it was suggested to consider stabilization observed in the work [14] as resulting from interference of different-order multiphoton transitions in the continuum. Probably, for clear explanation of the experiment [14] this interpretation, as well as all others, require further analysis and investigation.

Finally, the third experimental work to be discussed here is that by Talebpour et al. [15]. Methodically, this work differs significantly from [13] and [14]. The work [15] was devoted to a traditional investigation of multiphoton ionization of noble gas atoms. However, in contrast to earlier works, the dependence of the ion yield w_i on the peak laser intensity I_{peak} was measured very carefully, with excluded contributions from peripheral regions of the laser focus. The dependence $w_i(I_{\mathrm{peak}})$ was found to have regions of a decreasing slope. Such a behavior was interpreted as manifestation of a partial trapping of the electron population at Rydberg levels excited in the process of ionization. A mechanism of trapping was identified as the interference stabilization. This was a very interesting confirmation that the phenomenon of interference stabilization does exist and manifests itself in such an unexpected way as modulation of the ion yield dependence on the peak laser intensity.

1.5 Two-Color Interference Stabilization

As an extension of the described above effect of interference stabilization in a strong-field single-frequency laser field, let us consider here the phenomenon of two-color interference stabilization [17]. This will be a direct generalization of the simplest two-level model considered above in the very beginning of the section on interference stabilization.

In a single-frequency field there are two conditions under which the two-level model can be considered as rigorously applicable: (i) the levels must be close to each other and sufficiently far from any other atomic levels and (ii) bound-free dipole matrix elements of transitions to the same states of the continuum from these levels must be approximately equal to each other. Both of these two conditions hardly can be satisfied in any realistic atoms. By extending the consideration to the case of a two-color laser field and by manipulating with laser frequencies and intensities, we can easily make the two mentioned conditions or their analogs satisfied.

So, let us take the field strength of a laser field in the form

$$\varepsilon(t) = \varepsilon_1(t)\cos(\omega_1 t) + \varepsilon_2(t)\cos(\omega_2 t) \tag{1.16}$$

and let us consider a scheme of transitions shown in Fig. 1.10, where $\varepsilon_1(t)$ and $\varepsilon_2(t)$ are pulse envelopes and the frequencies ω_1 and ω_2 are assumed to satisfy approximately the Raman-type resonance transitions for two arbitrary selected atomic levels E_1 and E_2. In other words, the resonance detuning

$$\Delta = E_2 + \omega_2 - E_1 - \omega_1 \tag{1.17}$$

is assumed to be small compared to ω_1 and ω_2 and to $|E_n - E_{1,2}|$, where E_n are energies of atomic levels different from E_1 and E_2.

A scheme of transitions of Fig. 1.10 has been widely discussed in literature in connection with theoretical [18–24] and experimental [25–28] investigation of the phenomenon known as the Light-Induced Continuum Structures (LICS). This phenomenon is associated usually with the case when only one of the two fields (ε_2) is strong, whereas the other one (ε_1) is weak and is considered as a probe field. The probe field ionizes an atom, and the main

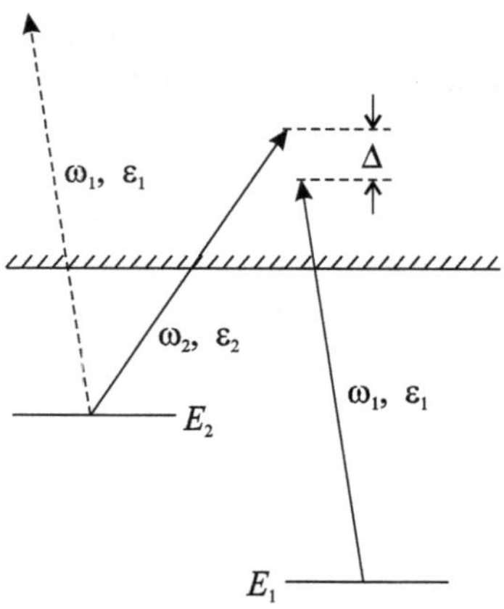

Fig. 1.10. A scheme of two atomic levels under the conditions of a Raman-type resonance in a two-color field (1.16)

investigation subject is the spectral dependence of the ionization yield on the probe-field frequency ω_1. The final result is a Fano-type curve $w_i(\omega_1)$ with parameters depending on the strong-field amplitude and frequency. This formulation of the problem differs from that on the strong-field stabilization, and the latter consists of the following. Let the field (ε_1) be strong enough to ionize an atom completely during the pulse duration. Let then the second field (ε_2) be added and let us look for conditions under which this addition suppresses ionization rather than accelerate it. Such a suppression can arise only owing to interference of direct transitions to the continuum from the level E_1 and transitions via the level E_2. Hence, if the effect takes place, this is a clearly understandable manifestation of interference stabilization in a rather simple scheme of levels and transitions. In such a formulation, the electron yield depends mainly on two parameters: the ratio of laser intensities $x \equiv I_2/I_1 = \varepsilon_2^2/\varepsilon_1^2$ and the resonance detuning Δ (1.17). The main task of a theoretical analysis is optimization of the stabilization effect with respect to a choice of these parameters.

Not dwelling upon any details of such an analysis (see [17]), we reproduce here only some most important results. By using again the rotating wave approximation and adiabatic elimination of the continuum, one can find two complex quasienergies of the system under consideration, $\gamma_\pm$. Their imaginary parts determine widths of quasienergy levels, $\Gamma_\pm = -2\mathrm{Im}\,(\gamma_\pm)$. A typical dependence of these widths on the resonance detuning is shown in Fig. 1.11. Specifically, this and all other results shown below were received for $1s2s$ (E_1) and $1s4s$ (E_2) states of a He atom and laser frequencies equal to $\omega_1 = 8.44\,\mathrm{eV}$ (the 2nd harmonics of a dye laser) and $\omega_2 = 1.17\,\mathrm{eV}$ (Nd:YAG laser).

In the picture of Fig. 1.11 both the widths and detuning are normalized by the first laser intensity I_1: $g_\pm = \Gamma_\pm/I_1$ and $\delta = \Delta/I_1$, where all quantities are in atomic units; the intensity ratio $x = I_2/I_1$ is kept constant. One of the quasienergy width ($g_+(\delta)$) has a clearly seen minimum. As usual, a small width of a quasienergy level is related to a long living part of the atomic population and stabilization of an atom. For this reason the detuning δ where

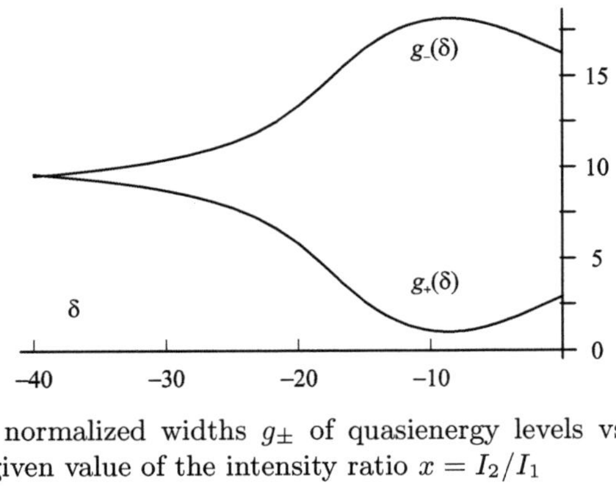

Fig. 1.11. The normalized widths $g_\pm$ of quasienergy levels vs. the normalized detuning δ at a given value of the intensity ratio $x = I_2/I_1$

the function $g_+(\delta)$ has a minimum, is interpreted as an optimal detuning δ_{opt}. An explicit analytical expression for δ_{opt} found in [17] is given by

$$\delta_{\text{opt}}(x) = \frac{1}{4}\left\{\alpha_2'(\omega_1) - \alpha_1'(\omega_1) - \frac{\alpha_{12}'}{\alpha_{12}''}\left[\alpha_2''(\omega_1) - \alpha_1''(\omega_1)\right]\right\}$$
$$+ \frac{x}{4}\left\{\alpha_2'(\omega_2) - \alpha_1'(\omega_2) - \frac{\alpha_{12}'}{\alpha_{12}''}\left[\alpha_2''(\omega_2) - \alpha_1''(\omega_2)\right]\right\}, \qquad (1.18)$$

where $\alpha_i(\omega_j)$ are the dynamical polarizabilities of the levels E_i, i, $j = 1, 2$

$$\alpha_i(\omega) = \int \mathrm{d}E\, |d_{i\,E}|^2 \left(\frac{1}{E - E_i - \omega - \mathrm{i}\delta} + \frac{1}{E - E_i + \omega}\right), \qquad (1.19)$$

α_i' and α_i'' are their real and imaginary parts, $\alpha_{12} = \alpha_{12}' + \mathrm{i}\,\alpha_{12}''$ is the cross-polarizability

$$\alpha_{12} = \int \mathrm{d}E\, d_{2\,E} d_{E\,1} \left(\frac{1}{E - E_1 - \omega_1 - \mathrm{i}\delta} + \frac{1}{E - E_1 + \omega_2}\right), \qquad (1.20)$$

and integrations over E include in themselves summations over discrete intermediate levels E_n different from E_1 and E_2.

If we change now the intensity ratio x and if we want to remain always at the minimum of the curve $g_+(\delta)$, we have to change the detuning synchronously with x in accordance with 1.18. The question is how does $[g_+(\delta)]_{\text{min}} \equiv g_+\left[\delta_{\text{opt}}(x)\right]$ (minimum of the lower curve in Fig. 1.11) change with a changing intensity ratio x? The answer is given by the calculated dependence [17] shown in Fig. 1.12. According to this result, in the framework of the two-level model under consideration, the width of the narrower quasienergy level optimized over detuning at all values of the intensity ratio x, falls unrestrictedly approaching zero at very large x. Limitations of this narrowing can be related only to the model applicability restrictions.

In any case, a possibility to get a very strong narrowing of the quasienergy level γ_+ can be considered as an indication that the achievable degree of

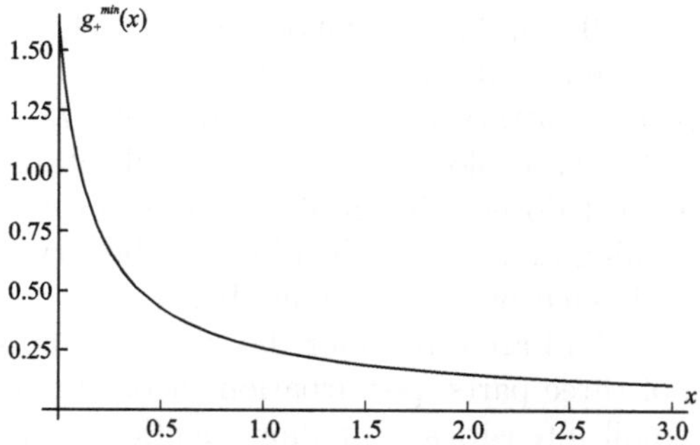

Fig. 1.12. The width of the narrower quasienergy level minimized with respect to the detuning δ vs. the intensity ratio x

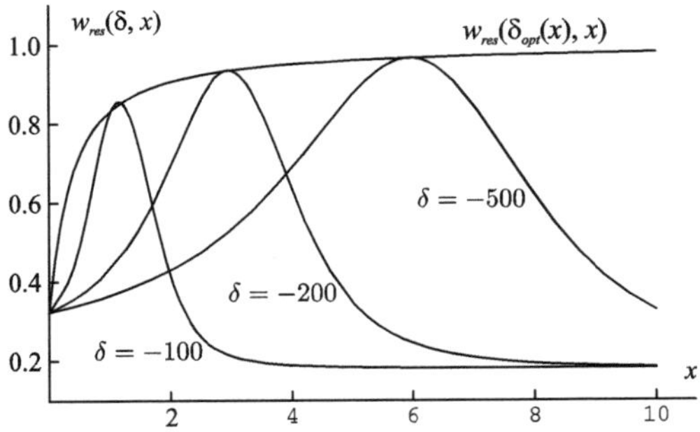

Fig. 1.13. The residual probability of finding an atom in bound states after the pulses have gone vs. the intensity ratio x at three given values of the detuning δ and at $\delta = \delta_{\rm opt}(x)$ (1.18)

stabilization in a two-color laser field can be very large. This assumption is confirmed by direct calculations of the ionization probability. Some results of such calculations are shown in Fig. 1.13. In this picture three resonance-like curves correspond to three different but given values of the detuning δ. The resonance-like behavior arises because at a given δ variations of x bring the system to the optimal conditions and then take it off the optimum. The envelope of these curves corresponds to the detuning tuned to the optimum at all values of x. The calculations are performed for coinciding rectangular envelopes $\varepsilon_1(t)$ and $\varepsilon_2(t)$. The normalized pulse duration $\theta = I_1 \times \tau$ is taken equal to 0.1. In this definition of θ both intensity I_1 and pulse duration τ are in atomic units.

If in calculations corresponding to Fig. 1.13, the residual probability $w_{\rm res}$ was considered in its dependence on the intensity ratio $x = I_2/I_1$, in other calculations of [17] this ratio was kept constant, and the ionization probability $w_i = 1 - w_{\rm res}$ was calculated in its dependence on the intensity of the field "1" normalized by an arbitrary constant intensity I_0. An example of such calculations is shown in Fig. 1.14. The curves $w_i(I_1/I_0)$ are calculated at $x = 3$ (i.e., $I_2 = 3I_1$) and constant detuning and pulse duration: $\Delta = -200\, I_0$ and $\tau = 0.1/I_0$ for the curve (a) and $= 1/I_0$ for the curve (b) (as usual, here Δ, τ, and I_1 are in atomic units).

Appearance of an arbitrary normalizing intensity I_0 reflect the scaling effect occurring in the two-color interference stabilization [17]: all pictures all features of the effect do not change if simultaneously both intensities I_1 and I_2 and the detuning Δ are multiplied by an arbitrary number λ and the pulse duration τ is divided by the same number λ.

The curves of Fig. 1.14 are typical for the stabilization phenomenon. The curve (a) consists of three parts: perturbation theory limit (a growing function $w_i(I/I_0)$ at small intensities), a falling function $w_i(I/I_0)$ in the stabilization region, and a growing function $w_i(I/I_0)$ at large values of intensity–destabilization region and quick complete ionization ($w_i = 1$). The case (b)

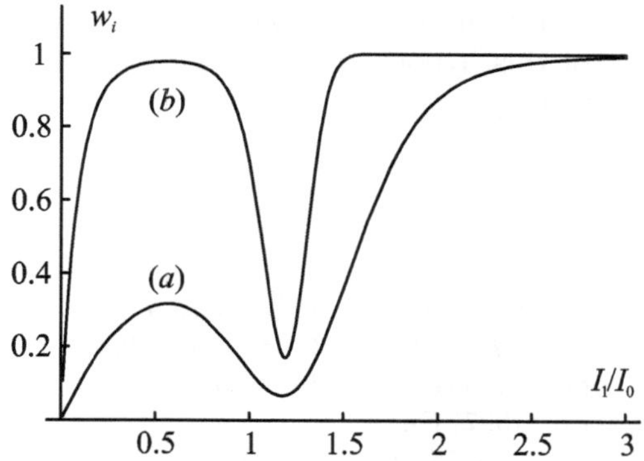

Fig. 1.14.

has an additional part: complete ionization ($w_i = 1$) between the perturbation theory and stabilization regions. This means that at the parameters corresponding to the curve (b), the synchronous increase of the laser intensities I_1 and I_2 at first increases the probability of ionization until it reaches the saturation value ($w_i = 1$). But then, an even further growth of the laser intensities gives rise to a very unexpected result: the probability of ionization falls and, in the specific case of the curve (b) in Fig. 1.14 it falls to the level as low as 0.2, which indicates a very high degree of stabilization. Note that, for example, at $I_0 = 3 \times 10^{10}\,\mathrm{W/cm}^2$ the maximal degree of stabilization (i.e., the minimum of curves in Fig. 1.14) is achieved at $I_1 \approx 4 \times 10^{10}\,\mathrm{W/cm}^2$ and $I_2 \approx 1.2 \times 10^{11}\,\mathrm{W/cm}^2$. Under these conditions the pulse duration, corresponding to the the curve (b) in Fig. 1.14 is $\tau \approx 3\,\mathrm{ps}$.

1.6 Conclusion

So, the provided overview above of the phenomena unified under the name of strong-field atomic stabilization shows that they are rather well studied theoretically. There are several experiments. One of them [13] confirms rather well the the main features of interference stabilization. On the other hand, interpretation of the second experiment discussed above [14], still, remains not completely satisfactory.

As shown above, two-color interference stabilization can increase significantly the achievable degree of stabilization. Note, that this result can open a rather important area of practical applications of this phenomena for the pulse propagation problem. Indeed, let a strong laser pulse be propagating in an atomic gas medium and let the length of its propagation be limited by ionization of atoms. Then it seems that by sending to the same medium two pulses with appropriately chosen parameters we can increase significantly the propagation length owing to the two color stabilization effect described above. Verification of this hypothesis requires further theoretical work and experimental investigation.

References

1. M.V. Fedorov and A.M. Movsesian: *Phys. Rev. A* **21**, L155 (1988)
2. M. Pont, M. Gavrila: *Phys. Rev. Lett.* **65**, 2362 (1990)
3. M.V. Fedorov: *Atomic and Free Electrons in a StrongLight Field* (World Scientific: Singapore, 1997) Ch. 8
4. M. Gavrila: *Phys. Rev. A* **35**, R147 (2002)
5. A.M. Popov, O.V. Tikhonova, E.A. Volkova: *Phys. Rev. A* **36**, R125 (2003)
6. H.A. Kramers: in *Rapport du 8eConseil Solvay, 1948* (Solvay Foundation, Paris, 1950) p. 241
7. W.C. Henneberger: *Phys. Rev. Lett.* **21**, 838 (1968)
8. R. Grobe and M.V. Fedorov: *Phys. Rev. Lett.* **68**, 2592 (1992)
9. N.B. Delone et al: *Phys. Rev. A* **16**, 2369 (1983); **22**, 2941 (1989); **27**, 4403 (1994)
10. M.V. Fedorov et al: *Phys. Rev. A* **29**, 2907 (1996)
11. M.V. Fedorov and O.V. Tikhonova: *Phys. Rev. A* **58**, 1322 (1998)
12. O.V. Tikhonova et al: *Phys. Rev. A* **60**, R749 (1999)
13. J. Hoogenraad et al: *Phys. Rev. A* **50**, 4133 (1994)
14. M.P. De Boer et al: *Phys. Rev. Lett.* **71**, 3263 (1993)
15. A. Talebpour, C.Y. Chien and S.L. Chin: *Phys. Rev. A*, **29**, 5727 (1996)
16. O.V. Tikhonova et al: *Phys. Rev. A* **65**, 053404 (2002)
17. M.V. Fedorov and N.P. Poluektov: *Phys. Rev. A* **69**, 033404 (2004)
18. L. Armstrong, Jr., B. Beers, and S. Feneuille: *Phys. Rev. A* **12**, 1903 (1975)
19. Yu.I. Heller and A.K. Popov: *Opt. Commun.* **18**, 449 (1976)
20. A.I. Andryushin and M.V. Fedorov: *Izvestiya Vuzov: Fizika*, #1, 63 (1978)
21. Bo-nian Dai and P. Lambropulos: *Phys. Rev. Lett.* **36**, 5202 (1987)
22. M.V. Fedorov and A.E. Kazakov: *Progress in Quantum Electronics* **13**, 97 (1989)
23. P.L. Knight et al.: *Phys. Rep.* **190**, 1 (1990)
24. A.I. Magunov, I. Rotter, and S.I. Strakhova: *J. Phys. B* **34**, 29 (2001)
25. M.H.R. Hutchinson and K.M.M. Ness: *Phys. Rev. Lett.* **60**, 105 (1988)
26. Y.L. Shao et al.: *Phys. Rev. Lett.* **67**, 3669 (1991)
27. S. Cavalieri and F. S. Pavone: *Phys. Rev. Lett.* **67**, 3673 (1991)
28. T. Halfmann et al.: *Phys. Rev. A* **58**, 46 (1998)

2 Creation of Novel Quasi-Bound States in High-Frequency Intense Laser Fields

Kiyohiko Someda and Tomokazu Yasuike

Department of Basic Science, Graduate School of Arts and Sciences,
The University of Tokyo, Komaba, Meguro-ku, Tokyo 153-8902
kiyon@isolde.c.u-tokyo.ac.jp

It has been shown that the photon field of intense infrared lasers works as a static electric field for electrons in atoms and molecules [1,2]. The primary process for the molecules placed in such a field is field-ionization. The photon field is, however, an alternating field. It forces electrons to quiver rather them kick away. It gives electrons a chance to travel around a molecule. The use of high frequency and intense lasers emphasizes the change of motion of electrons bound in atoms and molecules. With increasing laser frequency, the ionization rate is apt to be suppressed [3–5], and one can discuss quasi-bound electronic states of atoms and molecules in intense fields.

In high-frequency intense laser fields, electron wavefunctions are deformed, and electronic states of molecules come to have different characters. Energy levels are shifted by the ac Stark effect, and acquire an energy width due to ionization. This kind of change is, in a sense, gradual and continuous. If we gradually increase the laser intensity, the electronic states change their characters continuously, in other words, adiabatically. In addition to such adiabatic changes, it is known that an abrupt change occurs on atomic and molecular electronic states. New electronic states can be created by the effect of intense fields. Those states are called light-induced states (LIS) [6]. In Sect. 2.2, we discuss the origin of LIS and clarify the condition, i.e., the combination of laser intensity and frequency, for the formation of LIS.

Due to the changes in the electron wavefunction, atoms acquire different chemical properties. The nature of the chemical bond is modified in intense fields. For instance, bond hardening and softening in the H_2^+ molecular ion have been observed experimentally [7–9, 11]. Deformation in molecular structure has also been reported [12–14]. In a recent study, the creation of a covalent bond between two helium atoms in intense fields is theoretically predicted [15]. We discuss laser-induced chemical bonding in Sect. 2.3.

2.1 Motion of Electrons in High-Frequency Photon Fields

In this section, basic concepts and terminologies are summarized. The Hamiltonian for electron motion in a photon field is written as

$$H = \frac{1}{2}\left(\boldsymbol{p} + \frac{1}{c}\boldsymbol{A}(t)\right)^2 + V(\boldsymbol{x}) , \tag{2.1}$$

where $\boldsymbol{p}$ is the electron momentum, $\boldsymbol{A}(t)$ is the vector potential, $V(\boldsymbol{x})$ is the scalar potential, c is the light velocity, and the atomic units are employed. In order to simplify the problem, we restrict ourselves in the case that the laser is linearly polarized and its intensity is stationary. With this simplification, the vector potential can be written as

$$\boldsymbol{A}(t) = \frac{cF}{\omega}\boldsymbol{e}\cos\omega t \,, \tag{2.2}$$

where $\boldsymbol{e}$ is the polarization vector, and ω is the angular frequency of laser. The intense lasers working in laboratories are all pulsed lasers. The assumption that the laser intensity is stationary is justified when the duration of the laser pulse is sufficiently longer than the typical time scale of electron motion under interest. If we are interested in transient electron dynamics triggered by an intense laser pulse, the time-dependent treatment should be employed. In this chapter, however, our attention is focused on quasi-bound states having a certain lifetime in intense laser fields. The stationary field assumption helps to clarify the physics of creation of new quasi-bound states.

2.1.1 Ponderomotive Radius

If the electron is not bound in the scalar potential, i.e., $V(\boldsymbol{x}) = 0$ in (2.1), the motion of electron is simple. If we treat H in (2.1) as the classical Hamiltonian, we can readily integrate the equation of motion. The electron coordinate $x(t)$ along the laser polarization is given by

$$x(t) = \frac{F}{\omega^2}\sin\omega t \equiv \alpha\sin\omega t \,. \tag{2.3}$$

Here, we choose the initial conditions as $x(0) = p(0) = 0$. Equation (2.3) indicates that the electron oscillates with the amplitude $\alpha \equiv F/\omega^2$. This amplitude α is called ponderomotive radius. The ponderomotive radius can be understood as a measure for the effect that the laser field causes on electrons. With fixed laser intensity, the ponderomotive radius decrease with increasing angular frequency ω.

Another measure which describes the behavior of electrons in intense fields is the Keldysh parameter [16] defined by

$$\gamma \equiv \frac{\omega}{F}(2E_{\mathrm{b}})^{1/2} \,, \tag{2.4}$$

where E_{b} is the binding energy of an electron in atoms and molecules. The Keldysh parameter is a measure for the field ionization. When the Keldysh parameter is sufficiently small, the electrons bound in atoms and molecules feel the photon field as a static field. In this case, the field ionization is the dominant process.

By the combination of the Keldysh parameter γ and the ponderomotive radius α, the behavior of electrons in intense fields is roughly classified.

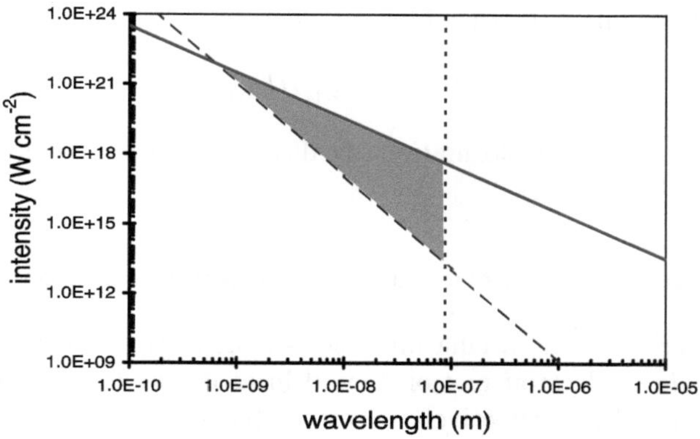

Fig. 2.1. Phase diagram of intense fields. The *horizontal axis* represents the wavelength of the laser, and the *vertical axis* represents the laser intensity. The *solid line* corresponds to the condition that the Keldysh parameter $\gamma = 0.1$. The *dash line* indicates the line $\alpha = 0.1$. The *dot line* indicates the ionization threshold for the hydrogen atom. The *shaded triangle* roughly indicates the region, where the non-trivial electron dynamics is expected. The *lines* do not indicate clear phase separation, but represents rough estimation

When α is sufficiently small the process can be well described by the perturbation treatment. In the case of large α and small γ, the field ionization prevails. Intriguing phenomena, such as creation of new quasi-bound states, is restricted in the case of large α and large γ.

We have to add one more condition for the high-frequency fields. The period of the electric field oscillation should be faster than the typical time scale of electron motion bound in atoms and molecules. Such a condition is fulfilled when the photon energy $h\nu$ exceeds the electron binding energy E_{b}.

Figure 2.1 shows something like a phase diagram. The light field of large α, large γ and high-frequency corresponds to the triangle region in the figure, where non-trivial electron dynamics is expected to occur.

2.1.2 Kramers–Henneberger Frame

The transformation to the moving frame that follows the quiver motion, (2.3), of an unbound electron gives us an insight. By this transformation, the Hamiltonian in (2.1) becomes

$$H_{\mathrm{KH}} = \frac{1}{2}\boldsymbol{p}^2 + V(\boldsymbol{x} + \alpha \boldsymbol{e} \sin \omega t) . \tag{2.5}$$

The moving frame is called Kramers–Henneberger (KH) frame [17, 18]. In quantum mechanics, the corresponding transformation is expressed by the unitary operator

$$U_{\mathrm{KH}} \equiv \mathrm{e}^{\mathrm{i}\alpha p \sin \omega t} . \tag{2.6}$$

The KH Hamiltonian H_{KH} is obtained as

$$H_{\mathrm{KH}} = U_{\mathrm{KH}}^{\dagger} H U_{\mathrm{KH}} \tag{2.7}$$

The frame after the transformation is called acceleration gauge or Kramers–Henneberger (KH) gauge.

2.1.3 Kramers–Henneberger High-Frequency Approximation

When the period of the oscillating field is sufficiently shorter than the typical time scale of electron motion bound in atoms and molecules, the time-averaged Hamiltonian effectively describes the electron dynamics [19]. The time-averaging leads to

$$\bar{H}_{\mathrm{KH}} = \frac{1}{2}\boldsymbol{p}^2 + V_{\mathrm{KH}}(\boldsymbol{x}) \ , \tag{2.8}$$

where

$$V_{\mathrm{KH}}(\boldsymbol{x}) = \frac{\omega}{2\pi} \int_0^{2\pi/\omega} \mathrm{d}t V(\boldsymbol{x} + \alpha \boldsymbol{e} \sin \omega t) \ , \tag{2.9}$$

The potential part V_{KH} is called KH potential. It is the KH approximation that helps the first discovery of LIS.

2.2 Light-Induced States

The first discovery of LIS (exactly speaking, it is a theoretical prediction) was reported in 1988 [6], and the studies on LIS have already history, as will briefly be summarized in Sect. 2.2.1. The first theoretical prediction is based on the high-frequency approximation applied to a simple one-dimensional (1D) model. Advances have been made in two directions: to treat realistic atomic systems having plural number of electrons and to lift the high-frequency approximation. More studies have been reported in the latter direction. A detailed analysis on a simple system [20–22] reported recently is reviewed in Sect. 2.2.2 and following subsections.

2.2.1 History of the Studies on Light-Induced States

The studies on LIS have started in 1988 [6]. In the early stage, the high-frequency limit is studied. In that limit, the Kramers–Henneberger (KH) high-frequency approximation is known to give us a good picture as mentioned in Sect. 2.1.3.

A simple example for the KH potential is shown in Fig. 2.2. There is a one dimensional potential well. When the laser is switched on, the KH potential gradually deforms. It becomes shallow, and eventually results in a double well

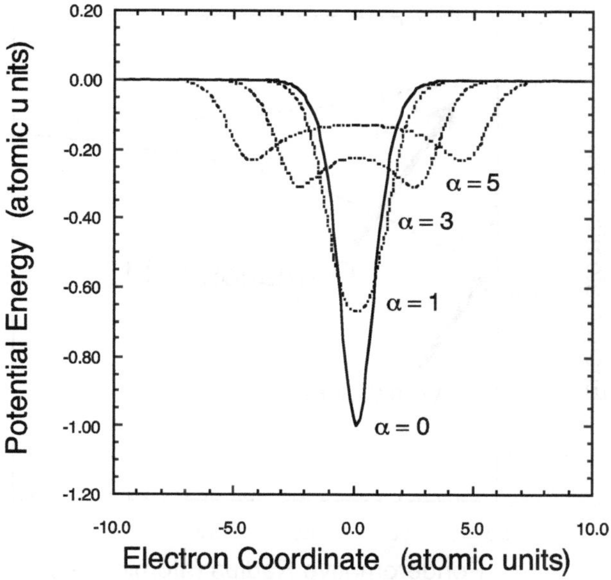

Fig. 2.2. An example of Kramers–Henneberger potential. The potential shape changes with increasing ponderomotive radius $\alpha \equiv F/\omega^2$, where F is the amplitude of the electric field and ω is the angular frequency of the photon

potential. The double well potential is a reflection of quiver motion of electron driven in the oscillating electric field. Changes in the energy levels held on the KH potential is also shown in Fig. 2.3. In the case of zero fields, the potential holds only one bound state. With increasing laser intensity, the energy level gradually shifts upwards. At the critical laser intensity, a new bound state is created, as shown in Fig. 2.3. This is an LIS. In the KH picture, LIS can be understood as excited states held on the deformed effective potential.

Later, Bardsley and Comella shoewed that the high-frequency limit is not necessary condition for the formation of LIS on the basis of a 1D model [23]. Dörr discussed LIS in H atom (in the three dimensional space) [24]. The analysis of a 1D model was reported also by Yao and Chu [25]. The first report on the LIS in the atom with plural number of electron was reported by Muller and Gavrila [26]. They calculated the LIS of H^- in the high-frequency limit. Fearnside et al. also calculated the the LIS of H^- [27]. Possibility of detection of LIS in photoelectron spectra measured in intense fields of finite frequency has been pointed out by Wells et al. on the basis of a 1D model [28].

No experimental observation of LIS has been reported so far. Theoretical study is indispensable in order to predict the condition, i.e., the combination of laser frequency and intensity, that gives rise to the formation of LIS. Such prediction requires theoretical technique which allows us to change the laser frequency and intensity freely without loos-

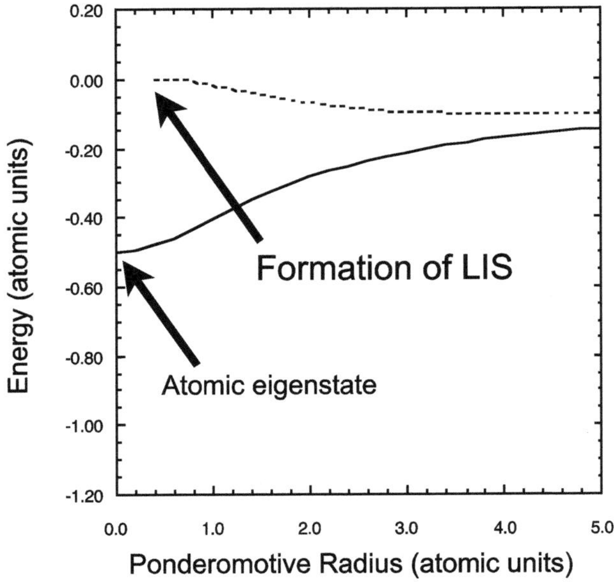

Fig. 2.3. Energy levels held on the Kramers–Henneberger potential shown in Fig. 2.2. The *horizontal axis* represents the ponderomotive radius α, which is proportional to the *square root* of the laser intensity. A new level appears at a threshold $\alpha \simeq 0.4$ (atomic units). This is an LIS

ing the accuracy of calculation. Researchers who tend to carry out exact calculation on a simple systems find their theoretical basis in the quantum scattering theory. Under laser field of a finite intensity with a finite frequency, all the atomic and molecular states are subject to photoionization. In this sense, all the states are resonances from the scattering theoretical point of view. The resonance states are identified as poles of resolvent, $G(E) = (E - H)^{-1}$, where H is the Hamiltonian of the system [29]. The resonance poles are located in an appropriate domain on an appropriate sheet of the complex-energy Riemann surface. The behavior of LIS can be explained by the motion of poles on different Riemann sheets of the complex-energy plane. Based on this idea, Fearnside et al. have analyzed a 1D model mimicking the Cl^- ion under irradiation of ArF laser [27]. Recently, the pole trajectories on the complex-energy plane are analyzed in a wide range of laser intensity and frequency, and the origin of LIS is clarified [20–22].

The framework of the quantum scattering theory is based on the time-independent Schrödinger equation. On the other hand, atoms and molecules in laser fields are described by time-dependent Schrödinger equation in the first principle. If the laser intensity is stationary, the problem is reduced to time-periodic system. In such a case, the Floquet formalism enables us to apply the quantum scattering theory.

2.2.2 Floquet Formalism

Under the stationary field assumption, (2.2), the Hamiltonian in (2.1) is time-periodic. According to the Floquet theorem [30,31], the solution of the time-periodic Schrödinger equation,

$$\left(H(t) - \mathrm{i}\frac{\partial}{\partial t}\right) \Psi(\boldsymbol{x}, t) = 0 , \tag{2.10}$$

can be expressed in the form

$$\Psi(\boldsymbol{x}, t) = \mathrm{e}^{-\mathrm{i}Et}\Phi(\boldsymbol{x}, t) , \tag{2.11}$$

where $\Phi(\boldsymbol{x}, t)$ is a periodic function with respect to t,

$$\Phi\left(\boldsymbol{x}, t + \frac{2\pi}{\omega}\right) = \Phi(\boldsymbol{x}, t) . \tag{2.12}$$

By substituting (2.11) into the Schrödinger equation, (2.10), it is shown that $\Phi(\boldsymbol{x}, t)$ satisfies

$$\left(H(t) - \mathrm{i}\frac{\partial}{\partial t}\right) \Phi(\boldsymbol{x}, t) = E\Phi(\boldsymbol{x}, t) , \tag{2.13}$$

which indicates that $\Phi(\boldsymbol{x}, t)$ is the eigenfunction of the Floquet–Hamiltonian operator [32],

$$H_{\mathrm{F}} \equiv H(t) - \mathrm{i}\frac{\partial}{\partial t} . \tag{2.14}$$

Due to the time-periodic property, (2.12), the Floquet eigenfunction $\Phi(\boldsymbol{x}, t)$ can be expanded in the Fourier series,

$$\Phi(\boldsymbol{x}, t) = \sum_{n=-\infty}^{\infty} \mathrm{e}^{-\mathrm{i}n\omega t}\Phi_n(\boldsymbol{x}) . \tag{2.15}$$

By substituting (2.15) into (2.13), we obtain the Floquet-coupled equation

$$(H_0 + n\omega - E)\,\Phi_n(\boldsymbol{x}) - \sum_m H_{m-n}\Phi_m(\boldsymbol{x}) = 0 , \tag{2.16}$$

where H_n is defined by

$$H_n \equiv \frac{\omega}{2\pi}\int_0^{2\pi/\omega} H(t)\mathrm{e}^{\mathrm{i}n\omega t}\mathrm{d}t . \tag{2.17}$$

We can thus convert a time-periodic problem, (2.10), into a stationary one, (2.16), based on the Floquet theorem. In photon fields, electrons are not bound in atoms and molecules, but ionization takes place eventually. In this sense, the Floquet coupled equation, (2.16), has no bound state eigenfunction, and should be treated by multichannel scattering theory.

It is noteworthy that the Fourier representation, (2.15), allows us to use the notion of photons on appropriate occasions, although the laser fields are treated as classical electromagnetic fields. Each Fourier component can be interpreted as the photon-number state [30]. But we have to be careful because we cannot define the absolute number of photons.

2.2.3 Trajectories of Poles and the Origin of Light-Induced States

Quasi-bound states of atoms and molecules in laser fields are identified by poles of the resolvent $G_{\mathrm{F}}(E) = (E - H_{\mathrm{F}})^{-1}$. The resolvent has branch points at every channel thresholds, and we need to consider Riemann sheets. The physical meaning of the pole depends on the sheet which it inhabits. In single-channel problems, the complex-energy Riemann surface is composed of two sheets, as shown in Fig. 2.4 [29].

The resolvent $G(E)$ is a single valued function of the momentum p. Due to the relation $E = p^2/2$, we need two Riemann sheets of complex energy in order to represent the value of $G(E)$ uniquely. Each sheet is specified by the sign of the imaginary part of momentum p. The physical sheet is specified by $\mathrm{Im}p > 0$, while the second sheet by $\mathrm{Im}p < 0$. The pole corresponding to bound states is placed on the negative part of the real E axis on the physical sheet, as indicated by the point “B” in Fig. 2.4. Due to the positive $\mathrm{Im}p$, the wavefunction of bound states goes to zero in the asymptotic region, $x \to \infty$. The sheet specifies the boundary condition of wavefunctions. The poles on the negative part of real E axis on the second sheet are called anti-bound states. With the negative $\mathrm{Im}p$, the wavefunctions of antibound states (indicated by the point “AB” in Fig. 2.4) diverge in the asymptotic region. The resonance poles should be in the fourth quadrant on the second sheet, as illustrated in Fig. 2.4. The resonance should be decaying with time and satisfy the Siegert boundary condition. It requires $\mathrm{Re}E > 0$, $\mathrm{Im}E < 0$ and $\mathrm{Im}p < 0$.

For N-channel problems, we need to specify all the signs of $\mathrm{Im}p$ in open and closed channels, and consider 2^N Riemann sheets. Each sheet can be identified by the set of signs $\Sigma = (\sigma_1, \sigma_2, \ldots, \sigma_N)$ [20]. The resonances have to satisfy the Siegert boundary condition [33], i.e., they should behave, in the asymptotic region $|x| \to \infty$, as outgoing wave in the open channels and should vanish in the closed channels. It follows that resonance poles should be on the sheet with $\sigma_{\mathrm{open}} = -$ and $\sigma_{\mathrm{closed}} = +$. Such a region of Riemann sheets is called resonance sector. The Riemann sheets for the three-channel problem is illustrated in Fig. 2.5.

The bound states of atoms and molecules become resonance states with finite lifetimes in light fields due to ionization. The bound states corre-

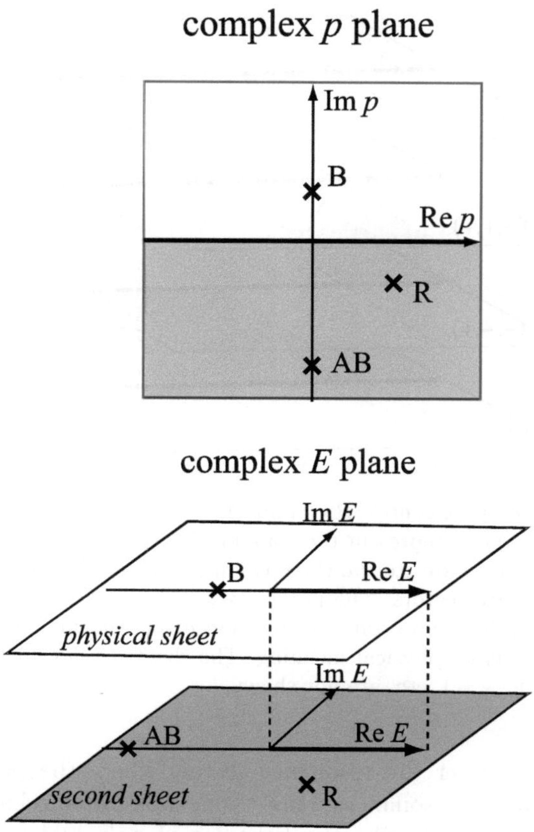

Fig. 2.4. The complex-momentum plane and the corresponding complex-energy Riemann sheets in the single-channel case. The *upper half* of the complex p-plane corresponds to the physical sheet of the complex E-plane, while the *lower half* to the second sheet. The points "B", "R", and "AB" indicate typical locations of a bound state, a resonance, and an anti-bound state, respectively

spond to poles on the real axis of the complex-energy plane. Irradiation of light pushes those poles into the resonance sectors. The situation is illustrated in Fig. 2.6. The changes in the characters of electronic states in laser fields can be understood by trajectories of poles with changing laser intensity. The irradiation of laser gives rise to another phenomena: The presence of light field couples the different Floquet channels, and generates poles on the other Riemann sheets. The circumstance is illustrated in Fig. 2.6. New poles are located at almost the same position as the original bound state but on different sheets. Such a pole is called shadow pole [34]. The eigenfunction for the shadow pole does not satisfy the physical boundary condition, and corresponds to a virtual state without any physical meaning. With increasing laser intensity, however, the

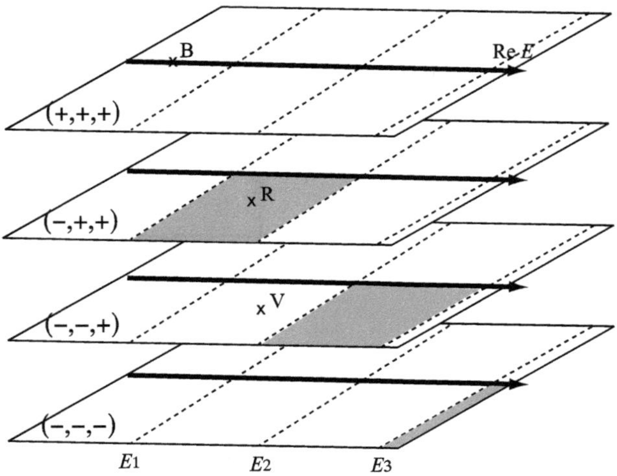

Fig. 2.5. The complex-energy Riemann sheets for the three channel problem. $(+,+,+)$, $(-,+,+)$ etc. represent the combination of signs specifying the Riemann sheet. E_1, E_2 and E_3 indicate the three channel thresholds. The *shaded region* indicates the resonance sectors. The point "B", "R" and "V" represent the typical position of poles of the bound state, the resonance, and the virtual pole, respectively. The virtual pole has no physical meaning. The sheets irrelevant to the resonance, e.g., $(+,-,+)$, $(+,+,-)$, etc. are not shown

shadow pole can enter the resonance sector, where the pole acquires the physical meaning of resonance. This is one possible mechanism of formation of LIS [35]. We describe the behavior of pole trajectories for simple 1D models.

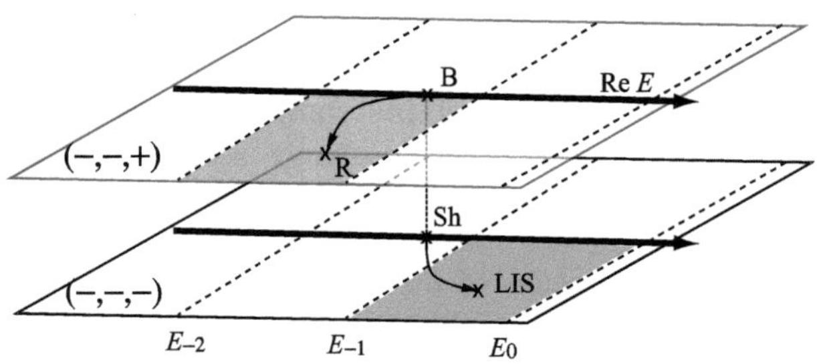

Fig. 2.6. Part of the complex-energy Riemann sheets for the Floquet coupled equation. Irradiation of laser pushes the bound state pole "B" into the resonance sector, and the bound state becomes resonance "R" embedded in the ionization continua. The irradiation of laser generates a shadow pole "Sh" at almost the same position as the bound state but on a different sheet. With increasing laser intensity, the *shadow pole* enters the resonance sector and becomes resonance. This is an LIS

Pöshl–Teller Potential Model

Behavior of pole trajectories in various combinations of laser intensity and frequency is reported in literature [20] on the basis of a 1D model with a simple scalar potential. Needless to say, the scalar potential which binds electrons on an atom should be the Coulomb potential. Exact calculation for the electron scattering by the Coulomb potential in laser fields is possible. However, simpler model potentials are object of many theoretical studies from the following reasons: The Coulomb potential has singularity at the origin, and it is liable to arise difficulty in numerical computation. In addition, the Coulomb potential holds an infinite number of bound states, the Rydberg states, and this forces us to pay much effort for keeping the numerical accuracy. Here, we restrict ourselves to a simple model potential, Pöshl–Teller potential,

$$V(x) = -\frac{1}{\cosh^2 x} . \tag{2.18}$$

The parameter is chosen so that there is only one bound state at $E = -1/2$. Here, the atomic units are used.

Pole Trajectories on the Complex-Energy Plane

The pole trajectories calculated on the basis of the complex-scaling method are shown in Figs. 2.7 and 2.8. Details of the method of calculation is described in [20] and references cited therein. Positions of poles on the complex-energy plane are calculated with many different laser intensities and fixed frequency. Trajectories are drawn by connecting these pole positions. They indicate the adiabatic changes of each resonance state with the adiabatic increase of laser intensity. In Fig. 2.7, the symbol T_n indicates the threshold of the nth Floquet channel, $E = n\hbar\omega$. We specify each trajectory by the notation $\Phi_N^{(\sigma_{-3},\sigma_{-2},\sigma_{-1})}$. The superscript $(\sigma_{-3}, \sigma_{-2}, \sigma_{-1})$ indicates the sign of Imp and specifies the sheet on which the pole is located. The subscript N indicates that the pole trajectory stems from the bound state at $E = E_0 + N\hbar\omega$. For example, $\Phi_0^{(---)}$ is on the $(---)$ sheet and correlates to the bound state at $E = E_0$ in the limit of zero fields. Due to the time periodicity of the Hamiltonian, the energy spectrum has a translational symmetry with the period of $\hbar\omega$ in the direction of the real axis. The translational symmetry can be derived from Eqs. (2.11) and (2.15). In Fig. 2.7, we display only Φ_0^{Σ}, and the other trajectories which can be derived from the translational symmetry are omitted for the sake of simplicity. For instance, the trajectories of $\Phi_{-1}^{(--+)}$ and $\Phi_{+1}^{(---)}$ exist in the region of $T_{-2} < \mathrm{Re}E < T_{-1}$ and $T_{-1} < \mathrm{Re}E < T_0$, respectively, but they are not displayed. The trajectories of $\Phi_{-1}^{(--+)}$ and $\Phi_{+1}^{(---)}$ have the same shape as $\Phi_0^{(---)}$ and $\Phi_0^{(--+)}$, respectively.

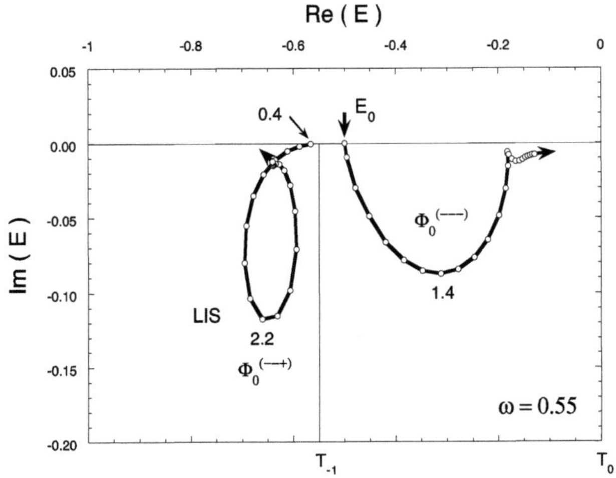

Fig. 2.7. Pole trajectories on the complex-energy plane. The trajectories are drawn with changing α from 0.0 to 5.0 with fixed laser angular frequency, $\omega = 0.55$. The resonance $\Phi_0^{(---)}$ stems from the original bound state at $E = E_0(= 0.5)$. The resonance $\Phi_0^{(--+)}$ is an LIS. The *symbols* are defined in the text

Behavior of Poles for $\hbar\omega > E_0$

In Fig. 2.7, the trajectories in the case of $\hbar\omega = 0.55$ are shown. The angular frequency $\hbar\omega = 0.55$ corresponds to slightly above the one-photon ionization threshold. The original bound state at $E = E_0(= 1/2)$ comes down to the lower half-plane and becomes the resonance $\Phi_0^{(---)}$. In the case of $\alpha < 1.4$, the width of resonance increases with α, namely, the lifetime becomes shorter with increasing laser intensity. When the laser intensity exceeds the critical value $\alpha_c \simeq 1.4$, the pole trajectory goes upwards, i.e., the lifetime increases with increasing laser intensity. Such a phenomenon is known as stabilization [3–5].

On the other hand, the resonance $\Phi_0^{(--+)}$ appears on the resonance sector at $\alpha = 0.4$. When $\alpha < 0.4$, it lies on the $(--+)$ but outside the resonance sector. This resonance is an LIS.

Behavior of Poles for $\hbar\omega < E_0$

Figure 2.8 shows the pole trajectories for the case of $\hbar\omega = 0.45$, i.e., slightly below the one-photon ionization threshold. We see a pair of pole trajectories having the similar shape as in Fig. 2.7. Although the shape is similar, their role is exchanged. The resonance correlating to the original bound state in the zero field is not $\Phi_0^{(---)}$ but $\Phi_0^{(--+)}$. In this case, $\Phi_0^{(---)}$ is an LIS, which enters the resonance sector at about $\alpha = 0.8$.

For the both cases shown in Figs. 2.7 and 2.8, a pair of pole trajectories has essentially the same shape. The pair, however, exchanges their role

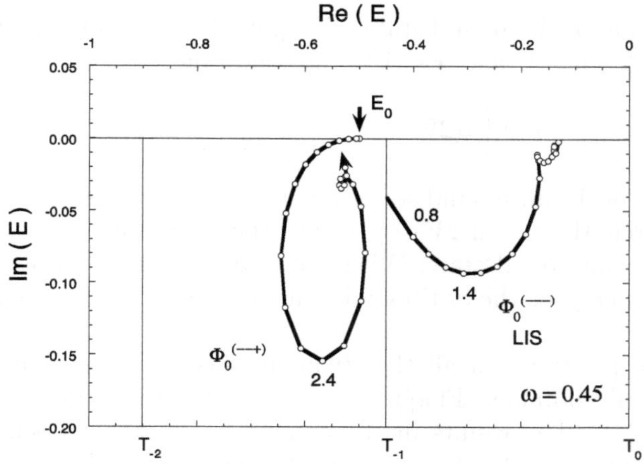

Fig. 2.8. Pole trajectories on the complex-energy plane for the case of $\omega = 0.45$. The resonance $\Phi_0^{(--+)}$ stems from the original bound state at $E = E_0(= 0.5)$, while the resonance $\Phi_0^{(---)}$ is an LIS

with each other when $\hbar\omega$ crosses the one-photon ionization threshold E_0. This phenomenon can be intuitively understood by the structure of Riemann sheets.

Origin of LIS

The study reported in [20] is based on the complex-scaling method, and the pole positions outside the resonance sector are not determined. The question arises: Where does the LIS pole arrive at in the limit of zero field? Stroe and Boca directly calculated the pole positions outside the resonance sector [22]. In the case of $\omega = 0.55$, the LIS pole $\Phi_0^{(--+)}$ reaches the threshold $E = T_{-1}$. Stroe and Boca pointed out that the Pöshl–Teller potential has an antibound state at zero energy. At each Floquet threshold $E = T_n$, there is an antibound state pole. The LIS pole $\Phi_0^{(--+)}$ originates from that anti-bound state pole at the threshold $E = T_{-1}$. The coincidence of the anti-bound state pole and the Floquet threshold is, in a sense, a very special case. Namely, the Pöshl–Teller potential, (2.18), with the given parameters is a singular case.

In the case of $\omega = 0.45$, the LIS pole $\Phi_0^{(---)}$ comes to the original bound state at $E = E_0$ but on the different Riemann sheet in the limit of zero fields. In this case, the origin of the LIS is a shadow pole associated with the original bound state. The situation is already illustrated in Fig. 2.6 schematically.

Square-Well Potential Model

In order to lift up the special coincidence of the anti-bound state and the threshold, the case of a 1D square-well potential is examined [21]. By adjusting the depth V and the width a of the square well, we are able to control the energy position of the bound state. The potential has only one bound state at $E_0 = -1/2$ when V and a satisfy the condition

$$\tan\left\{a(2V-1)^{1/2}\right\} = \frac{1}{(2V-1)^{1/2}} \, . \tag{2.19}$$

The energies of the antibound states are determined by the parameter, called potential strength, $\gamma = a(2V)^{1/2}$. Under the condition $\gamma < \pi/2$, there exists only one antibound state. When we change V and a so as to maintain (2.19), the energy of the antibound state E_a varies, while E_0 remains unchanged.

The pole positions on all the Riemann sheets are calculated based on the analytically obtained Floquet S-matrix with truncation of the number of channels [21]. The results in the case of $\hbar\omega = 0.55$ are summarized as follows: In the case of $-V = E_a < E_0$, i.e., the antibound state is located at the bottom of the potential, it is found that the LIS originates from a shadow pole of the original bound state. The pole starting from the antibound state travels outside the resonance sector, and does not play any role with physical meaning. The situation is schematically illustrated in Fig. 2.9.

When the energy of the antibound state comes to above E_0, i.e., $E_0 < E_a < 0$, the pole originating from the antibound state enters the resonance sector as schematically illustrated in Fig. 2.10. In this case shadow poles associated with the bound state does not enter any resonance sectors.

Origin of LIS

In summary, the origin of LIS switches between the shadow pole of the bound state and the antibound state depending on the relative energy positions of

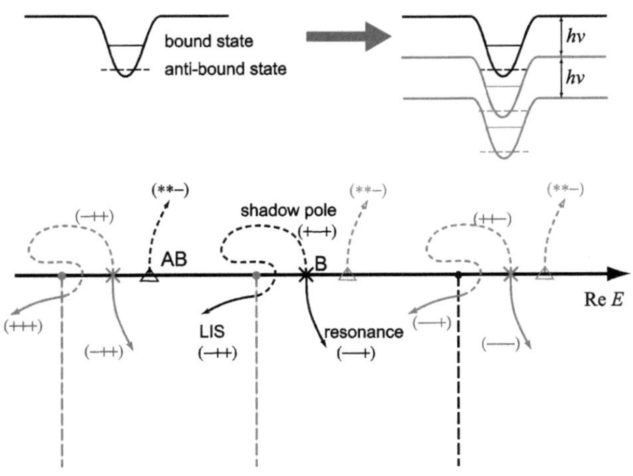

Fig. 2.9. Schematic illustration of pole trajectories indicating the origin of LIS. The case of $E_a < E_0$ and $\hbar\omega > |E_0|$

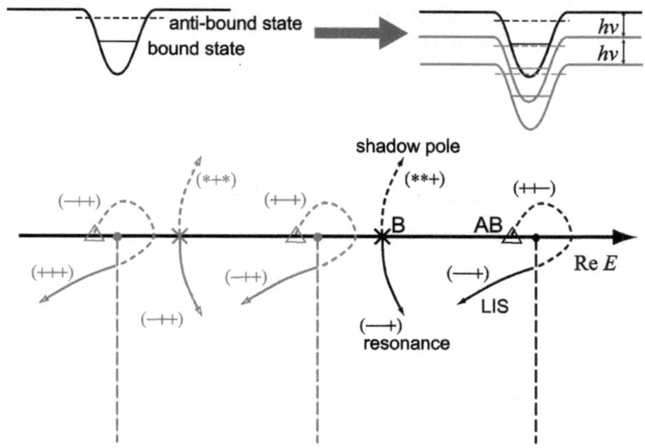

Fig. 2.10. Schematic illustration of pole trajectories indicating the origin of LIS. The case of $E_0 < E_a$ and $\hbar\omega > |E_0|$

the bound state and the antibound state. The numerical results imply that the origin of LIS is the shadow pole when the antibound state is located deeply in the well. This situation is realized when the condition $E_a < E_0 + \Delta$ is fulfilled, and the value of Δ is estimated to be in the range from 0.0167 to 0.0439 for the square-well potential model under question.

In the case of $\hbar\omega < |E_0|$, the numerical results imply that the origin of LIS is always the shadow pole of the bound state regardless to the energy position of the antibound states. Schematic illustrations are displayed in Fig. 2.11 for the case of $E_a \ll E_0$ and in Fig. 2.12 for the case of $E_0 \ll E_a < 0$.

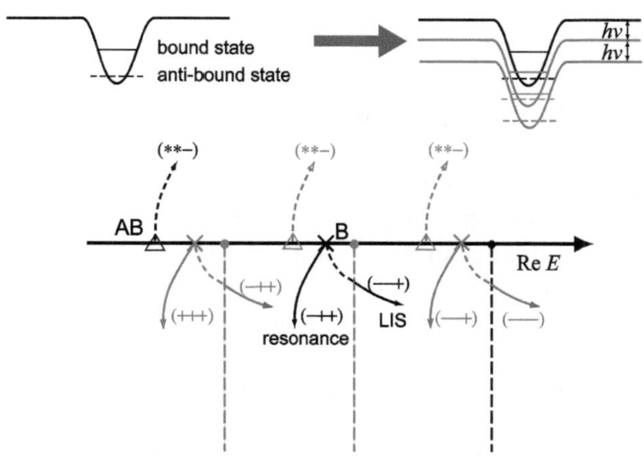

Fig. 2.11. Schematic illustration of pole trajectories indicating the origin of LIS. The case of $E_a < E_0$ and $\hbar\omega < |E_0|$

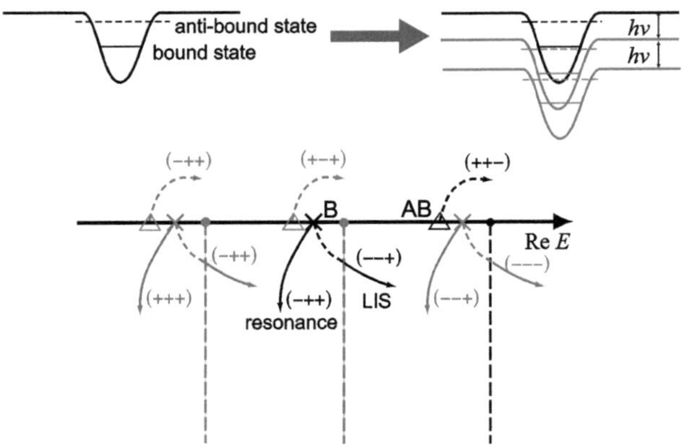

Fig. 2.12. Schematic illustration of pole trajectories indicating the origin of LIS. The case of $E_0 < E_a$ and $\hbar\omega < |E_0|$

2.2.4 Possibility of Experimental Observation

If we could measure photoelectron spectra in intense fields, LIS should be observed as peaks at the corresponding electron binding energy. Several authors have presented numerical simulation for the photoelectron spectra of LIS [20, 28]. In order to have sharp peaks, LIS should live sufficiently long.

On the other hand, a low appearance intensity for LIS is desirable from a practical point of view. The condition for formation of long lived LIS under a low appearance intensity can be inferred on the basis of the general behavior of pole trajectories. A longer-lived LIS appears in the case of $\hbar\omega > E_0$ than in the case of $\hbar\omega < E_0$. The appearance intensity increases with the detuning from the one-photon ionization threshold $|\hbar\omega - E_0|$. Consequently, the desirable experimental condition is obtained when one tune the laser frequency so that $\hbar\omega$ slightly exceeds E_0.

2.3 He–He Chemical Bonding in Intense Laser Fields

A class of phenomena originating from changes in the nature of chemical bonding in intense fields has been reported. Representative examples are the bond hardening/softening in H_2^+ [7–11], and the structural deformation of linear molecules such as CO_2 and CS_2 [12–14]. These phenomena have successfully been explained by the dressed-state picture. Based on the Floquet formalism, one considers the energy-shifted adiabatic potentials. The avoided crossing among them eventually causes the bond hardening/softening and the structural deformation. Such phenomena should be primarily ascribable to deformation of molecular orbitals in intense fields. Deformed atomic orbitals in intense fields are expected to form unusual molecular orbitals. A modified molecular orbital theory helps us to predict and understand the changes in chemical characters of molecules in intense fields. In this section, we discuss the chemical bonding of an He dimer in intense fields based on the molecular orbital calculation [15].

2.3.1 Molecular Orbital Calculation with the Kramers–Henneberger High-Frequency Approximation

With the Kramers–Henneberger (KH) high-frequency approximation, the effects of light fields caused on the electron motion are described by the KH effective potential. As described below, the effective potential can easily be included in the conventional quantum chemical computation.

Modification of the Ordinary Quantum Chemical Computation

Within the Born–Oppenheimer approximation, the electron wavefunction can be obtained as the eigenfunction of the Hamiltonian in (2.1), where $V(\boldsymbol{x})$ consists of the Coulomb interaction among electrons and nuclei,

$$H_{\text{el}} = \frac{1}{2}\sum_j \left(\boldsymbol{p}_j + \frac{F}{\omega}\boldsymbol{e}\cos\omega t\right)^2 - \sum_j\sum_a \frac{Z_a}{|\boldsymbol{r}_j - \boldsymbol{R}_a|} + \sum_{j<k}\frac{1}{|\boldsymbol{r}_j - \boldsymbol{r}_k|} \tag{2.20}$$

where $\boldsymbol{R}_a$ and Z_a are the position and charge of the ath nucleus, respectively. The use of the KH high-frequency approximation leads to the effective Hamiltonian

$$H_{\text{KH}} = \sum_j \frac{1}{2}\boldsymbol{p}_j^2 - \frac{\omega}{2\pi}\int_0^{2\pi/\omega} \sum_j\sum_a \frac{Z_a}{|\boldsymbol{r}_j - \boldsymbol{R}_a + \alpha\boldsymbol{e}\sin\omega t|}\mathrm{d}t + \sum_{j<k}\frac{1}{|\boldsymbol{r}_j - \boldsymbol{r}_k|} \tag{2.21}$$

The KH effective Hamiltonian differs from the usual molecular electronic Hamiltonian only in the nuclear attraction terms. The quantum chemical self-consistent field (SCF) calculation can be carried out by the use of usual quantum chemical computation codes with the nuclear attraction integrals modified. Great care is needed in preparation of basis set functions. Basis functions should be placed at each well of the KH effective potential as well as at the position of atoms. Details about the numerical technique and basis functions are presented in the literature [15].

Adiabatic Potential Curve of He–He

The adiabatic potential curves of He–He in intense fields calculated on the basis of the quantum chemical SCF method are shown in Fig. 2.13.

The values of the ponderomotive radius are $\alpha = 0$, 1.0 and 2.0. These values correspond to the laser intensities of 0, 35 and 140 PW cm^{-2}, respectively, in case that $\hbar\omega$ is set equal to 1. The polarization vector was set parallel to the molecular axis. The origin of energies is defined as the dissociation limit of each case.

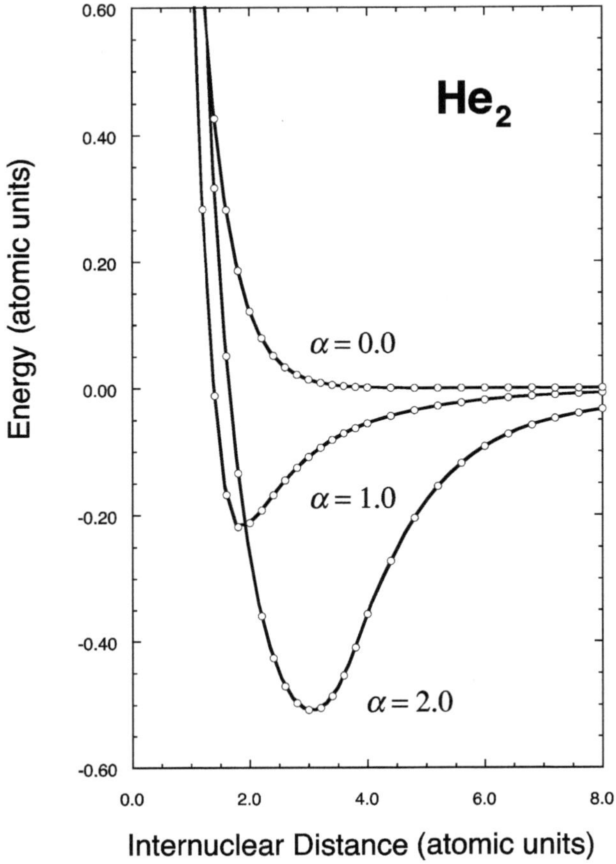

Fig. 2.13. Adiabatic potential curves for the He dimer in intense fields

The potential curves in intense fields are found to possesses a deep well as shown in Fig. 2.13. Since the SCF calculation does not reproduce any dispersion force, the observed attractive potential indicates that a chemical bond is created between two He atoms. It is noteworthy that the equilibrium distance is much longer than that of the ordinary H_2 molecule. In both the cases of $\alpha = 1.0$ and 2.0, the electron configuration of the ground state is $(1\sigma_g)^2(1\sigma_u)^2$ for any internuclear distance. This suggests that the creation of chemical bond is due to changes in the nature of molecular orbitals.

Changes in MO Characters

Figure 2.14 shows the changes in the shape of the occupied MO's with increasing laser intensity.

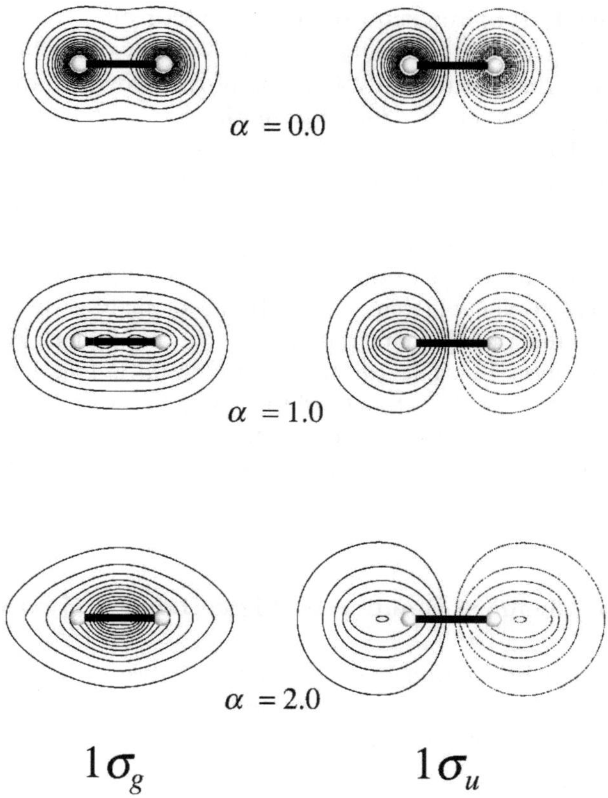

Fig. 2.14. Electron densities of occupied molecular orbitals. The contours of the electron density on the plane that includes the molecular axis are shown

The internuclear distance is fixed at $R = 3.0$. With increasing laser intensity, the maxima of the electron density of the $1\sigma_g$ MO shift towards the internuclear region, and the bonding character seems to be strengthened. On the other hand, the $1\sigma_u$ MO becomes diffuse with increasing laser intensity, and the anti-bonding character seems to be weakened. These change can be understood by hybridization of the atomic orbitals as discussed below.

2.3.2 Molecular Orbital Construction in Intense Fields

In general, the change in the nature of molecular orbitals can be discribed either by mixing among molecular orbitals or by changes in atomic orbitals. As discussed below, the latter description provides a simple picture for the issue under question.

Light-Induced Hybridization of Atomic Orbitals

The changes in the electron-wave function of the hydrogen atom have been investigated by Pont et al. [36]. Within the KH high-frequency approximation, the shape of the $1s$ wavefunction is prolonged along the polarization vector and its energy becomes higher. On the other hand, the energy of the $2p_z$ state becomes slightly lower due to the mixing mainly with the $2s$ wavefunction. The energy gap between $1s$ and $2p_z$ decreases, and this assists the s-p hybridization of these orbitals despite that these orbitals belong to the different shells. Detailed analyses of the contributions of these deformed atomic orbitals to the molecular orbitals are reported in reference [15]. The $1\sigma_g$ molecular orbital is found to consist not only of $1s$ but also of $2p_z$ atomic orbital (AO) of each He atom. The degree of the s-p hybridization reaches the maximum at around the equilibrium internuclear distance of He_2 molecule, implying that the s-p hybridization plays a significant role in the formation of the He–He chemical bond.

Creation of Chemical Bonding by Light-Induced Hybridization

The usual s-p hybridization, i.e., the mixing between $2s$ and $2p$ AOs, is known to strengthen σ bonds. Intense lasers are capable of inducing a different kind of s-p hybridization, i.e., the mixing between $1s$ and $2p$ AOs.

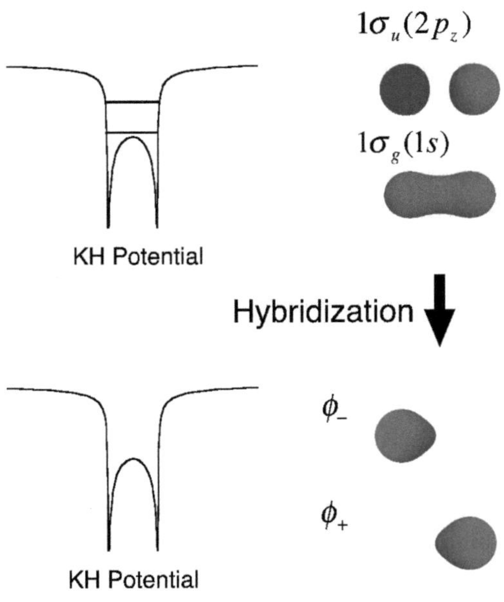

Fig. 2.15. Hybridization of the $1s$ and $2p_z$ atomic orbitals in intense fields

The mixing between $1s$ and $2p_z$ is a reflection of oscillation of the electron cloud along the laser polarization. The hybridization produces two lobes spanned toward the left and right sides of the He atom as illustrated in Fig. 2.15.

Molecular orbitals are constructed as the linear combination of the hybridized AOs as illustrated in Fig. 2.16. Two AOs from each He atom participate in the formation of MOs, and thus, four MOs, $1\sigma_g, 1\sigma_u, 2\sigma_g$ and $2\sigma_u$ are formed. These MOs possess following characters: The $1\sigma_g$ MO is a linear combination of the hybridized AOs spanning their lobes toward each other, and has high electron density concentrated in the internuclear region. Consequently, this MO has a strong bonding character. On the other hand, the electron distributions in the $1\sigma_u$ and $2\sigma_g$ MOs are diffuse, because these MOs are linear combination of AOs having their lobes outside the molecule, and consequently, these MOs have non-bonding character rather than anti-bonding one. The $2\sigma_u$ MO has a node in the middle of high electron-density region between the two atoms, and has anti-bonding character. In short, one strongly bonding MO, the two almost non-bonding MOs and one anti-bonding MO are formed as illustrated in Fig. 2.17. Four electrons occupy the bonding and one of the almost non-bonding MOs. In consequence, the He–He chemical bond is created.

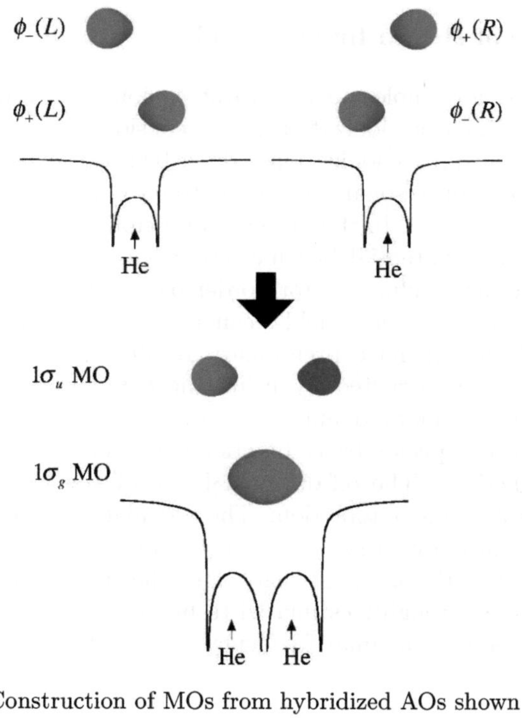

Fig. 2.16. Construction of MOs from hybridized AOs shown in Fig. 2.15

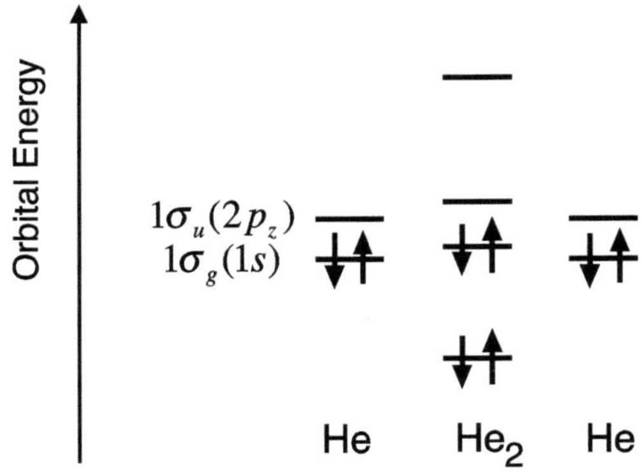

Fig. 2.17. The MO diagram for He molecule in intense fields

In the ordinary chemical bonding theory for the He–He system, two $1s$ AOs form one bonding and one anti-bonding MO. The anti-bonding MO is fully occupied, and the chemical bonding cannot be formed. It is, in some sense, the result of the shell structure of atoms. Intense fields destroy the shell structure and make $2p_z$ AO participate in. This causes the formation of non-bonding MOs and eventually leads to the creation of covalent bond between two He atoms.

2.3.3 Lifetime of He_2 in Intense Fields

In intense fields, any molecule is subject to ionization. In the KH high-frequency approximation, ionization is not considered. In the framework of the Floquet formalism, the ionization is described by the decay of resonances. The coupling between resonances and continua is ignored in the KH high-frequency approximation. In the presence of coupling, the bound states held on the KH effective potential become resonances with finite lifetimes. The KH bound states are useful as zeroth-order picture for resonances.

Lifetime of He_2 in intense fields, can be estimated on the basis of the Floquet formalism. We have preliminary results of such calculation. The Floquet matrix is constructed by using the configuration state functions generated by the ordinary quantum chemical computation. The basis set is augmented with special basis functions in order to describe the continuum wavefunctions. The off-diagonal block of the Floquet matrix contains the field-molecule interaction. The calculation shows that the lifetime of He_2 against ionization sensitively depends on the combination of the laser frequency, the laser intensity, and the internuclear distance. The maximum lifetime is roughly estimated to be around 1 picosecond, which is achieved when the laser angular frequency is $\omega \simeq 10$ and the ponderomotive radius $\alpha \simeq 1.0$.

2.3.4 New Chemical Bonding Theory in Intense Fields

Chemical characters of atoms are governed by the periodic law of elements. Two hydrogen atoms, when come close to each other, come to possess a chemical bond, but two helium atoms do not. The periodic law stems from the shell structure of atoms. Only the electrons from the outermost shell, i.e., the valence shell, participate in the formation of chemical bonding.

Intense fields are able to cause huge ac Stark effects, and eventually destroy the shell structure of atoms. Atoms come to have characters different from those expected from the periodic law. The bond creation between two helium atoms discussed above can be ascribed to the breakdown of the shell structure, i.e., the mixing between the K and L shells.

For the elements such as C, N, O, F and Ne, it is expected that $3s$, $3p$ and possibly $3d$ orbitals are allowed to participate in the formation of chemical bonds. Hybridization of these AOs into the valence $2s$ and $2p$ AOs gives rise to unusual chemical bonding and molecular structure. We emphasize that the scheme of deformed MO construction by light-induced hybridization helps to predict and understand possible unusual chemical species or unusual molecular structure in intense fields.

2.4 Concluding Remarks

High-frequency intense fields enable us to create novel quasi-bound states in atoms and molecules as discussed in Sect. 2.2. Creation of new states provides a novel way of thinking in the quantum control. The common paradigm of the quantum control stems from the π pulse-based coherent population transfer among given stationary states. Creation of new (quasi-)stationary states can add new solutions to the optimum control problems.

The optimum control scheme has successfully been applied to the chemical reaction control [37]. In the present stage, the control mainly concerns selective bond breaking in unimolecular dissociation. Control of bimolecular encounters and bond rearrangement should be challenging targets of the next phase of the research field [38]. The bond creation discussed in Sect. 2.3 can be a clue to open a new aspect of chemical reaction control. We emphasize the usefulness of molecular orbital consideration discussed in Sect. 2.3.2. Clarification of the scenario describing the construction of deformed molecular orbitals will be useful in the design of chemical reaction control by intense fields.

References

1. *Atoms in Intense Laser Fields*, ed by M. Gavrila (Academic, New York 1992)
2. *Molecules in Intense Laser Fields*, ed by A.D. Bandrauk (Dekker, New York, 1994)

3. M. Pont, N.R. Walet, M. Gavrila, C.W. McCurdy: Phys. Rev. Lett. **61**, 939 (1988)
4. M. Pont, M. Gavrila: Phys. Rev. Lett. **65**, 2362 (1990)
5. J.H. Eberly, K.C. Kulander: Science **262**, 1229 (1993)
6. R. Bhatt, B. Piraux, K. Burnett: Phys. Rev. A **37**, 98 (1988)
7. S.W. Allendorf, A. Szöke: Phys. Rev. A **44**, 518 (1991)
8. A. Zavriyev, P.H. Bucksbaum, J. Squier, F. Saline: Phys. Rev. Lett. **70**, 1077 (1993)
9. L.J. Frasinski, J.H. Posthumus, J. Plumridge, K. Codling, P.F. Taday, A.J. Langley: Phys. Rev. Lett. **83**, 3625 (1999)
10. P.H. Bucksbaum, A. Zavriyev, H.G. Muller, D.W. Schumacher: Phys. Rev. Lett. **64**, 1883 (1990)
11. A. Giusti-Suzor, F.H. Mies, L.F. DiMauro, E. Charron E, B. Yang: J. Phys. B **28**, 309 (1995)
12. A. Hishikawa, A. Iwamae, K. Yamanouchi: Phys. Rev. Lett. **83**, 1127 (1999)
13. H. Hasegawa, A. Hishikawa, K. Yamanouchi: Chem. Phys. Lett. **349**, 57 (2001)
14. A. Hishikawa, H. Hasegawa, K. Yamanouchi: Chem. Phys. Lett. **361**, 245 (2002)
15. T. Yasuike, K. Someda: J. Phys. B **37**, 3149 (2004)
16. L.V. Keldysh: Sov. Phys. JETP **20**, 1307 (1965)
17. H.A. Kramers: *Collected Scientific Papers* (North-Holland, Amsterdam 1956), p. 272
18. W.C. Henneberger: Phys. Rev. Lett. **21**, 838 (1968)
19. M. Gavrila, J.Z. Kaminski: Phys. Rev. Lett. **52**, 613 (1984)
20. T. Yasuike, K. Someda: Phys. Rev. A **66**, 053410 (2002)
21. T. Yasuike, K. Someda: Phys. Rev. A **71**, 017402 (2005)
22. M. Stroe, M. Boca: Phys. Rev. A **71**, 017401 (2005)
23. J.N. Bardsley, M.J. Comella: Phys. Rev. A **39**, 2252 (1989)
24. M. Dörr, R.M. Potvliege, D. Proulx, R. Shakeshaft: Phys. Rev. A **43**, 3729 (1991)
25. G. Yao, S.-I. Chu: Phys. Rev. A **45**, 6735 (1992)
26. H.G. Muller, M. Gavrila: Phys. Rev. Lett. **71**, 1693 (1993)
27. A.S. Fearnside, R.M. Potvliege, R. Shakeshaft: Phys. Rev. A **51**, 1471 (1995)
28. J.C. Wells, I. Simbotin, M. Gavrila: Phys. Rev. Lett. **80**, 3479 (1998)
29. J.R. Taylor: *Scattering Theory* (Wiley, New York 1972)
30. J.H. Shirley: Phys. Rev. B **138**, B979 (1965)
31. H. Sambe: Phys. Rev. A **7**, 2203 (1973)
32. U. Peskin, N. Moiseyev: Phys. Rev. A **49**, 3712 (1994)
33. A.J. F. Siegert: Phys. Rev. **56**, 750 (1939)
34. R.J. Eden, J.R. Taylor: Phys. Rev. **133**, B1575 (1964)
35. R.M. Potvliege, R. Shakeshaft: Phys. Rev. A **38**, 6190 (1988)
36. M. Pont, N.R. Walet, M. Gavrila: Phys. Rev. A **41**, 477 (1990)
37. R.J. Levis, G.M. Menkir, H. Rabitz: Science **292**, 709 (2001)
38. R. Itakura, K. Yamanouchi, T. Yasuike, K. Someda: Chem. Phys. Lett. **386** 208 (2004)

3 Multielectron Effects of Diatomic Molecules in Strong Laser Fields

Chunlei Guo

The Institute of Optics, University of Rochester, Rochester, NY 14627 USA
guo@optics.rochester.edu

Summary. The study of atoms and molecules in strong laser fields is one of the most active research areas in Atomic, Molecular, and Optical Physics. Following extensive study over the past two decades, most phenomena involving strong field-atom interactions have been relatively well understood. However, there have been a number of observations suggesting that molecules in strong fields behave significantly differently from what we have learned from atoms. Revised concepts are discussed in this article for phenomena as simple as single- and double-electron ionization in molecules. These findings provide a more uniform picture to understand behaviors of atoms and molecules in strong laser fields.

3.1 Introduction

With recent advances in high-intensity femtosecond (fs : $1\,\text{fs} = 10^{-15}\,\text{s}$) laser techniques, fs lasers can readily generate a high electric field with a strength comparable to or even exceeding the Coulomb binding potential that holds an atom or molecule together [1–6]. In this high field, light-atom/molecule interactions can be significantly different from those in weak electromagnetic fields and many interesting phenomena began to be observed, such as above-threshold ionization [7], non-sequential ionization [8,9], high harmonic generation [10], and attosecond pulse generation [11]. Fundamentally speaking, all the above-mentioned phenomena are directly or indirectly related to electron ionization since ionization almost always occurs when electrons in atoms or molecules are exposed to such a strong laser field.

Most effects in single-electron and sequential multiple-electron ionization of rare gas atoms can be relatively well understood by the single-active-electron (SAE) approximation [12]. Time-dependent quantum mechanical calculations using the SAE approximation have been shown to provide accurate single- and sequential multiple-ionization rates, above-threshold ionization yields, high-harmonic spectra, and angular distributions for rare gas atoms in strong laser fields [12–14]. In the tunneling regime [15], the SAE-based Ammosov–Delone–Krainov (ADK) tunneling model also provides an accurate fit to single- and sequential multiple-electron ionization rates of rare gas atoms [16]. The model is relatively simple: different atoms are characterized by their ionization potentials Ip and the effective quantum numbers obtained from weak field measurements; apparently, the details of the electronic structure are unimportant. However, most of the subjects studied extensively are the rare gas atoms that all have similar closed-shell electronic structures [17].

Molecules can potentially provide important tests of strong field dynamics as even easily studied diatomic molecules have a greater diversity of electronic structures. Despite the success in explaining strong-field ionization in rare gas atoms, the interpretation of these responses in molecules has encountered great difficulties. In this article, we will review some of our work in extending the fundamental understanding of strong-field phenomena from atoms to molecules. Multielectron collective effects are shown to play an important role in understanding these behaviors. Note, in this article, we will focus on discussing molecular phenomena that are electronic in nature and unrelated to internuclear dynamics. Molecular behaviors also related to nuclear motion have been studied extensively [18–22].

3.2 Technical Background

Strong-field experiments to be reviewed in this article are carried out using high-intensity femtosecond laser systems. An efficient way to generate the strong laser fields is to use ultrashort laser pulses. In principle, an unlimited temporal energy concentration can produce an unlimited electromagnetic field strength with a finite amount of input energy. With recent advances in laser technology, a tabletop laser source can readily generate an ultrashort laser pulse with a duration less than 10 fs [23, 24]. Since the spectral bandwidth and the duration of a pulse are related by the inequality $\Delta t \Delta \omega \geq 1/2$, a broad bandwidth lasing material is required to generate the ultrashort laser pulses. In the past fifteen years, one of the most spectacular advances in the field of ultrashort light pulse generation was the development of titanium-doped aluminum oxide (Ti:sapphire) as a laser gain medium [25]. The Ti:sapphire crystal is produced by substituting the Al^{3+} ion in the sapphire structure by the Ti^{3+} ion. The absorption band of Ti^{3+} ion is in the blue-green part of the visible spectrum, while the emission band of Ti:sapphire is shifted towards lower energies ($\sim$780 nm) and the width is as large as 200 nm. In principle, a $\sim$ 5 fs Gaussian-shape pulse can be generated with this bandwidth. Besides, synthetic sapphire also has a thermal conductivity as high as a metal, which allows for the high power pumping.

The way to obtain the ultrashort laser pulses is using mode-locking techniques. For Ti:sapphire laser, self mode-locking is the most commonly employed method. Self mode-locking is based on the Kerr lens effects, which induce a better spatial mode matching between the lasing beam and the pump beam inside the Ti:sapphire crystal at a higher intensity, thus providing a more favorable lasing condition for the short-pulse emission. Pulses generated from a femtosecond oscillator usually only have pulse energy of a few nJ. To achieve high field strength, the energy of a single pulse needs to be further amplified. To avoid damaging of the amplifier crystal and efficiently extract its stored energy without incurring serious nonlinear effects, a chirped-pulse-amplification technique is commonly employed [26]. The essence of the chirped-pulse-amplification technique is that the pulse duration of the ultrashort pulses from the oscillator is first broadened by a stretcher and then amplified to a sufficiently high pulse energy (e.g., mJ level) before compressed down to the original pulse duration with a compressor.

Light-molecule interactions are usually studied inside a vacuum chamber. In the work to be reviewed below, laser beam is focused inside an ultra-high-vacuum chamber filled with a certain dilute molecular gas. Molecular ions created at the laser focus are extracted by a dc field and accelerated towards a microchannel plate detector at the end of a standard time-of-flight (TOF) mass spectrometer [27]. The signal is further amplified, discriminated, and either integrated with a boxcar to produce ion yields or sent to a time digitizer to generate TOF mass spectra. Ions are separated by their different mass to charge ratio, m/q, in a TOF spectrum. For molecules, dissociating ions with the same m/q ratio can also be separated by their different initial kinetic energy due to the dissociation. The dissociating molecular spectra are usually symmetric around the arrival time of a zero-kinetic energy ion with the two peaks representing ions initially directed towards or away from the detector. To illustrate this, Fig. 3.1 shows a typical TOF mass spectrum containing metastable channel N_2^{2+} and the dissociating N^+ signals from both the $N^+ + N^+$ and $N^+ + N$ channels.

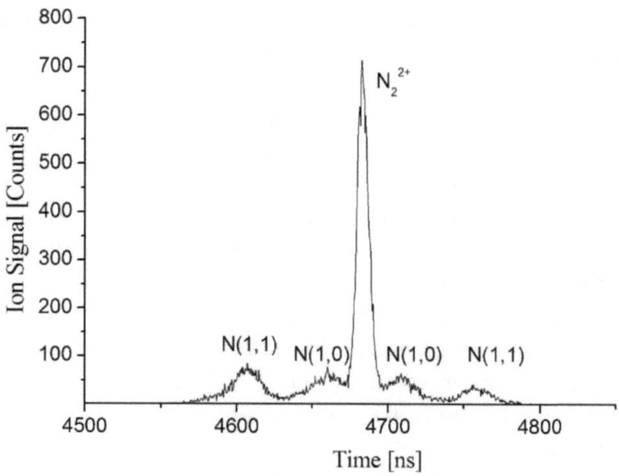

Fig. 3.1. A typical TOF mass spectrum showing metastable channel N_2^{2+} and the dissociating N^+ signals from both the $N^+ + N^+$ [labeled as N(1,1)] and $N^+ + N$ channels [labeled as N(1,0)]

3.3 Diatomic Molecules in Strong Fields

In this article, we will review some of our work in extending the fundamental understanding of strong-field phenomena from atoms to molecules. Some of the most fundamental questions in molecular responses will be addressed, such as the physical mechanisms of single and double ionization in molecules. Multielectron collective effects are shown to play an important role in understanding these behaviors. This article starts with the discussion of single electron ionization in molecules. Then both experimental and theoretical studies of nonsequential double ionization are reviewed. The article concludes with a discussion of multielectron effects in molecular dissociation.

3.3.1 Single Electron Ionization of Molecules

In the past, some studies have provided evidence that many aspects of strong field ionization in molecules are similar to those in atoms. In the multiphoton regime [15], photoelectron spectroscopy of N_2 showed atomic-like multiphoton resonance similar to Ar [28]. In the tunneling regime, several simple molecules have been studied with 10.6-μm CO_2 laser radiation, and the ion yields were found to agree with the ADK model [29]. This implies that in the absence of vibrational resonance with the CO_2 laser radiation, 'simple' molecules are ionized through tunneling as if they were structureless atoms with an ionization potential equivalent to that of the molecular ground state [29]. More recently, however, with 800-nm Ti:sapphire laser radiation, O_2^+ was found to have a significantly lower ion yield signal compared to Xe^+ [17,30], as shown in Fig. 3.2, although O_2 has virtually the same single-ionization potential as Xe ($\sim 12\,eV$). Furthermore, it has been further verified that Xe^+ behaves normally and agrees with the prediction of the ADK model, but O_2^+ has an anomalously low single ionization yield [17].

At first glance, the anomalous behavior of O_2 may be due to the fact that O_2 is a molecule and is different from rare gas atom Xe. However, in considering a list of general differences between atoms and molecules, such as dissociative recombination, other dissociative process, the Stark shift of the ground state, accidental resonances, and electron localization, no effect can explain the suppression of the O_2^+ signal [17]. Furthermore, comparison study was also carried out for diatomic molecule N_2 versus rare gas atom Ar as N_2 has the same single-ionization potential as Ar ($\sim 15.6\,eV$) [17]. The low ion signal is not observed in N_2, i.e., N_2 behaves like the rare gas atom Ar with the same single-ionization potential, as shown in Fig. 3.3 [17]. This indicates that the peculiar behavior of O_2 is not a simple consequence of its being a molecular. Rather, the fact that N_2 behaves like a structureless atom, while O_2 shows distinct non-atomic-like behavior, indicates that details of electronic structure play key roles in influencing strong field excitation

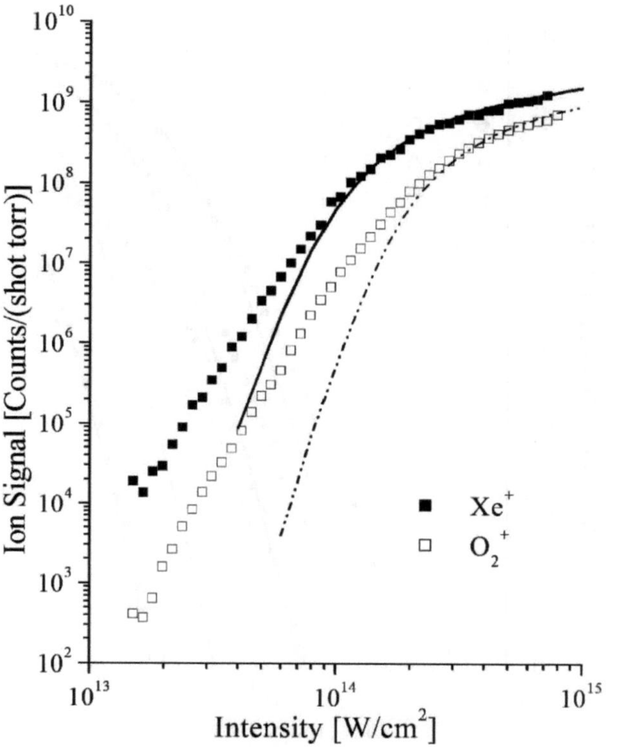

Fig. 3.2. Single ionization yields for Xe and O_2 in linearly polarization light (from [17]). The *full curve* is the ADK-model calculations using single ionization potential obtained in weak fields. The *dash dotted curve* is the ADK model calculations uses an effective charge z* of 1.4 and an effective ionization potential I* of 16.9 eV [33]

and ionization, since N_2 has a closed shell electronic structure, while the outermost orbital $(1\pi_g)^2$ of the ground state O_2 is only half filled [17].

Recently, extensive efforts have been devoted to developing suitable theoretical models to explain the anomalous ionization behavior in O_2 [30–35]. However, only a few models are able to quantitatively account for the observed suppression in O_2^+ [33–35], and these few models are all based on the original concept proposed in [17] that the detailed electronic structures play the dominant role in influencing electron ionization dynamics. Among the more quantitative models, an intuitive model explains the suppressed O_2^+ by taking into account subtle factors related to multielectron effects [33]. This concept and a corresponding model developed allow a generalization of the ADK model for predicting strong field ionization rates of more complicated atoms and molecules, including the suppression in O_2^+. As elaborated in [33], ionization in strong laser fields is a fundamentally different process compared to that in weak fields. The major difference, resulting from the distinctly dif-

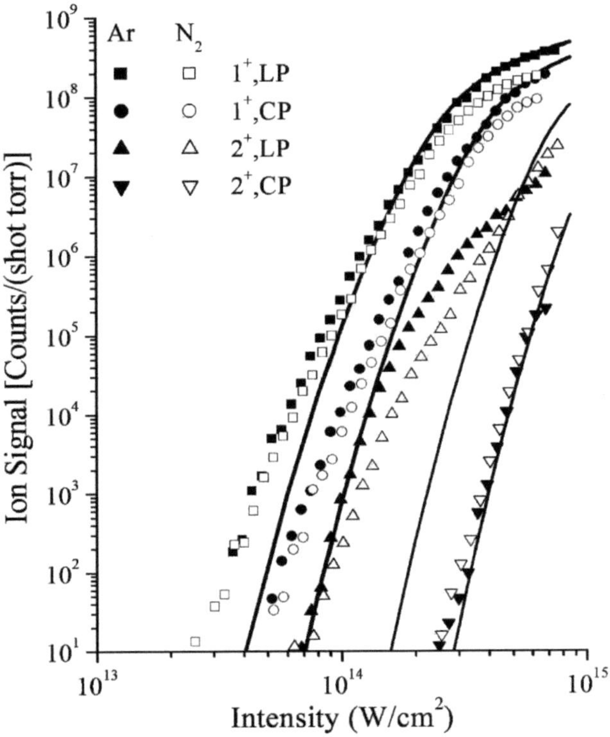

Fig. 3.3. Single- and double-ionization yields for N_2 and Ar in linearly and circularly polarized light (from [17])

ferent ionization process between strong-field versus weak-field experiments, is the strength of electron screening of the ion core. The difficulty in fitting O_2^+ using the ADK model comes from the need to introduce an effective charge and ionization potential for some atoms or molecules, like O_2, which are not necessarily the same as the weak field values due to the fundamentally different electron screening process. Again, the key factor here will be the detailed electronic structures. For example, N_2 and O_2 have different electronic structures and thus different values of effective charge and ionization potential. Using a deduced effective charge z^* of 1.4 and an effective ionization potential I^* of 16.9 eV, the ADK calculations show excellent agreement with the experimental observation in the tunneling regime, see Fig. 3.2 [33]. The uniqueness of O_2 stems from the fact that O_2 has an open-shell electronic structure. Thus, according to the model, we do not expect the same level of suppression for closed-shell molecules, such as N_2 and F_2, that have a more uniform distribution of outer electrons. Note, some previous interpretations [36] of this model on the closed-shell molecules were not always accurate .

3.3.2 Nonsequential Double Ionization of Molecules

A clear deviation from single-active-electron approximation in strong fields can be more easily seen from the so-called nonsequential double ionization (NSDI) [8, 9, 17]. NSDI of atoms and molecules is one of the most interesting problems in strong-field physics. NSDI normally involves an enhancement in the double ionization yields compared to those based on sequential rates. In a linearly polarized field, rare gas atoms and some molecules exhibit an unambiguous enhancement in double ionization yields compared to calculations based on the SAE approximation, which has accurately reproduced the single and sequential multiple ionization rates [8, 9]. This enhancement can be seen in Fig. 3.3: N_2 and Ar have almost the same single and sequential double ionization potential. Despite a good fit of single ion yield to both N_2 and Ar by the SAE-based ADK model, an enhancement (a knee structure) is clearly seen in the double ionization yields. It is commonly believed that the enhanced ionization is due to electron–electron correlation, which leads to nonsequential ionization, namely a simultaneous double electron emission instead of a stepwise process [8, 9].

Two theoretical models have been proposed in order to explain the experimental observations. On the one hand, the so-called shake-off model suggests that as one electron is ionized, another electron could be shaken off into the continuum due to the rearrangement caused by the sudden removal of the first electron [8]. However, this shake-off model is still not fully tested due to the extreme demands of a realistic calculation. On the other hand, the 'rescattering' model states that an electron can collide with its parent ion core one half cycle after being ionized by the laser field, and this e-2e rescattering could lead to an enhanced double ionization [37].

Similar to the case of single electron ionization discussed above, double ionization also exhibits a greater complexity when our study evolves from atoms to molecules. We will again discuss NSDI behaviors by studying diatomic molecules N_2 and O_2. Generally, NSDI is characterized by a knee structure in double ionization yield, but this cannot quantitatively measure the rate of NSDI. The intensity dependence of the ratio X^{2+}/X^{+} has been used to indicate NSDI [9, 17], where X^{+} and X^{2+} are the yields for single and double ionization of some atom or molecule, X. Non-sequential dynamics are most fundamentally related to this ratio: if this ratio is only weakly dependent on intensity, it shows that the precursor to X^{2+} is not X^{+}, and there should be a two-electron simultaneous emission component in X^{2+}.

Ratio curves of X^{2+}/X^{+} are plotted for N_2, O_2, and Ar in linearly polarized light in Fig. 3.4. N_2 and Ar have similar double ionization potential ($\sim 27\,\text{eV}$), and the ratio curves for these species are also very close. In the figure, the ratios for N_2 and Ar approach the sequential ion yield of X^{2+} when the average intensity is higher than $3 \times 10^{14}\,\text{W/cm}^2$, where the ionization to X^{+} has saturated. The weak dependence of the ratio on intensity in the intensity range of $1.2\times10^{14}\,\text{W/cm}^2$ to $3\times10^{14}\,\text{W/cm}^2$, reflects the NS aspect

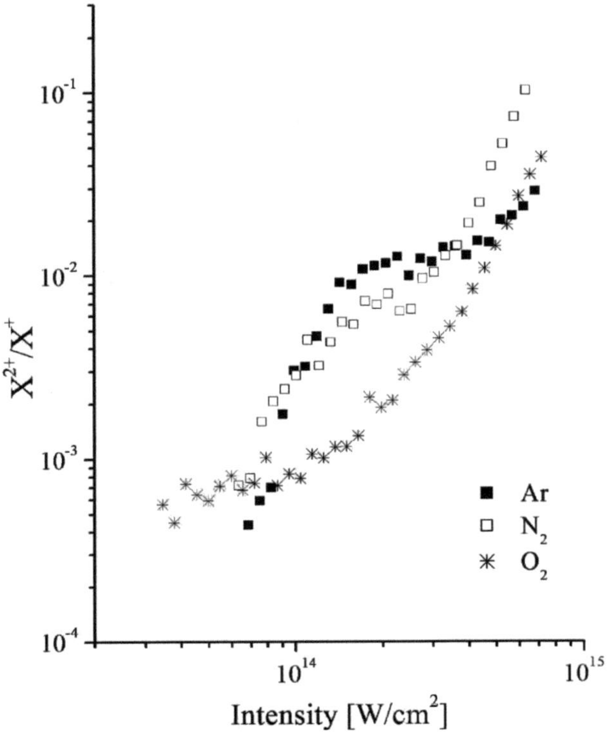

Fig. 3.4. Ratio curves of double/single ion yield for N_2, O_2, and Ar in linearly polarized light (from [17])

of double ionization. The overall agreement of single- and double-ionization between N_2 and Ar indicates that N_2 tunnel ionizes as if it is a structureless atom. From the ratio curve, we can see that O_2^{2+} has a distinctly reduced NSDI rate compared to N_2 and Ar (nearly an order of magnitude lower). The different NSDI rates in N_2^{2+} and Ar^{2+} vs. O_2^{2+} will again have to be related to the different electronic configurations in N_2 and O_2: N_2 has a closed shell electronic structure, while the outermost orbital $(1\pi_g)^2$ of the ground state of O_2 is only half filled [17]. NSDI has also been studied systematically in various dissociation channels of N_2 and O_2, and we refer the readers to [38] for detailed discussion.

All the above experimental evidence indicates that the detailed electronic structures play a key role leading to the different ionization behaviors in N_2 versus O_2 since N_2 is a closed-shell molecule, while O_2 is an open-shell molecule. Perhaps more significantly, the outermost two-electron wavefunction is spatially symmetric in N_2 but anti-symmetric in O_2. Motivated by these experimental observations, further investigation has also been performed theoretically. In a study reported in [39], two-electron model atoms are constructed with either a spatially symmetric electronic wavefunction (SSW) or a spatially anti-symmetric wavefunction (SAW) that correspond approximately to N_2 and O_2 molecules. The calculations are carried out by numerically solving the full time-dependent Schrödinger equation with the presence of a strong external laser field.

For a two-electron system, models are often used to study NSDI in a 1D or 3D atomic system [40, 41], but an exact treatment is always desirable in order to reveal the underlying physics without any assumptions [42, 43]. However, the *ab initio* calculations are mostly limited to 1D due to the heavy computational demands at higher dimensions. For example, it has been argued that 3D calculation on two fully correlated electrons is at the limit of tera-scale supercomputing power [44]. 1D calculation, on the other hand, has been routinely used because it represents well the physics of a 3D system. For example, a commonly used 1D soft-core potential eliminates the 1D singularity at the origin and also reproduces the 3D like Ryberg spacing of highly excited states [42]. In a 1D two-electron system, the total Hamiltonian including full electron–electron interaction and an external laser field is [in atomic units, (a.u.)],

$$
\begin{aligned}
H = & -\frac{1}{2}\frac{\partial^2}{\partial x^2} - \frac{1}{2}\frac{\partial^2}{\partial y^2} - \frac{z}{\sqrt{x^2+a^2}} - \frac{z}{\sqrt{y^2+a^2}} \\
& + \frac{1}{\sqrt{(x-y)^2+a^2}} + E(t)(x+y)
\end{aligned}
\tag{3.1}
$$

where x and y are the coordinates of electron 1 and 2, respectively. z is the charge of the atomic core, and the parameter 'a' controls the width of the potential well. $E(t)$ is the external laser field strength, and the interaction between the atom and external field is based on the dipole coupling. Evolution dynamics of the two-electron wavefunction exposed to an external field can be obtained by solving the Schrödinger's equation with the above Hamiltonian.

For the first time, an *ab initio* two-electron quantum mechanical calculation is performed in 1D on both SSW and SAW atoms [39]. In these calculations, single, sequential double, and nonsequential double ionization rate can be each obtained for both SSW and SAW atoms. Figure 3.5 shows a plot of the non-sequential component versus total double ionization production. Below the intensity where the single ion yield starts to saturate, NSDI component is higher in the double ionization for the SSW atom than SAW atom. This is consistent with the previous experimental observation that N_2 has a higher NSDI rate than O_2 [17].

In the past, extensive study of NSDI in rare gas atoms has also established a variety of properties for NSDI. One of the signature properties of NSDI is its ellipticity effect, where the NSDI rate is relatively high for linear polarization (LP), but appears to be completely suppressed with circularly polarized light (CP) [9, 17]. This can be seen in Fig. 3.3: despite the significant 'knee' structures in N_2^{2+} and Ar^{2+} with LP characterizing the NSDI component,

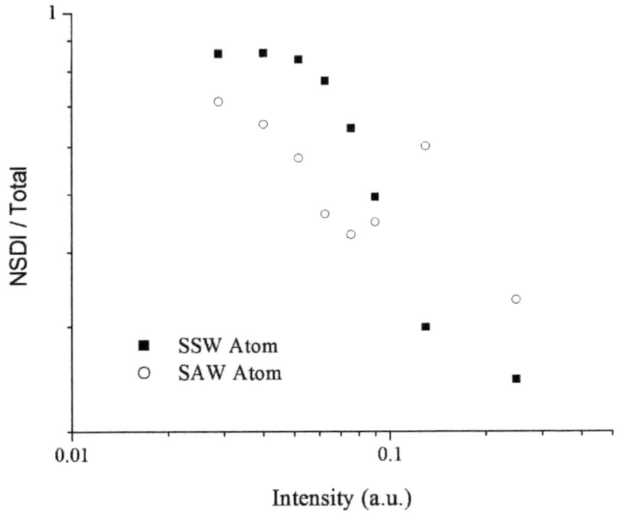

Fig. 3.5. The fraction of the nonsequential component in the total double ionization for both the SSW and SAW atoms from numerical results (from [39])

the ion yields with CP drop off rapidly with decreasing intensity showing an excellent agreement with the ADK-model calculations [16]. However, most NSDI properties, including the ellipticity effects, are established based on the extensive study of rare gas atoms that all have similar closed-shell electronic structures. When the study of NSDI is extended to molecules, some of these properties may need further examination. Recently, an anomalously high NO^{2+} signal with circularly polarized light has been observed. This indicates, for the first time, a non-sequential ionization signature with CP [45]. This can be seen in Fig. 3.6: the NO^{2+} signal with CP does not fall off rapidly with decreasing intensity as in N_2^{2+} and Ar^{2+}; instead, the ion yield of NO^{2+} with CP is relatively high at lower intensities showing the knee structure commonly seen in LP yields. As mentioned above, the rescattering model has been used to explain NSDI. In the rescattering model, the probability for an electron to rescatter off its atomic parent core will be greatly reduced in a CP field since there will be a significant transverse motion ($\sim$ 50 a.u. [46]), leading to a nearly vanishing possibility for e-2e rescattering even for a large atom or molecule [45,46]. Therefore, the observation of a clear NSDI component in NO with CP seems to be at odds with the rescattering model. The details of the experimental observation are discussed in [45], and the physical mechanism behind this observation still needs further investigation.

3.3.3 Molecular Dissociation

In strong fields, unlike an atom that only undergoes excitation and ionization, the interactions in molecules can also lead to dissociation fragments.

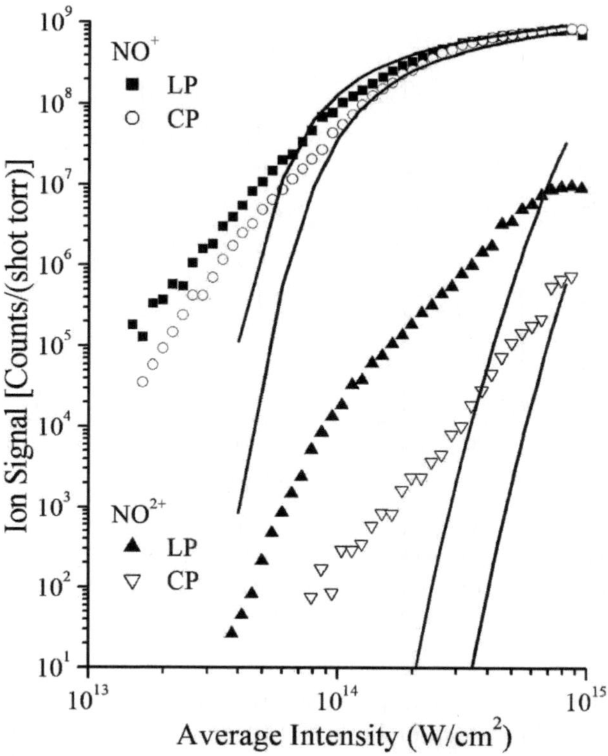

Fig. 3.6. Single- and double-ionization yields for NO in linearly and circularly polarization light. Calculations based on the ADK model are also shown by the *full curves* (from [45])

Most of the previous work in studying multielectron effects in molecules has mainly focused on electron ionization [17,33–36,39]. In this section, we extend our discussion to the subsequent dissociation channels in molecules. A heteronuclear diatomic molecule, nitric oxide (NO), will be used in the following experiments. Three chemical pathways exist following single electron ionization of NO: one ionic channel (NO^+) and two dissociation channels, $N^+ + O$ [labeled as N(1,0)], and $N + O^+$ [labeled as O(1,0)].

Figure 3.7 shows a TOF mass spectrum taken at different intensities across the delay range corresponding to the N(1,0) and O(1,0) channels. We can see that both N(1,0) and O(1,0) have a double peak structure corresponding to the ions with initial velocity towards and away from the MCP detector at the end of the TOF drift tube. Laser polarization in the experiment is parallel to the TOF axis. The clear double-peak structure at lower intensities indicates that the molecules oriented along the laser polarization have the highest ionization and dissociation probability since electron ionization needs to overcome the least amount of potential energy along the molecular axis,

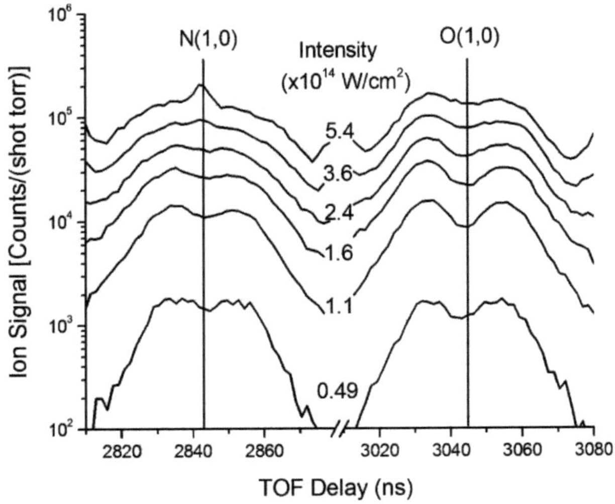

Fig. 3.7. TOF mass spectra of the N^+ + O and N + O^+ channels in linearly polarized laser light at different intensities. The laser intensity increases by a constant ratio of 1.5 from the *second lowest curve* (1.1×10^{14} W/cm^2) to the *highest one* (5.4×10^{14} W/cm^2) (from [47])

due to a double-well molecular potential structure [18–22]. The central regions between the double peaks of both N(1,0) and O(1,0) channels correspond to ions with nearly zero initial kinetic energy along the TOF axis. These regions have been shown resulting from molecules oriented perpendicular to the laser polarization [47]. From Fig. 3.7, one observation is that the central part of N(1,0) grows faster than that of O(1,0), despite the fact that both channels are from the same precursor NO^+. Therefore, for molecules oriented perpendicular to the laser polarization, it appears that NO^+, if it dissociates, will preferably dissociate into N(1,0) rather than O(1,0). This is interesting, because the energy level of the former channel is actually higher than the latter channel by about 1 eV. To understand this, a more careful examination of strong-field induced dissociation process for NO^+ is necessary.

As discussed in [47], the formation of the dissociation channels following single electron ionization involves removing an electron from the inner $(1\pi)^4$ binding orbital. Pictorially speaking, if we remove an electron closer to N, the N(1,0) channel should be more likely to result; if we remove an electron closer to O, the O(1,0) channel should be more likely to result. It is well known that the O atom has a greater electron affinity than the N atom and therefore, in NO molecules, the electrons in the 1π binding orbital between the N and O cores should shift towards the O atom. Thus, a 1π electron closer to the N atom is less bound to the overall core potential than a 1π electron closer to the O atom. Therefore, a 1π electron closer to the N atom should be ionized more easily, and the removal of this N side electron should result in the N(1,0) channel, as discussed earlier. This explains why the N(1,0) channel is easier to reach than the O(1,0) channel in strong fields. This binding energy difference mainly influences electron ionization perpendicular to the molecular axis since, unlike the ionization along the molecular axis, the electron does not need to move through a double-well molecular potential that will complicate the ionization process [18–22]. However, this observation appears to differ drastically from what we may expect based on energy level analysis, where the N(1,0) channel is supposed to be harder to reach than the O(1,0) channel.

This apparently anomalous charge separation following single-electron ionization in NO again may be explained by considering the distinctly different ionization dynamics in strong fields compared to in weak fields. For strong-field ionization, the initial electronic structure plays a more crucial role. The energy level of different channels, on the other hand, is based on experimental observation in weak fields under conditions where ionization typically is a relatively slow process, as discussed in detail in [33]. This different ionization dynamics between strong fields versus weak fields lead to the above discussed observation: the O(1,0) channel is more energetically favorable in the weak field case, but the reversed case seems to be more applicable in strong-field ionization [47]. This experiment suggests another evidence to reinforce our understanding of the fundamental differences between ionization dynamics in strong fields versus weak fields [33].

Molecular dissociation has also been studied following double ionization in NO molecules. One interesting study is a recent observation of channel competition between the doubly charged metastable state NO^{2+} and a dissociation channel $N^+ + O^+$ [labeled as N(1,1)]. The dissociation channel N(1,1) requires more energy to reach than the metastable channel NO^{2+}. Experimentally, the branching ratio of N(1,1)/[NO^{2+} + N(1,1)] versus intensity is measured for both LP and CP, as shown in Fig. 3.8 [48]. We can see that the ratio curve for CP follows our analysis: as intensity increases, the branching ratio for the dissociation channel N(1,1) steadily increases, in consistent with the fact that the dissociation channel requires more energy to reach than the metastable channel. In linearly polarized light, however, this branching ratio shows a significantly different pattern. Although the ratio curve shows the same intensity dependence as CP at higher intensities, it shows the opposite dependence for intensities below 6×10^{13} W/cm^2: the branching ratio of N(1,1) decreases dramatically with intensity. This observation deviates from what we would expect from the energy analysis of NO^{2+} and N(1,1) channels. The physical mechanism of this observation again relates to multielectron effects in double ionization of NO. At lower intensities, nonsequential double ionization plays a more significant role in linearly polarized light. As elaborated in [48], nonsequential double ionization contributes more in producing dissociation channel $N^+ + O^+$ than metastable channel NO^{2+}. Thus, we see a reversed intensity dependence of the LP branching ratio at lower intensities. This LP branching ratio curve deviates from the simple energy

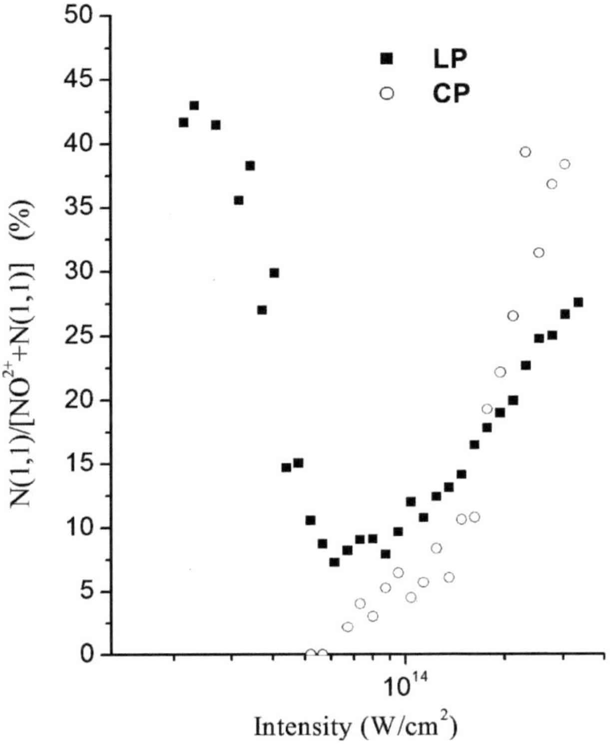

Fig. 3.8. Intensity dependent branching ratios of N(1,1)/[NO^{2+} + N(1,1)]. Note, N^+ data in the N^+ +O^+ dissociation channel is used to represent the N(1,1) channel although O^+ can be equally used as well. LP and CP denote linear and circular polarization, respectively (from [48])

analysis and is the result of two-electron collective effects. The study presented in this subsection demonstrates that multielectron collective effects also play an important role in molecular dissociation.

3.4 Summary

During the past two decades, most phenomena involving strong field-atom interactions have been relatively well understood. However, molecules in strong fields show a greater level of complexity and challenge what we have learned from atoms. In this article, we review some of the most fundamental aspects when we extend our understanding of strong-field phenomena from atoms to molecules. Experimentally, 1-kHz high-repetition-rate Ti:sapphire femtosecond laser systems are employed for various experimental studies reviewed in this article. Theoretically, *ab initio* quantum mechanical calculations are performed to investigate the key factors in nonsequential double ionization.

Following these studies, a general picture is formulated in understanding small molecules in strong laser fields and multielectron collective effects are shown to play an important role in various molecular behaviors. Despite the progresses we have made, strong field-molecule interactions are a complicated problem and further experimental and theoretical investigations are needed for us to fully understand the behaviors of atoms and molecules in strong laser fields.

References

1. M.D. Perry and G. Mourou: Science **264**, 917 (1994).
2. L.F. DiMauro and P. Agostini: in Adv. At. Mol. Opt. Phys. **35**, edited by B. Bederson and H. Walther (Academic, San Diego, 1995).
3. P. Salieres, A. L'Huillier: Ph. Antoine and M. Lewenstein: Adv. At, Mot. Phys. **41**, 83 (1999).
4. Edited by M. Gavrila: *Atoms in Intense Laser Fields* (Academic, New York, 1992).
5. M. Protopapas, C.H. Keitel, and P.L. Knight: Rep. Prog. Phys. **60**, 389 (1997).
6. T. Brabec and F. Krausz: Rev. Mod. Phys. **72**, 545 (2000).
7. P. Agostini, F. Fabre, G. Mainfray, G. Petite and N. Rahman: Phys. Rev. Lett. **42**, 1127 (1979).
8. D.N. Fittinghoff, P.R. Bolton, B. Chang, and K.C. Kulander: Phys. Rev. Lett. **69**, 2642 (1992).
9. B. Walker, B. Sheehy, L.F. DiMauro, P. Agostini, K.J. Schafer, and K.C. Kulander: Phys. Rev. Lett. **73**, 1227 (1994).
10. A. McPherson, G. Gibson, H. Jara, T.S. Luk, I.A. McIntyre, K. Boyer and C.K. Rhodes: J. Opt. Soc. Am. B, **4**, 595 (1987).
11. M. Hentschel, R. Kienberger, Ch. Spielmann, G.A. Reider, N. Milosevic, T. Brabec, P. Corkum, U. Heinzmann, M. Drescher, and F. Krausz: Nature **414**, 509 (2001).
12. K.C. Kulander: Phys. Rev. A **35**, 445 (1987).
13. K.J. Schafer, B. Yang, L.F. DiMauro, and K.C. Kulander: Phys. Rev. Lett. **70**, 1599 (1993).
14. B. Yang, K.J. Schafer, B. Walker, K.C. Kulander, P. Agostini, and L.F. DiMauro: Phys. Rev. Lett. **71**, 3770 (1993).
15. L.V. Keldysh: Zh. Eksp. Teor. Fiz. **47**, 1945 (1964) [Sov. Phys. JETP **20**, 1307 (1965)].
16. M.V. Ammosov, N.B. Delone, and V.P. Krainov: Zh. Eksp. Teor. Fiz. **64**, 2008 (1986) [Sov. Phys. JETP **64**, 1191 (1986)].
17. C. Guo, M. Li, J.P. Nibarger, and G.N. Gibson, Phys. Rev. A (Rapid Comm.) **58**, R4271 (1998).
18. A.D. Bandrauk, *Molecules in laser fields* (M. Dekker Pub, N.Y. 1994); A.D. Bandrauk, Comments At. Mol. Phys., **22**, 325 (1989).
19. L.J. Frasinski, K. Codling, P. Hatherly, J. Barr, I.N. Ross, and W.T. Toner: Phys. Rev. Lett. **58**, 2424 (1987).
20. A. Cornaggia, J. Lavancier, D. Normand, J. Morellec, and H.X. Liu: Phys. Rev. A **42**, 5464 (1990).

21. T. Seideman, M.Y. Ivanov, and P.B. Corkum: Phys. Rev. Lett. **75**, 2819 (1995).
22. G.N. Gibson, M. Li, C. Guo, and J. Neira: Phys. Rev. Lett. **79**, 2022 (1997).
23. J.D. Simon: Rev. Sci. Instrum. **60**, 3597 (1989).
24. W.H. Knox: IEEE J. Quan. Elect. QE **24**, 388 (1989).
25. D.E. Spence, P.N. Kean, and W. Sibbett: Opt. Lett. **16**, 42 (1991).
26. D. Strickland and G. Mourou: Opt. Comm. **56**, 219 (1985).
27. W.C. Wiley and I.H. McLaren: Rev. Sci. Instrum. **26**, 1150 (1955).
28. G.N. Gibson, R.R. Freeman, and T.J. McIlrath: Phys. Rev. Lett. **67**, 1230 (1991).
29. S.L. Chin, Y. Liang, J.E. Decker, F.A. Ilkov, M.V. Ammosov: J. Phys. B **25**, L249 (1992); T.D.G. Walsh, F.A. Ilkov, J.E. Decker, and S.L. Chin: J. Phys. B **27**, 3767 (1994).
30. A. Talebpour, C.Y. Chien, and S.L. Chin: J. Phys. B **29**, L677 (1996).
31. M.J. DeWitt and R.J. Levis: J. Chem. Phys. **108**, 7739 (1998).
32. A. Saenz: J. Phys. B **33**, 4365 (2000).
33. C. Guo: Phys. Rev. Lett. **85**, 2276 (2000).
34. J. Muth-Bohm, A. Becker, and F.H.M. Faisal: Phys. Rev. Lett. **85**, 2276 (2000).
35. X.M. Tong, Z.X. Zhao, and C.D. Lin: Phys. Rev. A **66**, 033402 (2002).
36. M.J. DeWitt, E. Wells, and R.R. Jones: Phys. Rev. Lett. **87**, 153001 (2001).
37. P.B. Corkum: Phys. Rev. Lett. **71**, 1994 (1993).
38. C. Guo, M. Li, J.P. Nibarger, and G.N. Gibson: Phys. Rev. A **61**, 33413 (2000)
39. C. Guo, R.T. Jones, and G.N. Gibson: Phys. Rev. A **62**, 15402 (2000).
40. J.B. Watson, A. Sanpera, D.G. Lappas, P.L. Knight, and K. Burnett: Phys. Rev. Lett. **78**, 1884 (1997).
41. D.G. Lappas and R. van Leeuwen: J. Phys. B **31**, L249 (1998).
42. J. Javanainen, J.H. Eberly and Q. Su: Phys. Rev. A **38**, 3430 (1988).
43. Q. Su and J.H. Eberly: Phys. Rev. A **44**, 5997 (1991).
44. J. Parker, K.T. Taylor, C.W. Clark and S. Blodgett-Ford: J. Phys. B **29**, L33 (1996).
45. C. Guo and G.N. Gibson: Phys. Rev. A **63**, 040701(R) (2001).
46. M. Protopapas, D.G. Lappas, and P.L. Knight: Phys. Rev. Lett. **79**, 4550 (1997).
47. C. Guo: Phys. Rev. A **71**, 021405(R) (2005).
48. C. Guo and K. Wright: Phys. Rev. A **71**, 021404(R) (2005).

4 Strong-Field Correlation Imaging: Revealing Molecular Geometries, Orientation and Dynamics

Wendell T. Hill, III[1,2], Kun Zhao[1], Lee N. Elberson[1], and Getahun M. Menkir[1]

[1] Institute for Physical Science and Technology, University of Maryland, College Park, MD 20742
[2] Department of Physics, University of Maryland, College Park, MD 20742

Summary. A review of several correlation techniques used to extract quantitative information from two-dimensional images obtained with fast-frame CCD cameras is presented. Three examples are discussed that demonstrate the power of these techniques for revealing field-induced dynamics: (1) the relative probability for Coulomb explosions of CO_2^{6+} as a function of the angle between the molecular and polarization axes; (2) the relative probability for Coulomb explosion of CO_2^{6+} and NO_2^{6+} as a function of bond angle and (3) the kinetic energy release and its ability to distinguish enhanced ionization from screening.

4.1 Introduction

Imaging has become the method of choice to probe strong-field dissociative ionization (Coulomb explosion – the dissociation of multiply ionized molecules into energetic atomic ions). Visualization is enabled through imaging, revealing pre-explosion molecular geometries and orientation relative to some preferred direction in space, as well as field-induced structure and dynamics. There are essentially four distinct flavors of approaches. The first is based on one-dimensional time-of-flight (TOF) detectors while the others exploit two-dimensional position sensitive detectors. In the one-dimensional approach, pseudo-images are constructed as composite polar plots of TOF waveforms obtained at different orientation angles of the polarization axis relative to the TOF axis. For best temporal resolution (of order nanosecond), the detector subtends a solid angle $\ll 4\pi$ sr. This technique has been used successfully by several groups to extract information regarding structure and dynamics. (See, for example, [1–3].)

While pseudo-imaging has been used to study three-atom explosions, it is most appropriate for two-atom or linear explosions. As shown in Fig. 4.1, TOF detection cannot capture all explosion partners simultaneously for non-linear explosions because of the small acceptance angle. Thus, much of the dynamics even for linear molecules, such as CO_2, will be missed. One of the dominant explosion channels, for example, is $CO_2^{6+} \rightarrow O^{2+} + C^{2+} + O^{2+}$ with the bond angle varying from 145° to 180° [4].

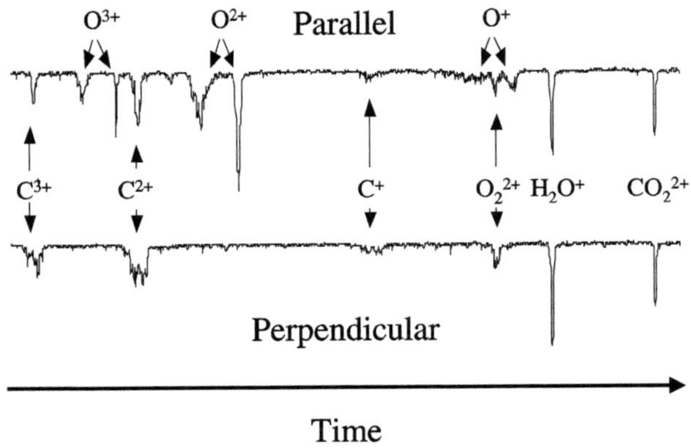

Fig. 4.1. Time-of-flight waveforms resulting from CO_2 Coulomb explosions obtained with 100 fs pulses focused to 10^{15} W/cm^2 when the polarization axis is parallel and perpendicular to the TOF axis. Note, the C structure is different in the *upper* and *lower traces* while some O components are missing altogether from the *lower trace*

Two-dimensional position sensitive detectors collect ions over a large solid angle (typically 4π sr) and are more ideally suited for studying nonlinear dynamics. Two orthogonal components of the momentum for the ejection partners can be determined simultaneously from their strike points, as shown in Fig. 4.2. The approaches based on position sensitive detectors are distinguished by how the third, perpendicular component of momentum is treated. This component can be measured directly via the ion's time of flight, as done in *COLTRIMS* (cold-target recoil-ion momentum spectroscopy) [5] and similar methods, extracted statistically by isolating explosion partners with correlation techniques [4, 6] or eliminated with one of several possible back-

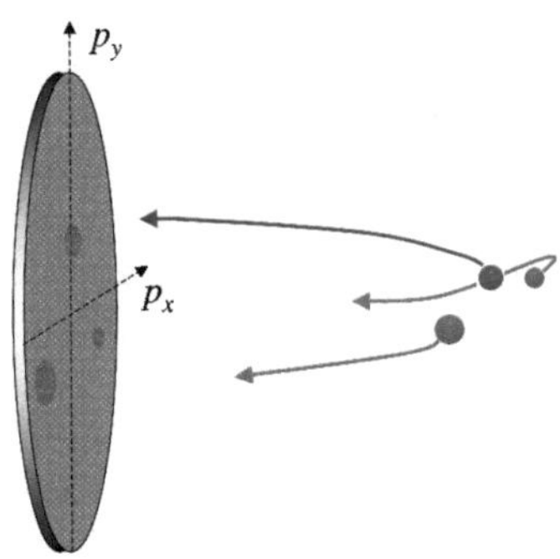

Fig. 4.2. Schematic of a position sensitive detector capturing all three ions ejected from the explosion of a three-atom system. Two components of the momentum are immediately determined from the strike point of each charge on the detector

plane projection procedures [7, 8]. There are advantages and disadvantages to each method.

The *COLTRIMS* family of approaches provide all kinematic information for all ejecta but are limited to only a few (typically of order ten) ions per laser pulse because the flight times ($\propto p_z$) are determined by delay-line anodes coupled to multi-hit time-to-amplitude converters, albeit with high resolution. Typically, the number of explosion events are held to one or fewer. Often, however, it is not necessary to measure this third component very accurately because an explosion event is constrained by momentum conservation. In some cases, once (p_x, p_y) are determined for all the explosion partners, p_z can be deduced after making a few assumptions. More generally, when the explosion partners are correlated, it is possible to select events in which p_z is known *a priori*, obviating the need to make these assumptions.

What turns out to be critical in these studies is an ability to distinguish charges with different masses and in different charge states. This discrimination often can be done with low temporal resolution. This allows CCD cameras to be used instead of delay-line readouts to determine (p_x, p_y), affording significantly higher event rates; several hundred ions per laser pulse can be handled in principle, which is limited by saturation of a particular channel. With CCD cameras, we use a technique we call *image labeling* [4] to select specific explosion events for detailed study. We point out, however, that both the delay-line anode and CCD approaches can exploit correlation and/or coincidence detection to isolate explosion events meeting predetermined characteristics.

When correlating explosion partners is not of interest, the backplane projection techniques can be used to eliminate charges with nonzero p_z from images obtained with CCD cameras or composite TOF waveforms. As shown in Fig. 4.3, if the perpendicular component of the momentum is not known, it is not possible to determine p^2 $(= p_x^2 + p_y^2 + p_z^2)$ accurately. Clearly, there are cases when $p_x = p_y = 0$ but $p^2 \neq 0$ – consider the center of the upper middle and the left hand corner of the lower left images. The upper and lower right images show that after the inversion is applied, p_z is eliminated and $p^2 = p_x^2 + p_y^2$. This approach is often useful for obtaining an overview of the structure of the explosion. After inversion,

$$p = r\sqrt{\frac{qEm}{2l}} \tag{4.1}$$

on the image, where p is measured from the center of mass (the center of the image). In (4.1), r is the radius of the rim, q is the charge, E is the strength of the static field, m the mass of the charge and l the distance between the interaction region and the detector. We emphasize that a correction related to an inversion to eliminate p_z contamination mentioned above should be applied to pseudo-images as well. As can be understood from Fig. 4.3, given two momenta $|\boldsymbol{p}_1| > |\boldsymbol{p}_2|$, there will be an angle θ where $|\boldsymbol{p}_1| \cos\theta = |\boldsymbol{p}_2|$.

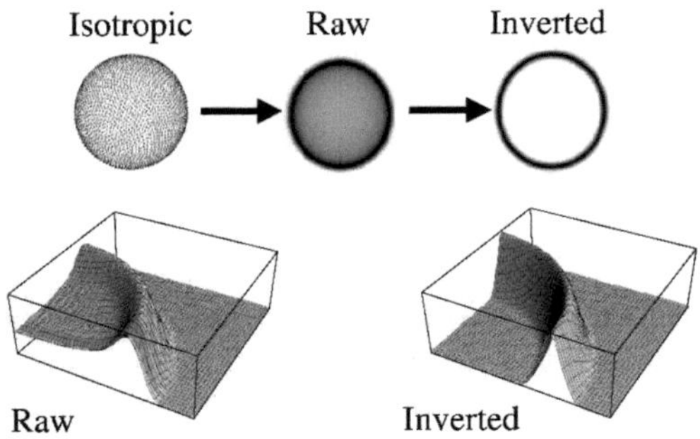

Fig. 4.3. Schematic of the backplane projection or inversion. *Top left* represents an isotropic distribution of charges ejected from the interaction region into 4π sr. The *upper middle* is a simulation of the CCD image of these ions striking the detector. The *lower left* shows that ions initially ejected parallel to the surface of the detector ($p_z = 0$) will land on the edge of the bowl-shaped distribution while those with $p_z \neq 0$ will land inside the rim of the bowl. The inversion techniques essentially remove the ions in the center leaving only the rim [8]. The *upper* and *lower right* images can be used to determine (p_x, p_y) directly from (4.1). The width of the rim in these simulations is associated with the energy spread of the distribution, which is not a delta-function, in general

Thus, the contributions from $\boldsymbol{p}_1$ and $\boldsymbol{p}_2$ will be commingled. Corrections have been discussed by a few authors [3,9].

All the approaches outlined above have been employed to study Coulomb explosions. In this chapter, however, we will concentrate on results obtained with *image labeling* and several correlation-imaging techniques stemming from it. In the Sects. 4.2 and 4.3 we will describe the experimental approach in more detail and in Sect. 4.4 we will look at three examples based on our ability to isolate specific explosion events in three-atom Coulomb explosions.

4.2 The 4π-Image Spectrometer

As mentioned above, *image labeling* requires simultaneous collection of all explosion partners ejected into 4π sr. The 4π-image spectrometer shown in Fig. 4.4, meets this need and was introduced by Zhu and Hill [10]. The spectrometer consists of a uniform extraction field, image quality microchannel plates (MCP) and a phosphor screen all housed in a vacuum chamber allowing measurements at pressures between 1×10^{-9} and 5×10^{-7} Torr. For the experiments discussed here, the sample gas was at room temperature and leaked into the chamber. Excitation was performed with 100 fs pulses generated with a Ti:sapphire oscillator/amplfier laser tuned near 800 nm and focused to intensities between 0.5 and 5×10^{15} W/cm^2. A typical image consisting of 500,000 laser shots is shown in Fig. 4.5 for a three-atom

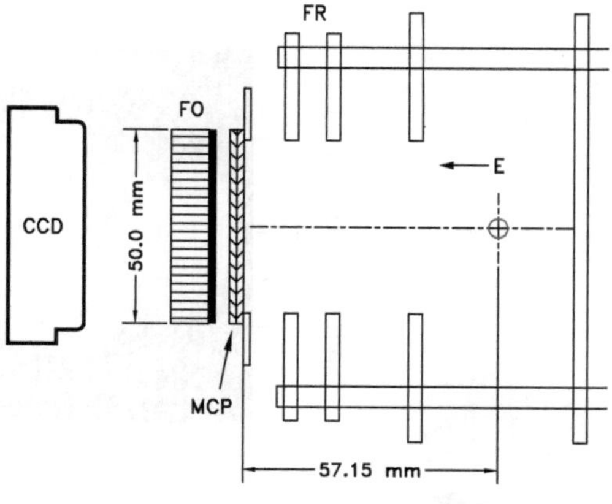

Fig. 4.4. Image spectrometer consisting of a uniform electric field (***E***) established by field rings (FR) and a pair of microchannel plates (MCP). Ejected charges are collected by the MCP (40 mm active area), where they are converted to a proportional number of electrons that are accelerated into a phosphor painted onto the MCP side of a fiber optic (FO) bundle. Two-dimensional images of the charges (see Fig. 4.5) are captured by a digital CCD camera with 128 × 128 elements and streamed to disk at up to 730 Hz. In atomic units (1.24×10^{-7} eV-s/cm), the maximum momentum, p_o, that can be captured is $\sim 4.86\sqrt{qEM}$, where q and M are the charge and mass in atomic units and E is the magnitude of the uniform electric field in V/cm. Typically, $E \sim 250$ V/cm leading to $p_o \sim 435$ au for O^{2+}. A narrow time slice of the charges striking the MCP (see Fig. 4.6) can be extracted by gating the potential across the MCP

explosion. The center of the image coincides with the center-of-mass of the explosion.

We have limited knowledge of p_z in the distributions in Fig. 4.5 so (4.1) cannot be used without first applying one of the backplane projection methods. Alternatively, (p_x, p_y) can be extracted when the explosion partners are correlated (image labeling). As we will discuss further in the next section, correlation is a statistical approach that requires each image frame to contain ions from one and only one laser pulse. Thus, 500,000 shots requires 500,000 image frames. To achieved this in a reasonable time (of order 30 to 60 min), a high frame-rate digital CCD camera is employed. We synchronized the camera to the laser repetition rate (500 Hz, typically) and streamed frames to a large SCSI disk. The fast data collection is further aided by the ability of this approach to handle multiple events simultaneously (many more than is possible with *COLTRIMS*).

The limit to the number of events per laser shot is set by two conditions. The first involves saturation where a specific channel (set of ions) exists in too many frames. The second is concerned with ions from two different channels landing at the same place on the MCP. The first constraint limits the dynamic range of cross sections we can study before false cor-

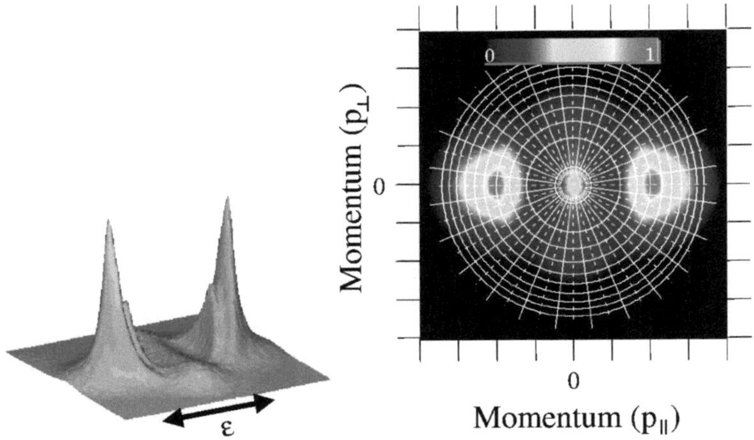

Fig. 4.5. Typical momentum distributions resulting from a Coulomb explosion of three-atom systems. Specifically, the three-dimensional image (*left*) corresponds to N^{2+} (central portion perpendicular to ε primarily) and O^{2+} ions resulting from the explosion of NO_2^{6+} while the contour plot corresponds to C^{2+} (*central, vertical* distribution) and O^{2+} ions resulting from the explosion of CO_2^{6+}. In both cases, the O^{2+} ions occupy the two end distributions parallel to ε (*horizontal axis* in the contour plot). The scale on the contour plot is in atomic units and varies from 50 to 80, depending on the mass of the ion

relations contaminate the resulting correlation image. The second can be minimized by keeping the sample pressure low and by gating the MCP so that it detects only a subset (time slice) of all the ions generated in one laser pulse. What turns out to be most important is an ability to segregate the charges by charge state. Gating takes advantage of the fact that ions with different charge-to-mass ratios arrive at the detector at different times. Gate widths of a few hundred nanoseconds are sufficient to isolate the first few charge states as shown in Fig. 4.6. We note that the images in Fig. 4.6 were taken with the same gas sample and with the same laser and detector conditions. It is important to recognize that in order not to disturb the static extraction field in the experimental region traversed by the ions, we change the back voltage of the MCP, which only affects the field between the MCP and the phosphor. It is only necessary to reduce the potential across the MCP below the threshold for signal generation, which is between 1 and 1.4 KV.

4.3 Correlation Imaging

Covariance mapping, introduced by Frasinski *et al.* [11], revolutionized the way Coulomb explosion experiments were performed as it is capable of reducing the ambiguity for correlating explosion partners in TOF waveforms.

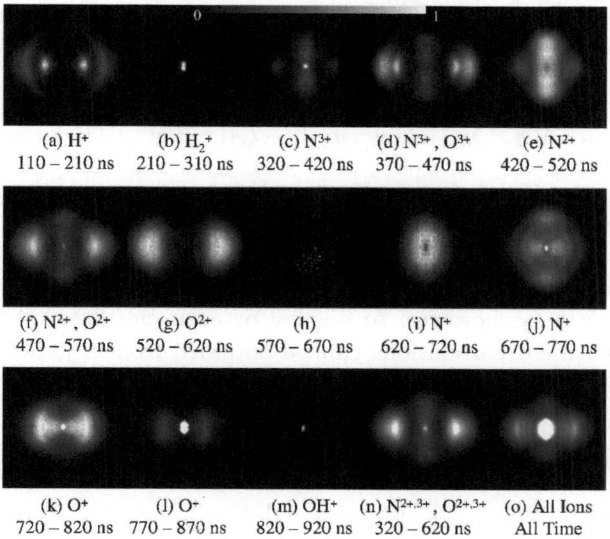

Fig. 4.6. Images of ions arriving in 100 ns time slices, panels **a**–**m**, relative to the laser pulse traversing the interaction region. Panels **n** and **o** were taken with a 300 ns gate and no gate at all (all ions), respectively. Panel **h** shows residual O^{2+} and N^+ ions in the time gap between panels **g** and **i**. The ions were induced by 100 fs, 10^{15} W/cm^2 pulses at 800 nm

The covariance matrix, which identifies the ions in the waveform (A or B) at times t_α and t_β that are explosion partners, is defined as

$$\xi_{\alpha\beta}(t_\alpha, t_\beta) = \langle A(t_\alpha)B(t_\beta)\rangle - \langle A(t_\alpha)\rangle\langle B(t_\beta)\rangle \,, \tag{4.2}$$

where

$$\langle A(t_\alpha)\rangle = \frac{1}{N}\sum_{i=1}^{N} A_i(t_\alpha) \tag{4.3}$$

$$\langle B(t_\beta)\rangle = \frac{1}{N}\sum_{i=1}^{N} B_i(t_\beta) \tag{4.4}$$

$$\langle A(t_\alpha)B(t_\beta)\rangle = \frac{1}{N}\sum_{i=1}^{N} A_i(t_\alpha)B_i(t_\beta) \,. \tag{4.5}$$

An actual correlation is given by,

$$C_{\alpha\beta} \equiv \frac{\xi_{\alpha\beta}}{\sigma_A \sigma_B} \,, \tag{4.6}$$

where

$$\sigma_\mu = \sqrt{\frac{1}{N}\sum_{i=1}^{N} \left[\mu_i(t_\alpha) - \langle \mu(t_\alpha)\rangle\right]^2} \tag{4.7}$$

and $\mu = A, B$. Zhu and Hill [10] were the first to extend this technique to two-dimensional images. The corresponding four-dimensional covariance matrix is

$$\xi_{\alpha_1\alpha_2\beta_1\beta_2} = \langle A(x_{\alpha_1}, y_{\alpha_2})B(x_{\beta_1}, y_{\beta_2})\rangle - \langle A(x_{\alpha_1}, y_{\alpha_2})\rangle\langle B(x_{\beta_1}, y_{\beta_2})\rangle \,, \quad (4.8)$$

with averages calculated as indicated above. This matrix identifies explosion partners at locations $(x_{\alpha_1}, y_{\alpha_2})$ and $(x_{\beta_1}, y_{\beta_2})$. With two-dimensional images, $\xi_{\alpha_1\alpha_2\beta_1\beta_2}$ has n^4 elements, which is awkward to display – 2.7×10^8 elements for an image generated with a CCD array with 128×128 pixels. It is not necessary, however, to display or even calculate all the elements because the salient correlations can be obtained from a judicious subset of two-dimensional projections as discussed in [10].

4.3.1 Image Labeling

While portions of $\xi_{\alpha_1\alpha_2\beta_1\beta_2}$ reveal correlation among the ions [4, 10], correlated partners can also be detected by a more visual approach we call *Image labeling.* To understand how it works, consider Fig. 4.7. On the left is an image generated by averaging all 500,000 frames. On the right is an image composed of a subset of the frames (a selective average of about 10,000) determined by the frames with contribution at the location indicated or labeled by the arrow. Only those ions generated at the same time are correlated with the ions labeled appear in this *correlation image.* These ions are explosion partners. *Image labeling* works well when each channel is only a few percent

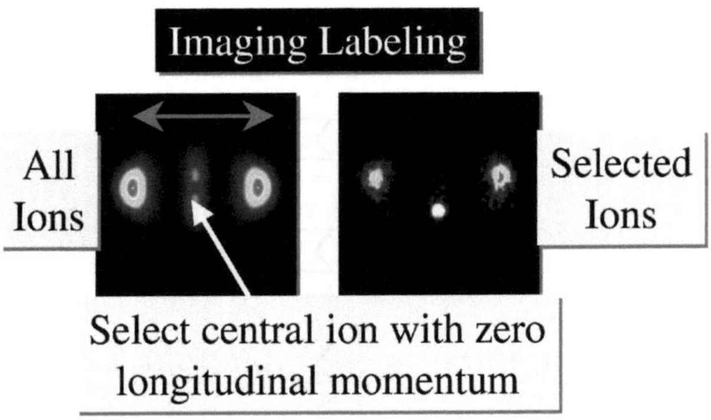

Fig. 4.7. Contours of the doubly charged ions resulting from the Coulomb explosion of CO_2^{6+}. The *left image* is the sum of 500,000 frames resulting from linearly polarized laser pulses (direction indicated by the *double-headed arrow*) focused to 10^{15} W/cm^2. The *single-headed arrow* indicates the labeled C^{2+} ions. The *right image*, containing about 10,000 images, is what we call a correlation image, constructed by summing only those frames having a nonzero contribution at the labeled location (a selective average). The ions in the correlation image are explosion partners. In this example CO_2^{6+} explodes with the bond angle of $\sim 163°$

of the total image. When any one channel appears too often relative to others, or conversely when one channel appears too infrequently compared with others, false correlations with the stronger channel occur. *Image labeling* is more sensitive to false correlation than $\xi_{\alpha_1\alpha_2\beta_1\beta_2}$ ((4.2) and (4.8)) but is both easier to run numerically and lends itself to a more graphical presentation.

4.3.2 Coincidence Imaging

While image labeling, described above, can isolate specific bond angles, extracting the relative strength of the explosion signal requires yet another step – coincidence imaging. This is because the correlated image contains events with a spread in p_z about $p_z = 0$, as indicated by the width of the correlated spots. We can reduce this spread by reducing the size of the correlated spots. Using the fact from Fig. 4.3 that the highest concentration of charges for a given momentum allows us to extract events with $p_z = 0$, if we focus on a narrow region about the peak point. Figure 4.8 shows an example of coincidence imaging for three-atoms. The right image is a subset of the ions appearing in the image on the left, consisting only of frames containing all three ions with $p_z = 0$ – a triple coincidence. The strength of the signal, or probability for explosion, is proportional to the number of frames that contribute to this image.

We carry out the signal strength determination in the following way. In general, each ion striking the MCP illuminates a cluster of adjacent pixels

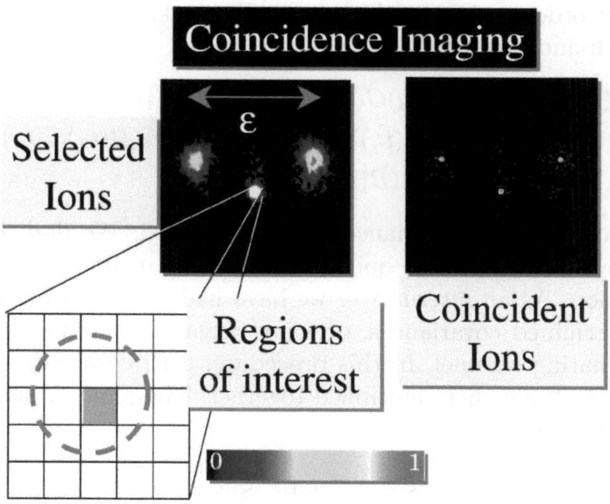

Fig. 4.8. Triple coincidence image construction. The *left image* is the same as the right image in Fig. 4.7. The *right image* is a subset of the ions in the left image. Contribution to the right image is made when an ion lands in all three circles of the regions of interest as described in the text. Note, although not shown, there are similar regions of interest for the two correlated spots as well

on the CCD camera. The number of pixels within a cluster follows a Poisson distribution and typically peaks between 1 and 5 pixels. To minimize the inclusion of clusters corresponding to two ions landing near each other, we restrict the clusters we consider to be valid to those that have a maximum number of pixels, typically between five and ten depending on the conditions of the experiment. There are two criteria a cluster must meet to be considered part of a *real* triple coincidence. The first signature is that the location of the centroid of a valid cluster must fall within circles of radius r_0 enclosing the final momentum of each ion (located at $\boldsymbol{p}_C$, $\boldsymbol{p}_{O_{\mathrm{left}}}$, and $\boldsymbol{p}_{O_{\mathrm{right}}}$). The size of r_0 is related to the uncertainty in determining the location of the centroid, which typically is $\pm$ 1 to 3 pixels. The second requirement for inclusion is that there can only be one centroid (cluster) within each circle. As shown in Fig. 4.8, it is also possible for a centroid to be only partially enclosed by the circle. Centroids falling entirely within the circle contribute unity to the count while those that are partially enclosed contribute fractional counts. The total contribution to the strength is given by the product of the count for each circle – if each centroid were only 50% enclosed, this frame would contribute 0.125 to the total count.

4.3.3 Higher Order Variance

When coincidence images are contaminated with false correlation because of one or more very strong channels, $\xi_{\alpha_1\alpha_2\beta_1\beta_2}$ does a better job of determining true ejection partners. Correlating three or more components, however, requires higher order variance calculations. Frasinski *et al.* [12] have explored this approach and suggested

$$\begin{aligned}\xi_{\alpha\beta\gamma} = &\langle A(t_\alpha)B(t_\beta)C(t_\gamma)\rangle - \langle A(t_\alpha)B(t_\beta)\rangle\langle C(t_\gamma)\rangle \\ &- \langle A(t_\alpha)C(t_\gamma)\rangle\langle B(t_\beta)\rangle - \langle A(t_\alpha)\rangle\langle B(t_\gamma)C(t_\gamma)\rangle \\ &+ 2\langle A(t_\alpha)\rangle\langle B(t_\beta)\rangle\langle C(t_\gamma)\rangle \,, \end{aligned} \tag{4.9}$$

as the appropriate triple variance to be calculated. We find, however, the image analogue to $\xi_{\alpha\beta\gamma}$ to be quite noisy, making it difficult to extract useful information. As an alternative, we have had success using a product of pairwise-determined covariances, or joint variance, to reduce the effects of the contaminating channel. In this procedure, each covariance is calculated using (4.2)–(4.5) and then multiplied together. The joint variance of B with A and C would be

$$\xi^J_{A\bar{B}C} = \xi_{AB} \cdot \xi_{BC} \,, \tag{4.10}$$

where the bar over B indicates the common entity in the calculation. Even though ξ_{AC} may be significantly larger than the other two, it plays a secondary role at best in $\xi^J_{A\bar{B}C}$. By contrast, it plays a primary role in the triple coincidence approach, skewing the results to higher values as we will show below.

4.4 Three-Atom Explosion Dynamics

We are now in a position to look at three examples of the implementation of the various correlation techniques discussed in the previous section. All of them exploit our ability to isolate specific explosion events. Our first example will focus on the measurement of the probability for linear explosion of CO_2. Next, will look at the probability for bent Coulomb explosions as a function of bond angle. Finally, we will return to linear explosions in CO_2 and probe the similarities between the screening and enhanced ionization models of strong-field induced Coulomb explosion.

4.4.1 Coulomb Explosions and the Molecular Axis

As we have seen, image labeling and coincidence imaging enables us to isolate explosion events for particular bond angles. We used this feature to measure the probability for Coulomb explosion as a function of the angle between the molecular axis and the polarization axis. Figure 4.9 is a graphical representation of the angular dependence for CO_2^{6+} explosions into C^{2+} and $2O^{2+}$ ions with bond angles of 180° (left) and 163°. The images in Fig. 4.9 are composites of coincidence images like the right image in Fig. 4.8. We point out that the angle measured in Fig. 4.8, θ_{CM}, is the center of mass angle between the momenta of the two outer ions in the far-field, which can be extracted from

$$\tan(\pi - \theta_{CM}) = \frac{|\boldsymbol{p}_\perp|}{|\boldsymbol{p}_\parallel|} , \tag{4.11}$$

where $\boldsymbol{p}_\parallel$ ($\boldsymbol{p}_\perp$) is the momentum of the ion parallel (perpendicular) to the polarization axis [13]. The center-of-mass angle is not the bond angle, θ_b, associated with the structure at the time of the explosion. However, for a given θ_b, θ_{CM} can be calculated numerically by integrating the equations of motion assuming nothing more than a pure Coulomb repulsion between the ions and an initial bond length that is the same for both bonds. The latter

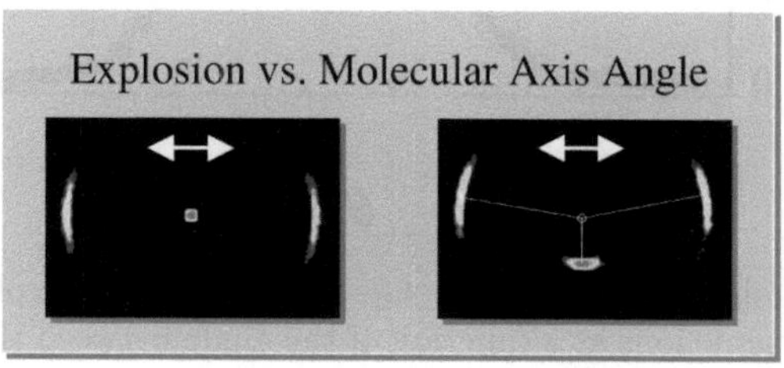

Fig. 4.9. These images show the distribution of linear (*left*) and bent (163°) CO_2^{6+} Coulomb explosions as a function of the angle between the molecular and polarization (indicated by the *arrow*) axes. These figures allow the bond-dependent angular distributions to be measured quantitatively

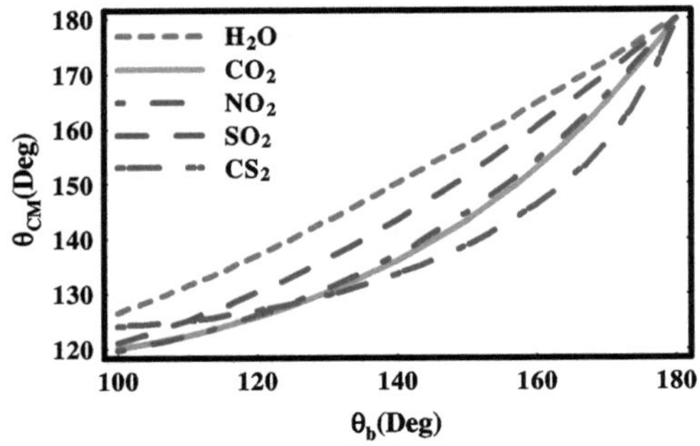

Fig. 4.10. The measured center of mass angle, θ_{CM}, as a function of bond angle, θ_b for the explosion of several three atom systems for $0.1\,\mathrm{nm} < R_C < 0.3\,\mathrm{nm}$

is true when the center ion has no momentum along the polarization axis, as is true for Figs. 4.8 and 4.9. The curves for $\theta_{CM}(\theta_b)$, which depend on the molecule (see Fig. 4.10), vary by less than 0.5% for bond lengths between 0.1 and 0.3 nm for a given molecule. Since every θ_b has a unique θ_{CM} the function can be inverted numerically to give $\theta_b(\theta_{CM})$.

The relative signal strength for a linear explosion is shown in Fig. 4.11. The curve obeys $\cos^{39}\theta$, which has a full width of 22°. The widths at other bond angles are about the same [6]. It is interesting to note that this width is considerably narrower than that for H_2, 31° [6]. The narrowness of the distribution has been linked to (1) the depth of the pendular potential [14,15]

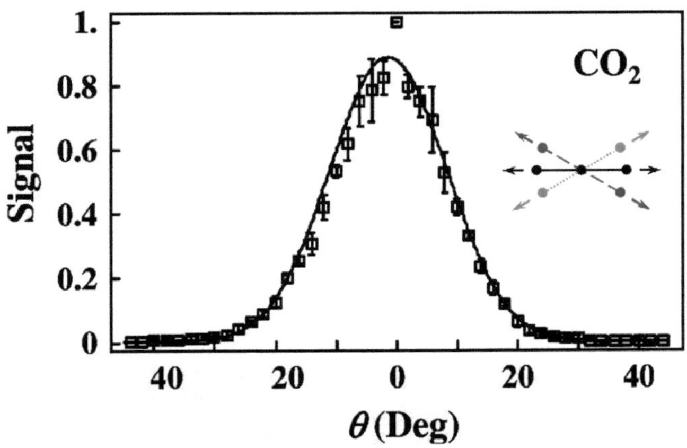

Fig. 4.11. The relative probabilities for linear Coulomb explosion of CO_2 into doubly charged atomic ions as a function of the angle between the polarization axis and the molecular axis. The *error bars* correspond to the standard deviation for four different runs. The *solid curve* is a $\cos^n\theta$ fit where $n = 39$ with a full width at half maximum of 22°

established by the field and (2) the time CO_2 spends in the laser field before undergoing enhanced ionization at the critical internuclear distance, R_c [6].

4.4.2 Bent Coulomb Explosion

It is well known that large amplitude vibrations are induced during Coulomb explosions [13, 16–18]. Figure 4.12 shows this for CO_2 and NO_2. Nitrogen dioxide is an interesting case because of the possibility of using the laser to straighten a bent system. Using pseudo-imaging, Hishikawa *et al.* [17] presented evidence for such straightening at 10^{15} W/cm^2. Since these measurements were done without the benefit of correlation, it is worthwhile re-examining this issue with the correlation imaging tools presented in this chapter.

In the lower panel of Fig. 4.12, we present the relative signal strength determined with triple coincidence (the triangles) and joint variance. The triple coincidence calculation agrees with with the conclusions of [17] while

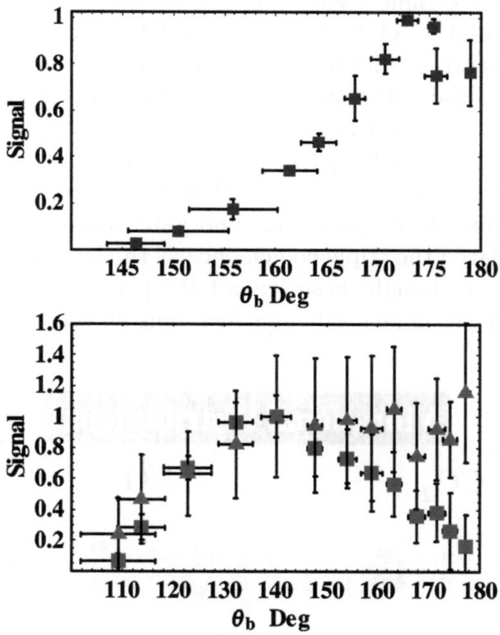

Fig. 4.12. Relative probability for CO_2^{6+} (*top*) and NO_2^{6+} explosions vs. bond angle (θ_b). For CO_2^{6+}, the probabilities were determined with triple coincidence while those for CO_2^{6+} were determined with both triple coincidence (*triangles*) and joint variance (*squares*). The *error bars* represent the standard deviation over four (seven) runs for CO_2 (NO_2). The anomalously high triple coincidence signal near 180° is due to contamination by a very strong two-body explosion channel as discussed in the text

the joint variance calculation implies a different story – the distribution peaks near the equilibrium bond angle and NO_2 never reaches the linear state.

To determine which is correct, we first recognize that the triple correlation calculation is contaminated by a very strong two-channel correlation. This can be seen in Fig. 4.13 where image labeling is used to map out the peak of the correlated distribution onto the uncorrelated image in the two left panels. The distribution for CO_2 is continuous, exhibiting a smooth arc across the most intense portion of the spectrum (the white circles), that for NO_2, however, is discontinuous near 180°. The momentum ($\propto r$ see (4.1)) suddenly drops and moves toward the peak of the uncorrelated distribution. The two-body channel in NO_2 (1) is about ten times stronger than the three-body channel and (2) has an O^{2+}–O^{2+} separation that is smaller than the corresponding three-body channel separation (see Fig. 4.14). Image labeling does a good job identifying where the ions are in CO_2 because it has no two-body channel. Image labeling is more problematic for NO_2 because contamination skews the three-body correlation signal toward smaller r near 180°. The precipitous drop near 180° is not observed in the right panel in Fig. 4.13 where the ions are located by joint variance. We point out that joint variance was calculated with (4.10) where $A = O^{2+}$, $B = N^{2+}$ and $C = O^{2+}$. The search space was limited to the areas identified by image labeling. Clearly, the peaks of the joint variance calculation and those of image labeling, and hence triple coincidence, do not necessarily coincide.

Image labeling not only gets the separation between the ions wrong near 180°, Fig. 4.12 shows it gets the relative probability wrong as well. To compare triple coincidence (triangles) and joint variance (squares) on equal footing, we calculated the triple coincidence at the locations identified by the joint variance. Even though these are not the peak triple coincidence values (the peaks follow the image labeling locations shown in the middle panel

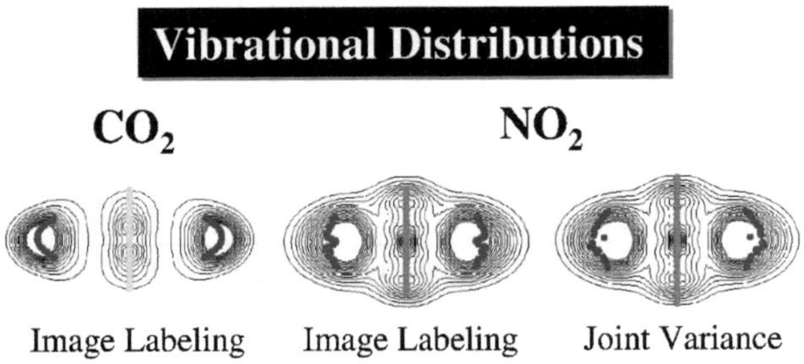

Fig. 4.13. Large amplitude vibration distribution determined by image labeling (CO_2 and NO_2) and joint variance (NO_2). The *dots* in the *vertical direction* across the center of each image indicate the location of C^{2+} or N^{2+} while those forming an arc indicate the location of O^{2+} as a function of bond angle. It should be understood that the central and outer atoms move vertically in opposite directions. The underlying contours correspond to the uncorrelated average images

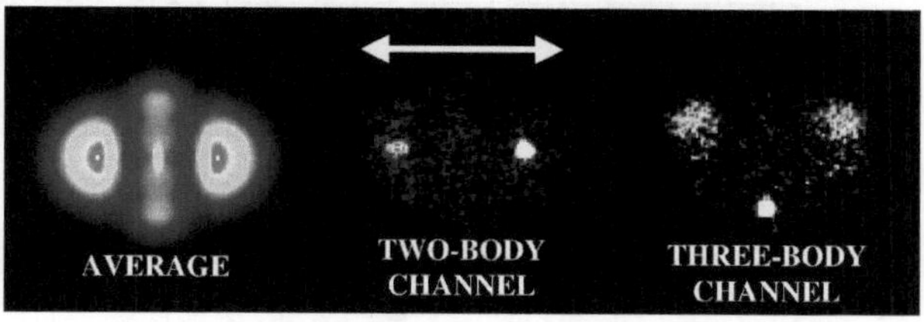

Fig. 4.14. Symmetric Coulomb explosion of NO_2^{6+}: uncorrelated image (*left*), two-body image labeling image (*middle*) and three-body image labeling image

in Fig. 4.13), they still over estimate the true signal strength. Again, the strength is enhanced by the concomitant two-body channel. The fact that the relative probabilities determined by joint variance and triple coincidence scale together at smaller bond angles ($< 140°$) gives credence to our believe that the probabilities it determines at larger angles are valid and NO_2 does not reach the linear state.

4.4.3 Enhanced Ionization vs. Screening

As our last example, we look at one of the hallmarks of strong-field induced Coulomb explosion, the tendency for the explosion energy (kinetic energy release) to be considerably smaller than the classical potential energy associated with the atomic ions separated by the equilibrium bond lengths of the parent molecule. Typically, the explosion energies are consistent with a separation, $R_c \simeq 2 - 3 \times R_e$, where R_c is the critical separation for enhanced ionization [19] and R_e the equilibrium separation of the neutral molecule. The primary models that have been suggested to explain the energy deficit are based on enhanced ionization at a critical separation (R_c-model) [19–21] and dynamic electron screening (DS-model) [22, 23]. The R_c-model assumes that the positively charged molecular ion explosion occurs at R_c. The value that R_c assumes is dictated by the observation that field ionization of the molecular precursor to the critical ionization stage (the stage where enough electrons have been removed that the system explodes) is far easier at R_c than at R_e. The DS-model accounts for the energy deficit by a reduced Coulomb potential during the explosion caused by the positive ions moving through the cloud of photoelectrons, some of which are trapped in the strong laser field, causing the ions to decelerate during the explosion. Thomas–Fermi calculations based on this model have shown good agreement with experiment [23]. Despite the fact that some of the electrons trapped in the laser field (Volkov states) will oscillate back and forth through the origin as the phase of the laser changes [24], Zhao *et al.* [13] showed the remarkable result that for linear explosions, treating the electron distribution as static leads to the same explosion energies as the more elaborate Thomas–Fermi calculations over several ionization stages (see in Fig. 4.15). Furthermore, a steady-state distribution

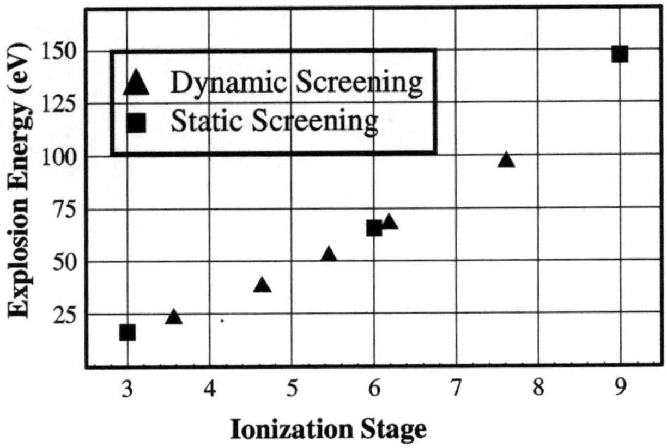

Fig. 4.15. A comparison between the linear explosion energies for CO_2^{q+} ($q = 3-9$) calculated by the Thomas–Fermi theory of [23] (*triangles*) and by σ (*squares*). Our experimental value for $R_c (= 0.22)$ nm measured for the 6-electron channel along with $R_e = 0.116$ nm were used to determine σ and Z_{eff} as discussed in the text

can be characterized by a single parameter we call the charge defect, σ, which produces an effective charge on each atomic ion, $Z^{\text{eff}} \equiv Z - \sigma$. If we let

$$\sum_{i<j} \frac{Z_i Z_i}{r_{ij}}\bigg|_{r_{CO}=R_c} = \sum_{i<j} \frac{Z_i^{\text{eff}} Z_i^{\text{eff}}}{r_{ij}}\bigg|_{r_{CO}=R_e} , \tag{4.12}$$

we determine

$$\sigma = Z\left(1 - \sqrt{R_e/R_c}\right) , \tag{4.13}$$

for symmetric explosions – all charges missing the same number of electrons. As a consequence, (4.13) provides a one-to-one mapping of R_c onto σ, which then varies from $0.54Z$ to $0.65Z$. While physically very different processes, kinetic energy release alone cannot distinguish the two models for symmetric explosions.

4.5 Conclusions

In conclusion, we have described three correlations techniques, image labeling, triple coincidence and joint variance, that can be used to reveal structure and dynamics in three-atoms Coulomb explosions. Image labeling can be used to isolate specific bond angles and molecular orientations, triple coincidence can be exploited for quantitative measurements while joint variance is useful for eliminating contamination from strong channels.

References

1. G.R. Kumar, P. Gross, C.P. Safvan, F.A. Rajgara, and D. Mathur, Phys. Rev. A **53**, 3098 (1996).
2. A. Hishikawa, A. Iwamae, K. Hoshina, M. Kono, and K. Yamanouchi, Chem. Phys. Lett. **282**, 283 (1998).
3. W.A. Bryan et al., J. Phys. B **33**, 745 (2000).
4. K. Zhao, G. Zhang, and W.T. Hill, III, Opt. Express **9**, 42 (2001).
5. J. Ullrich et al., J. Phys. B: At. Mol. Opt. Phys. **30**, 2917 (1997).
6. K. Zhao and W.T. Hill, III, Phys. Rev. A **71**, 013412 (2005).
7. C.J. Dasch, Appl. Opt. **31**, 1146 (1992).
8. K. Zhao, T. Colvin, Jr., W.T. Hill, III, and G. Zhang, Rev. Sci. Instrum. **73**, 3044 (2002).
9. A. Iwamae, A. Hishikawa, and K. Yamanouchi, J. Phys. B: At. Mol. Opt. Phys. **33**, 223 (2000).
10. J. Zhu and W.T. Hill, III, J. Opt. Soc. Amer. B **14**, 2212 (1997).
11. L.J. Frasinski, K. Codling, and P.A. Hatherly, Science **246**, 1029 (1989).
12. L.J. Frasinski, P.A. Hatherly, and K. Codling, Phys. Lett. A **156**, 227 (1989).
13. K. Zhao, G. Zhang, and W.T. Hill, III, Phys. Rev. A **68**, 063408 (2003).
14. B. Friedrich and D. Herschbach, Phys. Rev. Lett. **74**, 4623 (1995).
15. B. Friedrich and D. Herschbach, J. Phys. Chem. **99**, 15686 (1995).
16. C. Cornaggia, Phys. Rev. A **54**, R2555 (1996).
17. A. Hishikawa, A. Iwamae, and K. Yamanouchi, J. Chem. Phys. **111**, 8871 (1999).
18. A. Hishikawa, A. Iwamae, and K. Yamanouchi, Phys. Rev. Lett. **83**, 1127 (1999).
19. T. Seideman, M.Y. Ivanov, and P.B. Corkum, Phys. Rev. Lett. **75**, 2819 (1995).
20. J.H. Posthumus, L.J. Frasinski, A.J. Giles, and K. Codling, J. Phys. B **28**, L349 (1995).
21. T. Zuo and A.D. Bandrauk, Phys. Rev. A **52**, R2511 (1995).
22. M. Brewczyk, K. Rzążewski, and C.W. Clark, Phys. Rev. Lett. **78**, 191 (1997).
23. P. Hering, M. Brewczyk, and C. Cornaggia, Phys. Rev. Lett. **85**, 2288 (2000).
24. P.B. Corkum, Phys. Rev. Lett. **71**, 1994 (1993).

5 First-Principles Density-Functional Approach for Many-Electron Dynamics Under Intense Laser Fields

Kazuhiro Yabana, Tomohito Otobe, and Jun-Ichi Iwata

[1] Institute of Physics and Center for Computational Sciences, University of Tsukuba, Tsukuba 305-8571, Japan yabana@nucl.ph.tsukuba.ac.jp
[2] Kansai Photon Science Institute, Japan Atomic Energy Agency, Kizu, Kyoto 619-0215, Japan otobe.tomohito@jaea.go.jp
[3] Center for Computational Sciences, University of Tsukuba, Tsukuba 305-8571, Japan iwata@comas.frsc.tsukuba.ac.jp

5.1 Introduction

When the intensity of the laser field is comparable to the strength of the internal electric field that binds electrons in materials, an interesting interplay is expected between the external laser field and the bound electrons in materials. The relevant electron motion is characterized by a nonlinear and nonperturbative dynamics. For a pulse with a few tens of femto-second duration, only electron dynamics occur in the first stage of the response, while the ion wave packet is frozen during the pulse period. To understand the electron dynamics under the intense and short-pulse laser field, computational approaches directly solving the time-dependent Schrödinger equation are very useful [1–5].

Even though computational power is increasing rapidly, quantum dynamics calculations for systems with many degrees of freedom are still very demanding. Computational studies are thus limited to systems with several spatial degrees of freedom. For a single electron problem, one may simply solve the time-dependent Schrödinger equation in three spatial dimensions. For systems with more than one electron, there are two distinct directions in the computational approaches: One is to solve the many-electron Schrödinger equation exactly. This approach enables us to treat two-electron correlation accurately. However, to make the calculation feasible, the analysis is often limited to cases with axial symmetry. The other is a mean-field approach in which a many-electron problem is transformed approximately into an effective single-electron problem. In the mean-field approach, a part of the correlation effects are taken into account through the time-dependence of the mean-field Hamiltonian. It can be applied to systems with a number of electrons, from atoms to large molecules, solids, and so on. In this chapter, we present basic aspects and our recent progress on the latter approach.

We rely upon the density functional theory (DFT), which is now well recognized as a basic tool to investigate properties of materials without any adjustable parameters. The ordinary DFT describes the electronic ground state. An extension to the electronic dynamics is known as the time-dependent density functional theory (TDDFT) [6], and has been applied successfully to the electronic excitations and optical responses [7–12]. We have been developing a computational method capable of describing three-dimensional electron dynamics in which the orbitals are expressed on the grid points in the three-dimensional Cartesian coordinate [9,13]. The method can be applied for various systems and dynamics, including the linear optical response [9,11,12], nonlinear optical response [14], collision of highly charged ions [15], and the dielectric function of solids [16]. We are developing the method as a general simulation tool to describe electron dynamics in the first-principles level.

In the TDDFT approach, the most significant correlation effect that can be incorporated is the dynamical screening, which is one of the key quantities that bridge the optical response between small systems (atoms and molecules) and extended systems (solids). However, phenomena directly related to the two-electron correlation, such as the collision induced ionization, may not be easily incorporated if one employs a simple approximation such as adiabatic local density approximation for the exchange-correlation potential. We consider that the electron dynamics under intense laser field is a very good test-ground to examine usefulness and limitation of the TDDFT to describe many-electron dynamics.

In the following, we first present a brief explanation of the DFT and the TDDFT. Then we discuss the optical response of molecules to a weak field in the linear response approach which is the most successful application of the TDDFT. The framework is presented in a real-time formalism with an impulsive dipole field as a perturbation. The linear response results will demonstrate usefulness and accuracy of the TDDFT. It also clarifies how the physical quantities are linked between the original many-body dynamics and the auxiliary Kohn–Sham dynamics. We then discuss electron dynamics under intense laser field. We describe a tunnel ionization process under a static intense field. There we elucidate a significance of the molecular orbital properties and the screening effect [17]. We then report our current progress on the real-time many-electron dynamics under intense laser field, taking ionization of oriented nitrogen molecule as an example.

5.2 Static and Time-Dependent DFT

As the name suggests, the electron density $n(\boldsymbol{r})$ is a basic variable in the DFT. The energy of the system is expressed as a functional of the density, $E[n(\boldsymbol{r})]$, and the variation of it with respect to the density yields the equation that determines the electronic ground state. Fundamental theory of the DFT ensures the existence of such an energy functional [18], however, no known systematic procedure describes how to construct it. In the practical applications, one must introduce a local density approximation (LDA) or its variants at some stage of calculation. Most of the practical calculations are achieved in the Kohn–Sham framework in which the auxiliary single-particle orbitals $\phi_i(\boldsymbol{r})$ are introduced. With the introduction of the single electron orbitals, the kinetic energy term can be treated without resorting to the LDA, and the concept of the molecular orbitals comes in in the DFT framework [19].

The variation with respect to the orbitals, $\phi_i(\boldsymbol{r})$, instead of the density yields a self-consistent equation for $\phi_i(\boldsymbol{r})$.

$$h_{\mathrm{KS}}[n(\boldsymbol{r})]\phi_i(\boldsymbol{r}) = \epsilon_i \phi_i(\boldsymbol{r}) \; , \tag{5.1}$$

$$h_{\mathrm{KS}}[n(\boldsymbol{r})] = -\frac{\hbar^2}{2m}\nabla^2 + V_{\mathrm{ion}}(\boldsymbol{r}) + \int \mathrm{d}\boldsymbol{r}' \frac{e^2}{|\boldsymbol{r}-\boldsymbol{r}'|} n(\boldsymbol{r}') + \mu_{\mathrm{xc}}[n(\boldsymbol{r})] \; , \tag{5.2}$$

where the density is given by $n(\boldsymbol{r}) = \sum_i \phi_i^2(\boldsymbol{r})$. The hamiltonian is composed of the electron kinetic energy, the electron–ion potential for which we will employ the norm-conserving pseudopotential [20, 21], electron–electron interaction (Hartree potential), and the so-called exchange-correlation potential. This equation looks similar to the Hartree–Fock equation: the exchange potential in the Hartree–Fock theory, which is a non-local function in coordinate, is replaced with the local potential, $\mu_{\mathrm{xc}}[n(\boldsymbol{r})]$. The DFT ensures that there exists an exchange-correlation potential that describes the ground state exactly. In practice, the LDA (with gradient corrections) is usually employed and it gives results with reasonable accuracy.

The DFT describes the electronic ground state, and the TDDFT is applicable to the electron dynamics in the presence of the time-dependent external potential, $V_{\mathrm{ext}}(\boldsymbol{r}, t)$ [6]. The electron dynamics is described by the time-dependent Kohn–Sham equation,

$$\mathrm{i}\hbar \frac{\partial}{\partial t}\psi_i(\boldsymbol{r}, t) = \{h_{\mathrm{KS}}[n(\boldsymbol{r}, t)] + V_{\mathrm{ext}}(\boldsymbol{r}, t)\}\, \psi_i(\boldsymbol{r}, t) \; . \tag{5.3}$$

The exchange-correlation potential, $\mu_{\mathrm{xc}}[n(\boldsymbol{r}, t)]$, is now a functional of the time-dependent density, $n(\boldsymbol{r}, t) = \sum_i |\psi_i(\boldsymbol{r}, t)|^2$. In this equation, the many-body correlation effects are included in the change of the Kohn–Sham Hamiltonian $h_{\mathrm{KS}}[n(\boldsymbol{r}, t)]$ through the time-dependence of the density. The change of the Hartree term is known as the correlation effect in the random phase approximation. Other correlation effects are included in the exchange-correlation potential. In many applications, fortunately, a simple adiabatic approximation which utilizes the same exchange-correlation potential as the static one has been known to provide reasonable results for electronic excitations in molecules. However, it is necessary to pay careful attention to describe the ionization processes, as will be discussed later.

5.3 Linear Optical Response: Electron Dynamics Under Weak Impulsive Dipole Field

Before proceeding to the application of the DFT and TDDFT for the intense laser physics, we present a framework and an application of TDDFT for linear optical response in molecules. There are several equivalent formulations for the linear response theory. We adopt here a real-time one with a weak impulsive field. We often achieve practical calculations in this form [9,11]. The real-time formalism provides a good opportunity to understand the electron dynamics after the pulse external field is applied. Results of the calculation will also be useful to demonstrate usefulness and accuracy of the TDDFT for describing optical responses.

To extract physical quantities from the TDDFT calculations, we make a comparison of electron density between two systems: the original many-electron system with electron–electron interaction and the auxiliary Kohn–Sham system in which the electrons interact through the mean-field. Let us first examine how the system evolves after an irradiation of the weak impulsive dipole field.

We denote the static Schrödinger equation of the original system as $H\Phi_n = E_n\Phi_n (n = 0, 1, 2, \cdots)$, where H is the many-electron Hamiltonian. We apply an instantaneous perturbation at $t = 0$, $\hbar k\hat{D}\delta(t)$. $\hat{D}$ is a dipole operator $\hat{D} = \sum_i z_i$, z_i is a z coordinate of i-th electron. k is a small parameter that controls the strength of the perturbation. We assume that the system is in the ground state Φ_0 before the external field is applied at $t = 0$. The solution $\Psi(t)$ of this system follows the time-dependent Schrödinger equation,

$$\mathrm{i}\hbar\frac{\partial}{\partial t}\Psi(t) = \left\{H + \hbar k\hat{D}\delta(t)\right\}\Psi(t) \, . \tag{5.4}$$

The wave function of this system right after the external pulse field at $t = 0$ is expressed as $\Psi(t = 0_+) = \mathrm{e}^{-\mathrm{i}k\hat{D}}\Phi_0$, and the solution for $t > 0$ is formally given as

$$\Psi(t) = \mathrm{e}^{-\mathrm{i}Ht/\hbar}\mathrm{e}^{-\mathrm{i}k\hat{D}}\Phi_0 = \sum_i \mathrm{e}^{-\mathrm{i}E_i t/\hbar}\Phi_i\langle\Phi_i|\mathrm{e}^{-\mathrm{i}k\hat{D}}|\Phi_0\rangle \, . \tag{5.5}$$

We then take the expectation value of the dipole operator $\hat{D}$, retaining terms linear in the parameter k,

$$D(t) = \langle\Psi(t)|\hat{D}|\Psi(t)\rangle \simeq -2k\sum_i \sin[(E_i - E_0)/\hbar]|\langle\Phi_i|\hat{D}|\Phi_i\rangle|^2 \, . \tag{5.6}$$

This expression implies that the function $D(t)$ as a function of time includes various information related to the excited states of the system. Taking a Fourier transformation of $D(t)$, we obtain

$$\begin{aligned}\tilde{D}(\omega) &\equiv \int_0^\infty \mathrm{d}t\mathrm{e}^{\mathrm{i}\omega t}D(t) \\ &= \hbar k\sum_i\left(\frac{1}{\hbar\omega - (E_n - E_0)} - \frac{1}{\hbar\omega + (E_n - E_0)}\right)|\langle\Phi_i|\hat{D}|\Phi_i\rangle|^2 \, . \end{aligned} \tag{5.7}$$

Thus the time-dependent dipole moment $D(t)$ includes information on the excitation energy $E_i - E_0$ and the transition matrix elements. We can extract them from the dipole moment $D(t)$ by a Fourier transformation.

We next consider a system described by the time-dependent Kohn–Sham equation. A comparison of the function $D(t)$ between the fully many-body system described above and the TDDFT system leads us how the physical quantities are extracted from the TDDFT calculation. At $t < 0$, the system is described by a static Kohn–Sham solution which we express as $\phi_i(\boldsymbol{r})$. We apply the impulsive dipole field $\hbar k\hat{z}\delta(t)$. The perturbation modifies the orbital as $\psi(t = 0_+) = \mathrm{e}^{-\mathrm{i}kz}\phi_i$, a product of the plane wave and the static orbital. For this initial condition, we solve the time-dependent Kohn–Sham equation and calculate the density as a function of time, $n(\boldsymbol{r}, t) = \sum_i |\psi_i(\boldsymbol{r}, t)|^2$. The dipole moment is calculated from the time-dependent density,

$$D(t) = \int \mathrm{d}\boldsymbol{r} z n(\boldsymbol{r}, t) \,. \tag{5.8}$$

We compare the dipole moment $D(t)$ in the Kohn–Sham calculation with a fully quantum one, (5.6), and extract excitation energies and the transition matrix elements.

For a comparison with measurements, the oscillator strength distribution is an important quantity. It is related to the imaginary part of the Fourier transform of $D(t)$,

$$\frac{\mathrm{d}f}{\mathrm{d}\omega} = -\frac{2m\omega}{\pi\hbar k}\mathrm{Im}\int_0^\infty \mathrm{d}t\mathrm{e}^{\mathrm{i}\omega t}D(t) \,. \tag{5.9}$$

The oscillator strength distribution is proportional to the linear absorption cross section. It can be easily shown that the TKR sum rule, $\int_0^\infty \mathrm{d}\omega(\mathrm{d}f/\mathrm{d}\omega) = N_\mathrm{e}$ with N_e being the valence electron number, is satisfied in the TDDFT calculation.

To demonstrate the real-time response calculation, we show a result of linear optical absorption of C_{60} molecule as an example. We apply the instantaneous dipole field at $t = 0$. For this molecule of high symmetry, calculation for only one distortion direction is sufficient to obtain the dipole response. We show the dipole moment as a function of time, $D(t)$, in Fig. 5.1. The dipole distortion modifies the orbital wave function as $\psi_i(\boldsymbol{r}, t = 0_+) = \mathrm{e}^{-\mathrm{i}kz}\phi_i(\boldsymbol{r})$. For σ orbitals, the distortion induces the $\sigma - \sigma^*$ excitation, while for π orbitals, $\pi - \pi^*$ excitations are induced. Right after the dipole field is applied, all the electrons move coherently and give rise to a sharp rise of the dipole moment. The time derivative at $t = 0$, $\mathrm{d}D(t)/\mathrm{d}t|_{t=0}$, is proportional to the number of electrons and to the TRK sum rule value.

The σ^* components included in $\psi_i(t)$ has energies above the ionization threshold. These components in the wave function will not come back to the molecule once they are emitted outside the molecule. After first maximum of $D(t)$, only the bound excitations, $\pi - \pi^*$, contribute to the dipole moment.

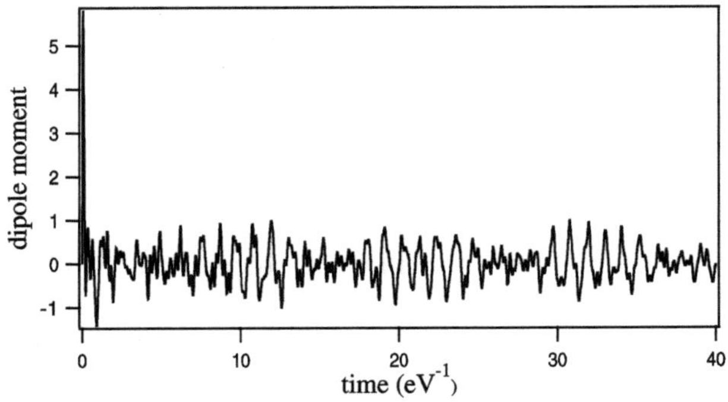

Fig. 5.1. The dipole moment $D(t)$ as a function of time is shown for C_{60} molecule

Taking the Fourier transform of the dipole moment, we obtain the dipole polarizability and the oscillator strength distribution, which are shown in Figs. 5.2 and 5.3, respectively. We note that the spectrum for a whole frequency region can be obtained from a single time-dependent calculation. Below the ionization threshold, the oscillator strength show discrete peaks which correspond to the electronic excitations of C_{60} molecule with dipole transition strength. In Fig. 5.3, the experimental spectrum is also indicated [22]. The position and the magnitude of the oscillator strength is described fairly well.

We summarize the real-time linear response theory presented in this section. The linear response calculation is the most successful application of TDDFT. The electronic excitation energies and the oscillator strength distribution for small and large molecules are described with a reasonable accuracy. Since there holds a principle of superposition for responses to weak perturbations, one may employ a distortion with any time profile in studying the linear response. We presented a formulation and a computation with an impulsive dipole field. This description is useful for gaining intuitive understanding on how the excited states comes in in the time evolution of the wave function.

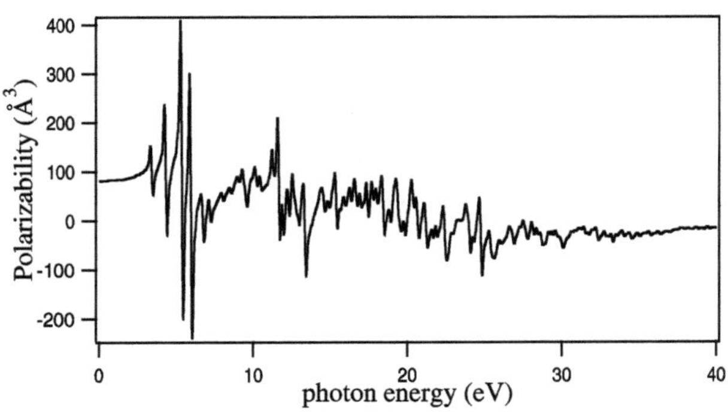

Fig. 5.2. The dipole polarizability of C_{60} molecule obtained from $D(t)$ by Fourier transformation

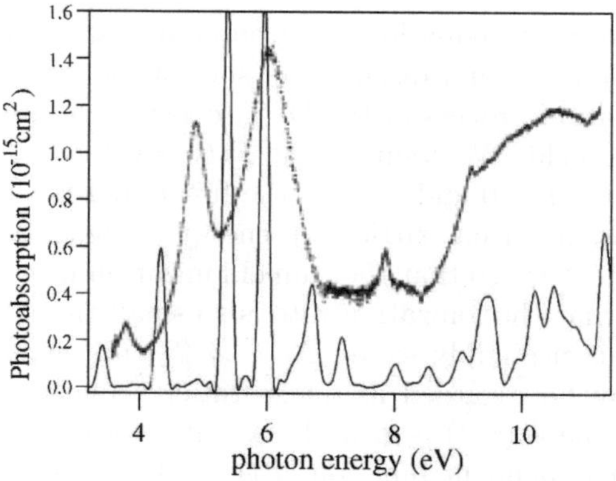

Fig. 5.3. The absorption cross section of C_{60} calculated from the Fourier transformation of $D(t)$. The measurement [22] is shown by *dots*

A single time-dependent calculation provides physical quantities, the dipole polarizability and the oscillator strength distribution, for a whole spectral region through the Fourier transformation.

5.4 Ionization Under Static Dipole Field: Tunnel Ionization

5.4.1 DFT Treatment for Static Ionization

We now turn to the electron dynamics under the intense laser field. The intense laser field is described by a dipole external potential. Taking the direction of the laser polarization as z-axis, the external potential is expressed as $V_{\mathrm{ext}}(\boldsymbol{r},t) = eF(t)z$, where the function $F(t)$ specifies the time profile of the electric field. It is characterized by a frequency ω and the envelope function $f(t)$ which is a smooth function in time, $F(t) = f(t)\sin\omega t$. The electron dynamics is described by the time-dependent Kohn–Sham equation,

$$\mathrm{i}\hbar\frac{\partial}{\partial t}\psi_i(t) = \{h_{\mathrm{KS}}[n(t)] + V_{\mathrm{ext}}(t)\}\,\psi_i(t)\ . \tag{5.10}$$

In this section, we consider an ionization process under a static external field. In most experiments, the laser frequency $\hbar\omega$ is much smaller than the excitation energy of atoms and molecules. More precisely, if the Keldysh parameter γ, which is a ratio between the period of the laser and the tunneling time, is much smaller than unity, the static treatment is considered to be appropriate.

To describe the ionization process under static intense field, we will solve the static Kohn–Sham equation. Unlike the vanishing boundary condition in ordinary Kohn–Sham equation for ground states, we impose a decaying boundary condition. We first discuss how such static treatment can be justified for the ionization process in the DFT, considering an adiabatic switching of the intense field. We assume that $F(t) = 0$ for $t < 0$ and $F(t) = F_0$ for $t > T_0$. Between $t = 0$ and $t = T_0$, the function $F(t)$ increases smoothly and slowly. We assume that barrier top energy is higher than the ionization potential I_{P}, $e^2F_0 < I_{\mathrm{P}}^2$, so that the tunnel ionization mechanism dominates. We also assume that the ionization rate is so small that the ionization during $0 < t < T_0$ is negligibly small. At $t \gg T_0$, the wave function around the molecule may be regarded as quasi-static. Namely, the wave function around the molecule $\psi_i(\boldsymbol{r}, t)$ is given by a static wave function, $\phi_i(\boldsymbol{r})$ multiplied with a time-dependent function $c_i(t)$, $\psi_i(\boldsymbol{r}, t) = c_i(t)\phi_i(\boldsymbol{r})$. The static function $\phi_i(\boldsymbol{r})$ is called the Gamow state. One should note that ϕ_i does not vanish asymptotically but is connected to the outgoing wave.

To define the ionization rate, we consider a volume V which encloses the molecule. We define a current $\boldsymbol{j}_i$ of the i-th orbital as usual,

$$\boldsymbol{j}_i = \frac{i\hbar}{2m}(\psi_i^* \boldsymbol{\nabla}\psi_i - \psi_i \boldsymbol{\nabla}\psi_i^*) \,. \tag{5.11}$$

The number of electrons that passes through the surface S of the volume V in a unit time is given by $\int_S \mathrm{d}S\boldsymbol{n} \cdot \boldsymbol{j}_i$ where $\boldsymbol{n}$ is the normal vector on the surface S. The number of electrons passing through the surface S divided by the number of electrons in V is the ionization rate of i-th orbital, w_i,

$$w_i = \frac{\int_S \mathrm{d}S\boldsymbol{n} \cdot \boldsymbol{j}_i}{\int_V \mathrm{d}V|\psi_i|^2} \,. \tag{5.12}$$

The total ionization rate is obtained by summing up all the orbitals, $w = \sum_i w_i$.

The Gamow state solution $\phi_i(\boldsymbol{r})$ satisfies the static Kohn–Sham equation,

$$\{h_{\mathrm{KS}}[n(\boldsymbol{r})] + eF_0 z\}\,\phi_i(\boldsymbol{r}) = \epsilon_i \phi_i(\boldsymbol{r}) \,. \tag{5.13}$$

Because the orbital ϕ_i does not vanish asymptotically, the orbital energy ϵ_i is complex. The imaginary part of the eigenvalue $\mathrm{Im}\epsilon_i$ is related to the ionization rate. To see it, we multiply ϕ_i^* to the static Kohn–Sham equation and ϕ_i to the complex conjugate of the equation. Then subtracting both sides, we have

$$-\frac{\hbar^2}{2m}(\phi_i^* \nabla^2 \phi_i - \phi_i \nabla^2 \phi_i^*) = 2i\mathrm{Im}\epsilon_i |\phi_i|^2 \,. \tag{5.14}$$

Integrating both sides over the volume V, and employing the Gauss theorem, we have

$$\int_S \mathrm{d}S\boldsymbol{n} \cdot \boldsymbol{j}_i = -\frac{2}{\hbar}\mathrm{Im}\epsilon_i \int \mathrm{d}\boldsymbol{r}|\phi_i|^2 \,. \tag{5.15}$$

We thus find that the ionization rate is related to the imaginary part of the energy eigenvalue, $w_i = -(2/\hbar)\text{Im}\epsilon_i$. The total ionization rate is given by the sum of the imaginary part of the orbital energies,

$$w = -\frac{2}{\hbar}\sum_i \text{Im}\epsilon_i \,. \tag{5.16}$$

5.4.2 Choice of the Exchange-Correlation Potential

In DFT calculations, electron correlation effects are incorporated in the exchange-correlation potential $\mu_{\text{xc}}[n(\boldsymbol{r})]$. Although the basic theorem assures the existence of an appropriate local potential, we must in practice employ an approximate one. The LDA with gradient corrections is a standard prescription. For a process accompanying ionization dynamics, however, the simple LDA significantly overestimate the ionization rate. This is because the absolute value of the Kohn–Sham orbital energy ϵ_i for the highest occupied orbital is much smaller than the ionization potential in the LDA, as well as the incorrect asymptotic behavior of the exchange-correlation potential. For an electron far outside the molecule, the electron interacts with the molecular ion with $-e^2/r$ potential, and the exact exchange-correlation potential should have such asymptotic behavior. In the LDA, however, the potential decays exponentially. We also note that the electron transfer probability in the highly charged ion shows a similar aspect, too large charge transfer probability if one employs the LDA [15].

In linear response calculations, a gradient correction that produces a correct asymptotic form of the potential has been proposed, which is often abbreviated as LB94 [23]. In the static calculations with LB94, the absolute value of the HOMO energy coincides with the ionization potential in good accuracy, at least for atoms and small molecules. The excitation energies of Rydberg states and the linear absorption spectra around and above the ionization threshold are well reproduced [10, 12]. However, in the intense field applications, the ionized electron density oscillates in time and space, and extends far-outside the molecule. The gradient correction, which produces a finite potential from an exponentially decaying density, is not designed to describe the potential induced from the emitted electrons. Therefore, the LB94 procedure will not be appropriate for the ionization dynamics induced by the intense field.

If one treats exchange potential exactly in nonlocal form, the correct asymptotic behavior of the potential is automatically satisfied. The self-interaction term included in the Hartree potential is cancelled by a term in the Fock potential. In our numerical scheme of three-dimensional grid representation, however, the exact evaluation of the nonlocal exchange potential requires heavy computational cost [15].

There is another variational approach for many-electron system, the optimized effective potential approach [24], in which the energy is considered as a functional of the local potential instead of the density in the DFT. The variation with respect to the local potential results in an integral equation. It has been known that the self-interaction problem is absent in this approach, the potential has a correct asymptotic behavior. An explicit construction of the optimized effective potential solving the integral equation is, however, a difficult numerical task. Fortunately, there has been proposed an approximate way of constructing the optimized effective potential by Krieger, Lee, and Iafrate (KIL) [25]. Their prescription allows us to construct the local potential of correct asymptotic behavior with a reasonable computational cost. In the static ionization calculations shown below, this prescription is employed.

5.4.3 Ionization Rate of Atoms and Small Molecules

Before showing results, we briefly mention about a computational procedure [17]. The tunnel ionization rate is calculated by solving the static Kohn–Sham equation

$$\{h_{\mathrm{KS}}[n(\boldsymbol{r})] + eF_0 z\}\,\phi_i(\boldsymbol{r}) = \epsilon_i \phi_i(\boldsymbol{r})\ . \tag{5.17}$$

In the decaying boundary condition, the electron density extends outside the molecule. However, the electrons already emitted to outside of the molecule should not influence the ionization process. In the actual calculation, we achieve the calculation in two steps. We first solve the self-consistent problem ignoring the ionization. We realize it by placing a potential wall at a certain radius. At this stage, we capture an important many-body correlation, the screening effect which is induced by the external dipole field, $eF_0 z$. After solving the self-consistent problem, we fix the Kohn–Sham Hamiltonian and solve the static Kohn–Sham equation again, with the outgoing boundary condition. In practice, we impose the outgoing boundary condition approximately by placing an absorbing boundary potential outside the molecule. We solve the equation expressing the wave function on grid points in a three-dimensional adaptive Cartesian coordinate. Further details of the computational procedure is found in [17].

We first discuss ionization rate of atoms and small molecules. We show in Fig. 5.4 the ionization rate of rare gas atoms, Ar and Xe. The calculated ionization rate is shown by open squares. The measurements are available with the linearly polarized, ultrashort (30 fs) laser at 800 nm [26]. In the measurements, relative ionization rates at various laser intensities are available. We plot the measured rates in the figure, normalizing the measurements to coincide with the calculation at the laser intensity of $9\times10^{13}\ \mathrm{W/cm^2}$ for Xe atom, and at the laser intensity of $1.3\times10^{14}\ \mathrm{W/cm^2}$ for Ar atom. We also plot the rate in the ADK theory [27] by the dotted curve. Since we show the rate for a static field, we compare the rate expression in the ADK theory for a static

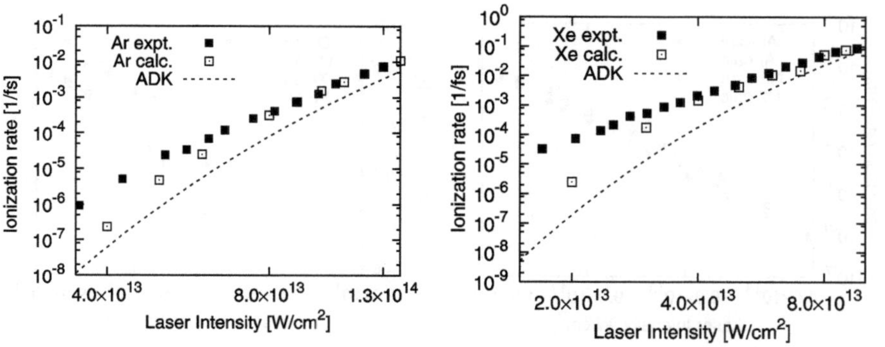

Fig. 5.4. The ionization rate of Ar (*left*) and Xe (*right*) as a function of the laser intensity. Calculated rates (*open squares*) are compared with the measurements (*filled squares*) [26] and the ADK theory (*dashed curve*)

field. The rate expression for the static field w_{static} is related to the rate expression for the periodic one w_{periodic} by $w_{\text{periodic}} = \sqrt{3F/\pi\beta^3}w_{\text{static}}$ where $\beta = \sqrt{2I_{\text{P}}}$.

As seen from the figure, the measured dependence on the laser intensity is described fairly well by the calculation. The calculated rate decreases much faster than the measurements when the laser intensity is weak ($< 3 \times 10^{13}\,\text{W/cm}^2$ for Xe and $< 6 \times 10^{14}\,\text{W/cm}^2$ for Ar). This trend may be attributed to the effect of the multiphoton process, which is not included in the present static treatment. Comparing our calculation with the ADK formula, the absolute values of the ionization rate are slightly higher (about a factor of two) for all the region of the laser intensity. Thus the laser intensity dependence is very similar between our calculation and the ADK results. This indicates the usefulness of the ADK theory in describing the relative ionization rates of atoms with many electrons for wide laser intensities.

We next show results of tunnel ionization rates for diatomic molecules, taking N_2 and O_2 as examples. We note that the properties of HOMO are different among molecules. The HOMO of the N_2 is the σ orbital, having zero angular momentum around the symmetry axis. On the other hand, the HOMO of O_2 is π^*, having one-unit angular momentum around the symmetry axis and having also the nodal plane perpendicular to the molecular axis. The O_2 molecule has a spin-triplet ground state. We take account of the spin unsaturated property in the KLI framework.

In the left panel of Fig. 5.5, we show the calculated ionization rate of N_2 as a function of the laser intensity. The ionization rate of Ar atom whose ionization potential is close to that of N_2 is also plotted. In the measurements, the absolute value of the rate is not available. As before, we scaled the measured rate of Ar atom to coincide with the calculation at laser intensity of $1.3 \times 10^{14}\,\text{W/cm}^2$. We note that the relative intensity between Ar and N_2 has been measured and reliable. The calculation nicely reproduces

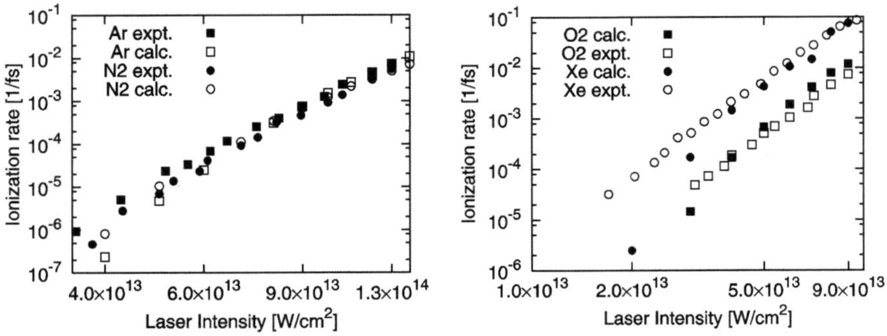

Fig. 5.5. The ionization rates of N_2 and Ar in the *left panel*, and O_2 and Xe in the *right panel*, are shown as a function of the laser intensity. Calculated rates are compared with the measurements [26]

the relative ionization rates between Ar and N_2 as well as the laser intensity dependence. The dependence on the laser intensity is very similar between N_2 and Ar.

We next turn to the ionization of O_2 molecules. In the right hand panel of Fig. 5.5, we compare the ionization rate of Xe atom and O_2 molecule. The measured ionization rate is also plotted. As before, we scaled the measured rate of Xe atom so that the calculated rate coincides with the measurement at the laser intensity of 9×10^{13} W/cm^2. The relative ionization rate of Xe atom and O_2 molecule shows marked difference, in spite of the similar ionization potential between two systems. Our calculation succeeds reproducing this difference. We have confirmed that this strong suppression of the ionization rate in O_2 molecule is related to the π^* character of the HOMO. Since the π^* orbital vanishes along the molecular axis, the ionization to the direction parallel to the molecular axis is strongly suppressed in O_2 molecule. In other words, the π orbital feels the centrifugal barrier along the molecular axis. This centrifugal barrier hinders the emission of the electrons along the molecular axis. The significance of the HOMO character on the ionization has been pointed out in other theories as well [28, 29].

We show in Fig. 5.6 the calculated ionization rate as a function of the angle between the laser polarization direction and the molecular axis. For N_2 molecule, the ionization rate is large when the direction of the electric field is parallel to the molecular axis, reflecting the σ character of HOMO. In the figure, the angle dependence with and without the screening effect is also plotted, which will be discussed in the next subsection. The dashed curve indicates recent measurements [30]. In contrast to the N_2 case, the ionization rate in O_2 molecule is suppressed either the electric field direction is parallel or perpendicular to the molecular axis, as in the right-hand panel of Fig. 5.6. This reflects the π^* character of the O_2 HOMO.

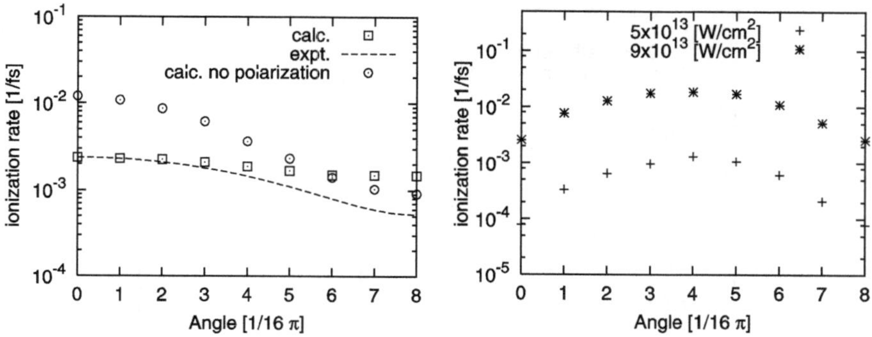

Fig. 5.6. The ionization rate of molecules as a function of the angle between the electric field and the molecular axis. The *left panel* show the ionization rate of N_2, and the *right panel* for O_2. See text for detail

The calculated suppression of the ionization rate for O_2 molecule is not enough to explain fully the measured suppression. One possible origin of this discrepancy is a limited accuracy of the calculated orbital energy in our approach. The ionization potential of O_2 molecule is 12.06 eV. However, in our calculation in the KLI treatment with the self-interaction correction, the orbital energy of the HOMO is −11.29 eV. Therefore, our calculated ionization rate of O_2 molecule may be overestimated to some extent. On the other hand, the HOMO energy of Xe is 12.07 eV in our calculation, close to the measured value, 12.13 eV.

5.4.4 Screening Effect on the Ionization Rate

As the molecular size increases, the electron polarization induced by the external field becomes significant. The polarized electrons screen the external electric field inside the molecule. In simplified model approaches such as the ADK and KFR theories, the screening effect is usually ignored. In our approach, the screening effect is taken into account through the construction of the self-consistent potential in (5.17). In this subsection, we will examine which molecular size the screening effect comes to show a sizable effect.

We show in Fig. 5.7 the calculated ionization rate for several systems of different sizes, Ar atom, ethylene, and benzene molecules. The ionization rates with and without screening effect are plotted. The ionization rate without the screening effect is calculated from the Kohn–Sham eigenvalue of Gamow state in which the Kohn–Sham Hamiltonian is constructed in the absence of the external dipole field. The ionization rate is shown when the electric field is parallel to the molecular axis for ethylene, and the electric field in the molecular plane for benzene.

For Ar atom, the screening effect suppresses the ionization rate, at most by a factor of two. This suppression rate is almost independent of the laser intensity. Therefore, for the dependence on the laser intensity which are often

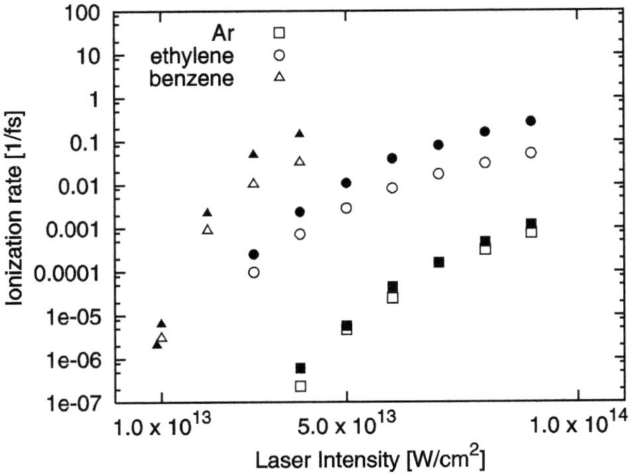

Fig. 5.7. Ionization rates of Ar, ethylene, and benzene are plotted as a function of the laser intensity. The electric field is applied parallel to the axis for ethylene and in the plane of molecule for benzene. The ionization rates are shown with screening by *open symbols*, while without the screening effect by *filled symbols*

analyzed, the screening effect is not significant. As the molecular size increases, the screening effect becomes more and more significant. For ethylene and benzene molecules, the screening effect suppress the ionization rate by factor of three and five, respectively, at the laser intensity of 3×10^{13} W/cm^2. We have also confirmed that the dipole moment induced by the intense laser field is almost linear to the applied external field. This indicates that the screening effect may be incorporated by considering the linear dipole polarization. Another interesting issue is the directional dependence of the screening effect. Indeed, as shown in the left panel of Fig. 5.6, the screening effect is quite large when the external field is parallel to the molecular axis.

We thus conclude that the screening effect substantially changes the absolute ionization rate, factor of two for rare gas atom (Ar) and five for a medium size molecules (benzene). This factor will increase more and more as the molecular size increases. Thus the screening effect is important to understand the system dependence of the ionization rate. In [31], a systematic suppression of the saturation intensity is reported as the molecular size increases. The screening effect will surely be one of the important factors that should be examined to understand this trend. We should, however, also keep in mind that the HOMO of the conjugated organic molecules are π orbitals. As discussed in the previous subsection, the HOMO character has a significant effects on the ionization rate. The ionization rate of O_2 molecule is, about an order of magnitude, suppressed, in comparison with Xe. One should also notice that the orbital level density around HOMO becomes larger as the molecular size increases. In our preliminary study, several orbitals with different orbital

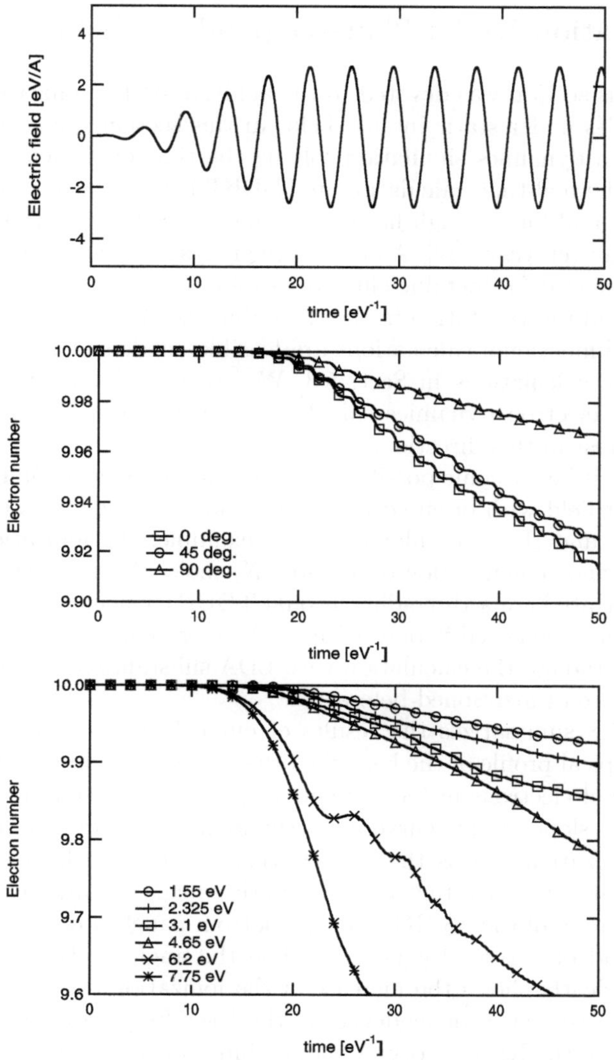

Fig. 5.8. Time-dependent calculation of ionization for N_2 molecule. (*Top*) The intensity of the laser electric field as a function of time (800 nm). (*Middle*) The number of electrons in N_2 as a function of time under an irradiation of 800 nm laser shown in the *top panel*. The angle between the electric field and the molecular axis is set at 0, 45, and 90 degrees. (*Bottom*) The number of electrons in N_2 as a function of time for various laser frequencies. The angle between the electric field and molecular axis is set at 45 degree

characters contribute to the total ionization rate in benzene molecule. For a unified understanding, these aspects should be taken into account in the ionization under static field.

5.5 Ionization Under Time-Dependent Field

The previous section was devoted to the analyses on the ionization of atoms and molecules under static intense field. In this section, we consider a real-time electron dynamics simulation solving the time-dependent Kohn–Sham equation. The real-time calculation in TDDFT has been achieved for ionization process and for the high harmonic generation of many-electron systems for more than ten years [32]. An extensive analysis on the electron dynamics has been achieved for metallic clusters as well [5].

In our grid representation in the three-dimensional Cartesian coordinate, fully three-dimensional calculations are feasible, as demonstrated in the linear response calculations in Sect. 5.3. We are now developing fully three-dimensional electron dynamics calculations. In this section, we report our recent progress in this direction.

Recently it has become possible to achieve measurements of interaction of intense laser field with oriented molecule, fixing the angle between the laser polarization and the molecular axis. For example, the ionization rate and the high harmonic generation of oriented N_2 molecule are measured [30, 33]. We demonstrate here a three-dimensional TDDFT calculation for ionization of N_2. We have achieved it only with LDA at present to save the computational time, though the calculation with LDA substantially overestimate the ionization rate as mentioned before.

Figure 5.8 summarizes the results of our calculation. The upper panel show the typical profile of the laser field that we employ. In the middle panel, the number of electrons inside a sphere of 5 Å radius is shown as a function of time. The slope of this curve gives the ionization rate. The figure shows that the ionization rate is the largest when the electric field is parallel to the molecular axis, consistent with the static calculation and reflecting the σ orbital character of the N_2 HOMO. The bottom panel shows the calculations for laser field of various frequencies while the angle is fixed at 45 degree. The figure clearly shows the increase of the ionization rate as the laser frequency increases. Such dependence on the laser frequency has been known analytically in the KFR theory. A sudden jump in the ionization rate is seen between the laser frequencies of 4.65 eV and 6.2 eV. We have not yet analyzed the origin of this behavior.

In the limit of low frequency calculation, the ionization rate obtained from the slope should coincide with the rate calculated from the Gamow state eigenvalue in the previous section. We numerically confirmed it in the ionization of the hydrogen atom.

5.6 Summary

The TDDFT has been developing as a general simulation tool to investigate electron dynamics in the first-principle level. The TDDFT has been most successful in the combined use with the linear response theory. In fact, the TDDFT has now come to be a standard quantum chemical tool to investigate electronic excitations in molecules. The application of the TDDFT for intense laser field has also been extensively investigated in the last decade. However, the application was mostly limited to problems with axial symmetry. We consider that the fully three-dimensional TDDFT calculations will be useful and fruitful to promote understanding for rich phenomena in the intense laser science.

We have been developing a computational approach in which the time-dependent Kohn–Sham equation is solved in the grid representation of three-dimensional Cartesian coordinate. It has been applied for linear optical responses of various molecules and solids. We are now proceeding to the electron dynamics induced by the intense laser field. Till now, we have analyzed a tunneling ionization rate of molecules under static intense field in the Kohn–Sham formalism. The calculations treat properties of the molecular orbitals appropriately, and include screening effect as a many-body correlation. We have also undertaken a real-time electron-dynamics simulation, taking an oriented nitrogen molecule as a first target. The three-dimensional real-time calculation requires large computational resources. It is a challenging subject in the field of large-scale computational physics.

This work is supported by NAREGI Nanoscience Project. Numerical calculations are achieved on the supercomputers at the Institute of Molecular Sciences, and Institute for Solid State Physics, University of Tokyo.

References

1. J.L. Krause, K.J. Schafer, K.C. Kulander: Phys. Rev. A **45**, 4998 (1992)
2. S. Chelkowski, T. Zuo, A.D. Bandrauk: Phys. Rev. A **46**, R5342 (1992)
3. I. Kawata, H. Kono, Y. Fujimura: J. Chem. Phys. **110**, 11152 (1999)
4. C.A. Ullrich, P.-G. Reinhard, E. Suraud: J. Phys. B **30**, 5043 (1997)
5. F. Calvayrac, P.-G. Reinhard, E. Suraud, C.A. Ullrich: Phys. Rep. **337**, 493 (2000)
6. E. Runge, E.K.U. Gross: Phys. Rev. Lett. **52**, 997 (1984)
7. A. Zangwill, P. Soven: Phys. Rev. A **21**, 1561 (1980)
8. W. Ekardt: Phys. Rev. Lett. **52**, 1925 (1984)
9. K. Yabana, G.F. Bertsch: Phys. Rev. B **54**, 4484 (1996)
10. M.E. Casida, C. Jamorski, K.C. Casida, D.R. Salahub: J. Chem. Phys. **108**, 4439 (1998)
11. K. Yabana, G.F. Bertsch: Int. J. Quantum Chem. **75**,55 (1999)
12. T. Nakatsukasa, K. Yabana: J. Chem. Phys. **114**, 2550 (2001)
13. J.R. Chelikowsky, N. Troullier, K. Wu, Y. Saad: Phys. Rev. B **50**, 11355 (1994)
14. J.-I. Iwata, K. Yabana, G.F. Bertsch: J. Chem. Phys. **115**, 8773 (2001)
15. R. Nagano, K. Yabana, T. Tazawa, Y. Abe: Phys. Rev. A **62**, 062721 (2000)
16. G.F. Bertsch, J.-I. Iwata, A. Rubio, K. Yabana: Phys. Rev. B **62**, 7998 (2000)
17. T. Otobe, K. Yabana, J.-I. Iwata: Phys. Rev. A **69**, 053404 (2004)

18. P. Hohenberg, W. Kohn: Phys. Rev. **136**, B864 (1964)
19. W. Kohn, L.J. Sham: Phys. Rev. **140**, A1133 (1965)
20. N. Troullier, J.L. Martins: Phys. Rev. B **43**, 1993 (1991)
21. L. Kleinman, D. Bylander: Phys. Rev. Lett. **48**, 1425 (1982)
22. H. Yasumatsu, T. Kondow, H. Kitagawa, K. Tabayashi, and K. Shobatake: J. Chem. Phys. **104**, 899 (1996)
23. R. van Leeuwen, E.J. Baerends: Phys. Rev. A **49**, 2421 (1994)
24. J.D. Talman, W.F. Shadwick: Phys. Rev. A **14**, 36 (1976)
25. J.B. Krieger, Y. Li, G.J. Iafrate: Phys. Rev. A **45**, 101 (1992)
26. C. Guo, M. Li, P. Nibarger, G.N. Gibson: Phys. Rev. A **58**, R4271 (1998)
27. A.M. Ammosov, N.B. Delone, V.P. Krainov: Sov. Phys. JETP **64**, 1191 (1986)
28. J. Muth-Böhm, A. Becker, F.H.M. Faisal: Phys. Rev. Lett. **85**, 2280 (2000)
29. X.M. Tong, Z.X. Zhao, C.D. Lin: Phys. Rev. A **66**, 033402 (2002)
30. I.V. Litvinyuk, K.F. Lee, P.W. Dooley, D.M. Rayner, D.M. Villeneuve, P.B. Corkum: Phys. Rev. Lett. **90**, 233003 (2003)
31. S.M. Hankin, D.M. Villeneuve, P.B. Corkum, D.M. Rayner: Phys. Rev. Lett. **84**, 5082 (2000)
32. E.K.U. Gross et al.: Topic in Current Chemistry **181**, 81 (1996)
33. J. Itatani et al.: Nature **432**, 867 (2004)

6 Plasma Physics in the Strong Coupling Regime: Intense VUV Laser–Cluster Interaction

Lora Ramunno[1], Christian Jungreuthmayer[1], Carlos F. Destefani[1], and Thomas Brabec[1]

Department of Physics and Center for Photonics Research, University of Ottawa, 150 Louis Pasteur, Ottawa, K1N 6N5 ON, Canada

6.1 Introduction

The first free electron laser (FEL) experiments in the VUV (vacuum-ultraviolet) regime, with photon energy of 12.7 eV and unprecedented peak intensities of nearly 10^{14} W/cm^2, were recently performed at DESY (Deutsches Elektronen-Synchrotron) [1]. Unexpectedly, when cold Xe clusters were exposed to such radiation, high ion charge states up to Xe^{8+} were observed. This has raised considerable interest, as it is difficult to explain the high charge states by conventional ionization or heating mechanisms.

The role of multi-photon ionization is currently under debate. More experiments will be needed to unambiguously determine the influence of nonlinear processes on the creation of high charge states [2]. Heating by a plasmon resonance was ruled out experimentally [3]. Macroscopic mechanisms such as surface absorption [4], responsible for light absorption of clusters exposed to near-infrared lasers, must be ruled out as well, as they require a free electron quiver motion amplitude that extends over a significant fraction of the cluster. The maximum quiver amplitude reached in the DESY experiment is below 1 Bohr.

A strong candidate for high charge state creation is inverse Bremsstrahlung (IB) heating of the valence electrons set free by single-photon ionization, and subsequent impact ionization of deeper bound electrons. However, conventional IB heating rates are an order of magnitude too low to account for the observed charge states. Currently, there exist three different approaches to explain the enhanced heating which are: atomic, non-Coulomb potential contributions to IB heating [6], charge enhanced single-photon ionization [5], and many-body (MB) recombination heating [7]. These different theoretical approaches highlight the fact that new physics is required to understand the so far unexplored area of intense VUV-light matter interaction. The classical molecular dynamics (MD) analysis presented here focuses on the MB recombination heating mechanism.

The key to understanding the DESY experiment is that the intense VUV pulse creates a strongly coupled electron–ion plasma [7]. A strongly coupled plasma appears in the limit of high density and low temperature [8–10]. In conventional weakly coupled plasmas, fluctuations resulting from the discrete nature of charges, and therewith collisions and Coulomb correlations, are kept at a low level through Debye shielding. In strongly coupled plasmas however, statistical mean field interactions are dominated by the microscopic nature of the Coulomb interaction. Strong coupling can exist between ions, between electrons, and between electrons and ions. In the VUV-FEL cluster experiments, strong coupling between electrons and ions (and electrons and electrons) is maintained over the laser pulse duration, as electron heating is offset by an increase in free electron density by ionization. The availability of the VUV-FEL thus opens a novel avenue for the controlled, laboratory-scale creation and investigation of strongly coupled plasma dynamics.

Our analysis shows that energy absorption in a strongly coupled plasma is enhanced, accounting for the missing order of magnitude in conventional IB heating theories [11]. The main enhancement of IB heating is due to a MB recombination heating, a schematic of which is depicted in Fig. 6.1. In a strongly coupled plasma, ion-electron correlation increases the probability of finding one or more electrons in the vicinity of an ion [12]. Due to the larger local density around the ions, many-body (i.e. multi-electron ion) collisions become much more likely. Many-body collisions result in an enhanced recombination of electrons to excited bound states, see Figs. 6.1(a) and (b). We call this process MB recombination. In a strongly coupled plasma, MB recombination is much more efficient than three-body recombination in a weakly coupled plasma. As Fig. 6.1 illustrates, the remaining free electrons absorb the energy set free during the electron transition to the bound excited state (b). Further collisions either reionize the recombined electron or scatter it from highly excited into deeper bound states. As a result, a broad range of the excited state spectrum is populated. The MB recombination heating-cycle is closed when the recombined electron is re-excited or re-ionized by photon absorption (c). Since the recombined electrons revolve continuously around an ion, light can be absorbed much more efficiently than in IB heating, where free electrons first need to find an ionic partner to absorb radiation.

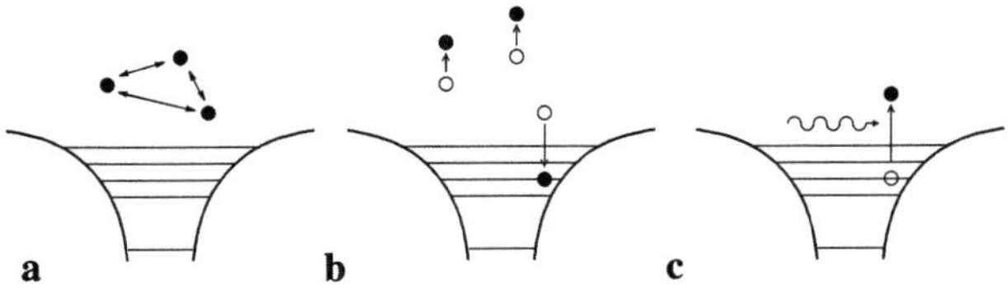

Fig. 6.1. Schematic of MB recombination heating, which takes place in a cycle consisting of three processes. Collisions of two or more electrons close to an ion (**a**) result in MB recombination to a highly excited bound state (**b**). Further collisions scatter the electron into deeper bound states, thus populating a broad range of the excited state spectrum. The total energy of the multi-electron-ion system remains conserved (**b**). Energy is absorbed when the recombined electron is re-ionized by photon absorption (**c**)

6.2 Theoretical Model

We model the ion and electron dynamics in the FEL cluster interaction through a tree-code [13], which scales more favorably ($\mathcal{O}(N \log N)$) with the number of particles N than a direct MD approach ($\mathcal{O}(N^2)$). The better scaling is achieved by approximating the force of a group of distant particles by a multipole expansion. Expansion terms are included up to quadrupole order. The distance at which the exact Coulomb interaction is approximated by the multipole expansion is determined by a tolerance parameter Θ, defined e.g. in [14]. In the limit of $\Theta = 0$, the tree-calculation becomes an exact MD analysis. To eliminate numerical heating, we use shielded electron and ion Coulomb potentials with a shielding radius of 1 Bohr. Whereas, the Coulomb potential for a radius greater than 1 Bohr remains unchanged, it is smoothed for distances below 1 Bohr. The results are insensitive to a further reduction of the shielding parameter. Convergence of the presented calculations is obtained for $\Theta = 0.3$ and a time step $\Delta t = 5 \times 10^{-20}$ s. The small time step is necessary to correctly calculate the bound state dynamics of the recombined electrons. A further decrease of Δt or Θ changes electron kinetic energies and ion charge states by less than 3%.

The simulation begins with a set of neutral Xe-atoms in fcc-structure with an interatomic distance $d = 8.2$ Bohr. Individual electrons and ions are created through ionization. Before the ionization event the electron does not exist in the code. When an electron is created by ionization, the charge of the ion is increased by one. In the DESY experiment, single photon ionization is the predominant process for the ionization of the first electron. Quantum mechanical single photon ionization rates for the Xe ground state are taken from [15]. The subsequent electrons bound in ionic ground states are ionized by impact ionization as determined by the Lotz cross section [16], where the single atom/ion ionization potential is used. This presents a lower bound to impact ionization in solid density plasmas, since in a positively charged cluster the net field of the ions reduces the ionization potential. The clusters investigated here are weakly charged, which makes the use of the unperturbed ionization potential a reasonable approximation.

Ionization does not only take place from the atomic or ionic ground state. Due to collisions, free electrons may recombine to ionic excited bound states, from where they can again be ionized. Once an electron is created initially, it remains in the calculation, even during recombination with an ion. Recombination and reionization are fully contained in the classical calculation. Note that our code models the description of the bound state dynamics by classical mechanics. We would like to emphasize that bound state dynamics in strongly coupled plasmas greatly differs from single atoms. The micro fields due to the strong charge fluctuations are of the order of 1 at.u. The resulting Stark shift becomes larger than the level spacing merging the excited states of even highly charged Xe ions into a quasi-continuum. An experimental proof of line merging in similar parameter ranges was given in [10]. For systems with high level density, such as Rydberg atoms, classical analysis is known to work reasonably. This validates our classical approach of modelling bound state dynamics in a strongly coupled plasma.

6.3 Discussion of Results

Our analysis focuses on a Xe_{1000} cluster. This is somewhat smaller than the Xe_{1500} clusters used in the DESY experiment [1] however, is at the limit of available computer capacity. Our parameters are close enough to the DESY experiment to allow a reasonable qualitative comparison. A quantitative comparison is difficult in any case due to the large uncertainty in the measurement of cluster size, FEL pulse shape, and peak intensity [17].

The Xe ion distribution for a Xe_{1000} cluster with intensities $1.5 \times 10^{12}\,\mathrm{W/cm^2}$, $1.5 \times 10^{13}\,\mathrm{W/cm^2}$, $7 \times 10^{13}\,\mathrm{W/cm^2}$, and for a Xe_{80} cluster with an intensity $2 \times 10^{13}\,\mathrm{W/cm^2}$, is plotted in Fig. 6.2(a)–(d) respectively. All calculations used $\lambda = 98\,\mathrm{nm}$, and an FWHM of 100 fs. The ion charge states are determined at a time after the laser pulse, when the free electrons

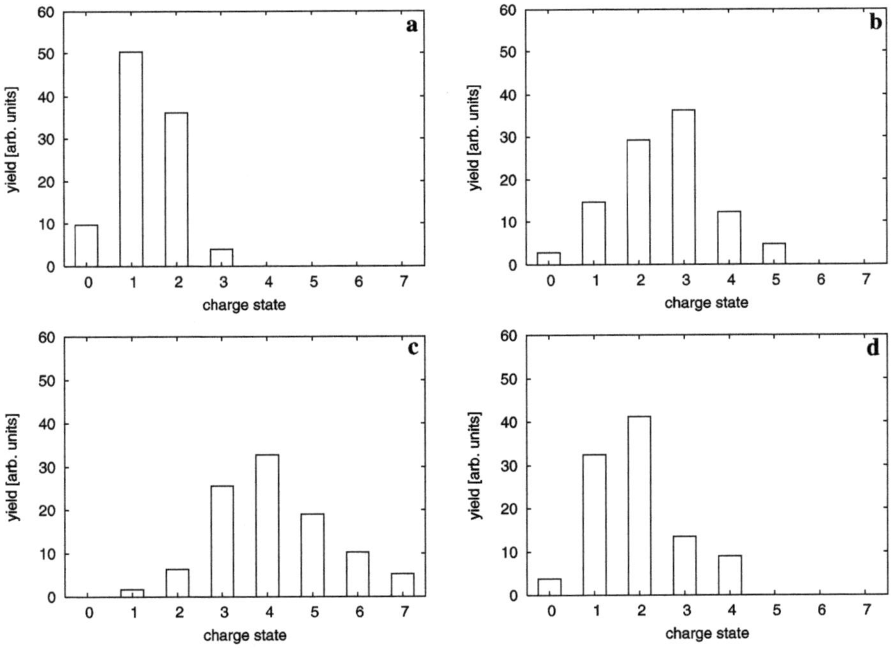

Fig. 6.2. Ion charge state distributions of a Xe cluster interacting with a FEL pulse with $\lambda = 98\,\mathrm{nm}$ and full width half maximum of 100 fs. A sin-squared pulse envelope was used; (**a**)–(**c**) Xe_{1000} and intensities of $1.5 \times 10^{12}\,\mathrm{W/cm^2}$, $1.5 \times 10^{13}\,\mathrm{W/cm^2}$, and $7 \times 10^{13}\,\mathrm{W/cm^2}$, respectively; (**d**) Xe_{80} and an intensity of $2.0 \times 10^{13}\,\mathrm{W/cm^2}$. No averaging over the transversal pulse profile was done

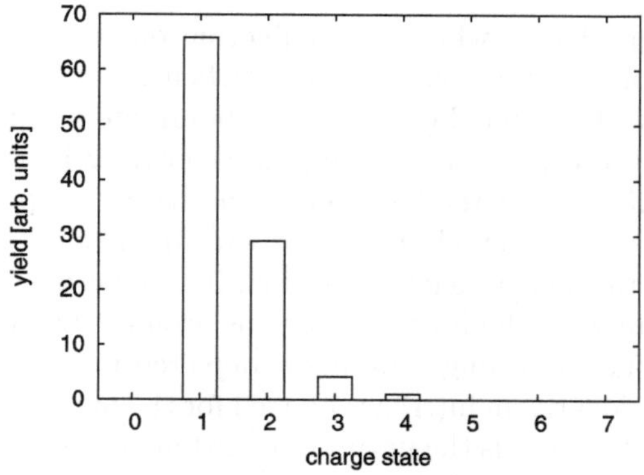

Fig. 6.3. Ion charge state distributions of a Xe cluster interacting with a FEL pulse, averaged over the transversal pulse profile, for the parameters of Fig. 6.2(d)

have escaped from the cluster. We find a reasonable agreement between our calculations and Figs. 1 and 2 of [1]. In (a) and (b) charge states up to 3 and 5 are created, respectively, in agreement with the DESY experiment. In (c) we observe ions up to Xe^{7+}, which is one charge state lower than observed in the experiment.

In general, the calculated distribution functions peak around higher charge states than in the experiment. This is because we did not average our calculations over the transversal intensity distribution of the pulse. We did however calculate the average distribution for (d), see Fig. 6.3. Averaging shifts the population towards lower charge states and decreases the population of the highest charge state, thus further improving the overall agreement.

To demonstrate that the FEL driven cluster dynamics takes place in the strongly coupled plasma regime, we plot in Fig. 6.4 the electron–electron,

$$\Gamma_{\mathrm{ee}} = \frac{V_{\mathrm{ee}}}{k_{\mathrm{B}} T_{\mathrm{e}}} , \tag{6.1}$$

and the electron–ion,

$$\Gamma_{\mathrm{ei}} = Z \Gamma_{\mathrm{ee}}^{3/2} , \tag{6.2}$$

coupling parameters as a function of time [12]. We use the parameters of 6.2(c). Here, Z denotes the average ion charge state, $k_{\mathrm{B}}T_{\mathrm{e}}$ the average thermal energy, and $V_{\mathrm{ee}} = e^2/(4\pi\epsilon_0 a)$ the average electrostatic energy between neighboring electrons. The average distance between electrons $a = (3/(4\pi n_e))^{1/3}$ is determined by the electron density n_e. The electron density is calculated by taking into account only the free electrons. The electrons bound by MB recombination are excluded. The parameter Γ_{ei} represents the ratio of ion charge to electron charge within the Debye-sphere. The Debye length

$$\lambda_d = \sqrt{\frac{\epsilon_0 k_{\mathrm{B}} T_{\mathrm{e}}}{e^2 n_e}} \tag{6.3}$$

determines the length over which charge fluctuations are screened by the electron plasma. A plasma is strongly coupled when $\Gamma_{\mathrm{ee}} \geq 0.1$ and $\Gamma_{\mathrm{ei}} \geq 1$ [12]. From Fig. 6.4, we find that both coupling parameters fulfill the conditions for a strongly coupled plasma. Although laser heating increases the electron temperature to 25 eV after the laser pulse, the strong coupling is maintained throughout, since a higher electron temperature results in an increase in electron density through impact ionization. The kinetic energy distribution is non-Maxwellian with the hottest electrons around 100 eV.

Further evidence of strong coupling is depicted in Fig. 6.5, where the λ_d and the Coulomb logarithm $\ln(\Lambda)$ are plotted for the parameters of Fig. 6.2(c). Here, $\Lambda = 4\pi n_e \lambda_d^3$, and ln is the natural logarithm. The Coulomb logarithm is smaller than 1 over most of the laser pulse and even becomes negative in the early stages of the laser-cluster interaction. The description of electron–ion collisional processes in conventional, weakly coupled plasma physics relies on a well-defined Coulomb logarithm. For negative values conventional plasma concepts become clearly meaningless. The Debye length remains between 1 and 3 Å over most of the laser pulse, which is smaller than the internuclear distance $d = 4$ Å. As a result, cluster charges are no longer shielded. Electrons and ions create strongly fluctuating micro-fields that determine the cluster dynamics. Such a scenario requires the full propagation of the equations of motion of all individual charges. Averaged mean field concepts used in weakly coupled plasmas lose their validity.

As a result of the strong electron–ion correlation, many-body collisions between several electrons and an ion are much more likely to occur, often

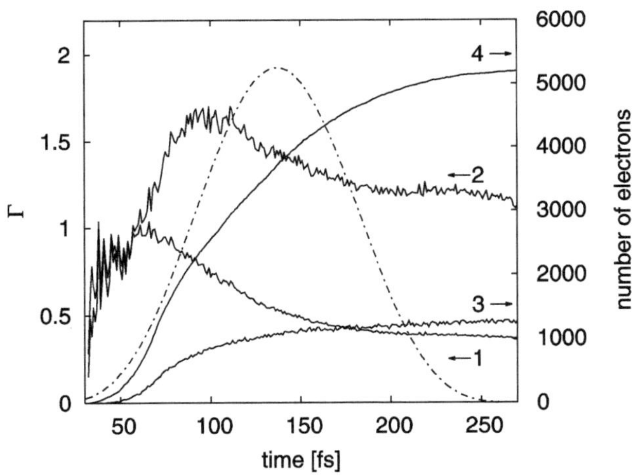

Fig. 6.4. Graphs (*1*) and (*2*) denote the electron–electron (Γ_{ee}) and electron–ion (Γ_{ei}) coupling parameter versus time, respectively, for the parameters of Fig. 6.2(c). Graphs (*3*) and (*4*) show the number of electrons recombined to an excited bound state and the total number of electrons created during the calculation, respectively. The *dash-dotted line* indicates the FEL pulse shape

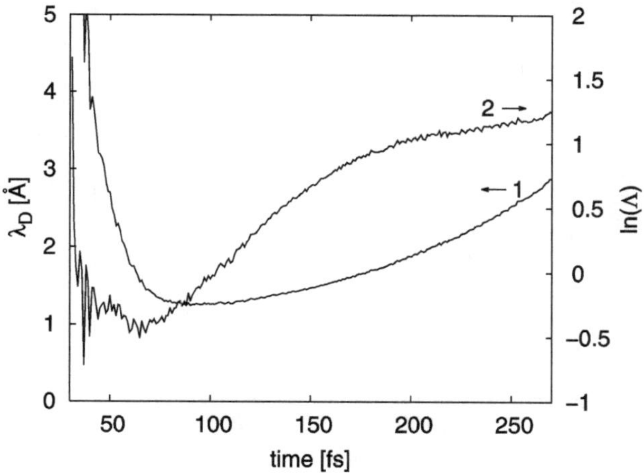

Fig. 6.5. Debye length λ_d (*graph 1*) and Coulomb logarithm $\ln(\Lambda)$ (*graph 2*) versus time for the parameters of Fig. 6.2(c)

leading to MB recombination to excited bound states. Graphs 3 and 4 in Fig. 6.4 show the number of electrons recombined to an excited bound state, and the total number of ionized (and of ionized-and-recombined) electrons as a function of time. At the peak of the laser pulse every fourth ionized electron is found to have recombined to an excited bound state. The distortion of the Coulomb potential by the rapidly fluctuating field of the adjacent electrons makes it difficult to identify an excited bound electron from its total energy. Therefore, they were identified in a two-step process. First, the total energy of an electron in the vicinity of an ion is calculated by using the single-ion Coulomb potential. Second, when the total energy is negative, the electron is observed over 1 fs. The electron is counted as a bound electron, when it remains in the vicinity of the original ion or its nearest neighbors during this time interval. This takes into account extended, molecule-like excited states that can exist due to the complex potential energy structure of the cluster. An electron with a thermal kinetic energy of 1 eV traverses the interatomic distance d in a time interval $\ll 1$ fs. The number of free electrons that remain in this vicinity for a time longer than 1 fs is negligible.

Figure 6.6 demonstrates that MB recombination to bound states and re-ionization is a major factor determining the electron dynamics in strongly coupled plasmas. From the numerical data we have calculated the $1/e$-life time τ_b and the MB recombination rate R_b as determined by the rate equation

$$\frac{\mathrm{d}N_b}{\mathrm{d}t} = -\frac{N_b}{\tau_b} + R_b(t) \; . \tag{6.4}$$

Here N_b is the number of MB recombination bound electrons. The MB recombination rate is a function of the free electron temperature and of the free electron density. The life time of the bound electrons is of the order of a few fs,

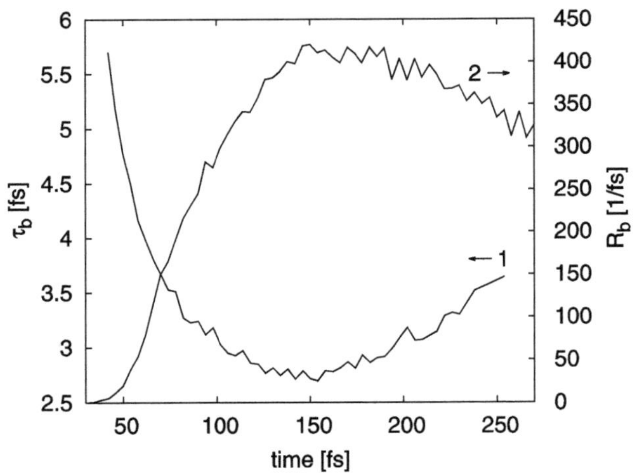

Fig. 6.6. $1/e$-lifetime τ_b of (MB recombination) bound electrons (*graph 1*) and MB recombination rate R_b (number of electrons recombining to a bound state per fs) (*graph 2*) as a function of time

decreasing at the beginning of the laser pulse, reaching its minimum around the peak of the laser pulse and then increasing again more slowly towards the trailing edge of the laser pulse. The asymmetry of τ_b with respect to the laser peak comes from the fact that two mechanisms contribute to re-ionization, which are photon absorption and electron–electron collisions. Whereas photon absorption has a maximum at the laser pulse peak and is symmetric with respect to the peak, electron collisions are weakest in the beginning and increase continuously towards the end of the laser pulse, as a result of the increasing electron density. The monotonic drop of the life time resulting from the increasing importance of collisions explains the asymmetry of τ_b.

The recombination rate in Fig. 6.6 reaches a maximum of 400 electrons per fs around the pulse peak demonstrating the profound influence of MB recombination. The MB recombination rate increases as a nonlinear function of the increasing free electron density in the beginning and starts to decrease after the laser pulse peak. The decrease is caused by the rising electron temperature, which reduces the plasma coupling parameters, see Fig. 6.4, and thus MB recombination.

The dominant contribution of MB recombination to the total energy absorption is shown in Fig. 6.7, where the energy absorption is plotted for the parameters of Fig. 6.2(c). Graph 4 shows the total absorbed energy obtained directly from our simulation. In order to identify the dominant light absorption process, we plot in graphs 1–3 the absorption from the individual heating processes that we found to be the most relevant. Graphs 1 and 2 show the energy absorption due to single photon ionization and IB heating, as calculated from the rates in [15, 18], respectively. In the calculation of IB heating

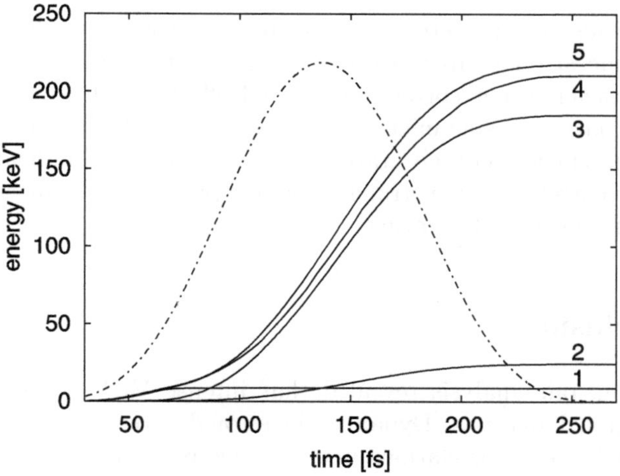

Fig. 6.7. Energy absorption versus time for various heating mechanisms for the parameters of Fig. 6.2(c); graph (*1*) single photon absorption, graph (*2*) IB heating, graph (*3*) MB recombination heating, graph (*5*) sum over graphs (*1*)–(*3*), graph (*4*) total energy absorption as calculated from the MD analysis. The *dash-dotted line* indicates the FEL pulse shape

we used the average ion charge state and the thermal velocity extracted from the simulation. The total light absorption is close to an order of magnitude more efficient than is predicted by conventional IB heating.

Graph (3) shows the energy absorbed by electrons recombined to excited states, which is about an order of magnitude larger than (1) and (2). This shows that MB recombination heating provides the bulk of the cluster heating. Graph (3) was determined in the following way. First, we obtained the number of electrons recombined to an ion with charge state Z as a function of time from our simulation. Next, the photo absorption rate as a function of Z, binding energy, and angular momentum was obtained from classical analysis of a single electron bound in a Coulomb potential. We averaged the rate over angular momentum and binding energies extending from the first excited state to 10% of the first excited state energy. The resulting rate depends only weakly on the boundaries of the energy interval. The energy absorption was calculated by multiplying the number of electrons bound in an excited state of a Z ion with the appropriate Z-dependent photo-absorption rate, then summing over all charge states. Finally, graph 5 plots the total absorbed energy obtained by adding graphs 1–3. The good agreement with graph 4 corroborates the fact that the three heating mechanisms depicted by graphs 1–3 largely determine the total energy absorption.

To our knowledge there exists currently no reliable approach to estimate the changes in IB heating induced by enhanced correlation in a strongly coupled plasma. It may be that the conventional rates underestimate IB heating, because of an increase in the free electron–ion collision frequency caused by the stronger free electron–ion correlation in a strongly coupled plasma. However, other correlation related effects, such as the strong and rapidly fluctuating plasma micro-fields, may dampen IB heating. In any case, the good agreement between graphs 4 and 5 in Fig. 6.7 indicates that free electron–ion correlation in a strongly coupled plasma has at most a moderate effect on conventional IB heating.

6.4 Conclusion

In conclusion, our analysis revealed that intense VUV sources can create strongly coupled plasmas. Dynamics in strongly coupled plasmas is dominated by high-order correlations between electrons and ions. Strongly coupled plasmas occur in diverse areas of physics, ranging from plasma dynamics in giant planets, white dwarfs and progenitors of super novae [19] to quark-gluon plasmas created in high energy ion collisions [20]. The recent availability of X-ray free electron lasers opens a new window of opportunity for the systematic investigation of strongly coupled plasmas. This is demonstrated by our finding that absorption of electromagnetic radiation is dominated by a new mechanism termed many-body recombination heating. The resulting enhanced energy absorption is an order of magnitude more efficient than conventional inverse Bremsstrahlung heating, and explains the observation of unexpectedly high charge states in recent intense X-ray cluster interaction experiments performed at DESY. Many-body recombination heating might also be a key mechanism in other wavelength regimes, determining e.g. the damage threshold of the near-infrared laser induced breakdown of solids. This will be subject to future investigations.

References

1. H. Wabnitz et al.: Nature **420**, 482 (2002).
2. H. Wabnitz et al.: Phys. Rev. Lett. **94**, 023001 (2005).
3. T. Laarmann et al.: Phys. Rev. Lett. **92**, 143401 (2004).
4. C. Jungreuthmayer, M. Geissler, J. Zanghellini, and T. Brabec: Phys. Rev. Lett. **92**, 133401 (2004).
5. C. Siedschlag and J.-M. Rost: Phys. Rev. Lett. **93**, 043402 (2004).
6. R. Santra and C.H. Greene: Phys. Rev. Lett. **91**, 233401 (2003).
7. C. Jungreuthmayer, L. Ramunno, J.Zanghellini, and T. Brabec: J. Phys. B **38**, 3029 (2005).
8. S. Ichimura: *Basic principles of plasma physics*, Addison-Wesley Publishing (1980); S. Ichimura and H.M. van Horn: *Strongly coupled plasma physics*, University of Rochester Press, Rochester (1993).
9. A.N. Mostovych et al.: Phys. Rev. Lett **75**, 1530 (1995).
10. M. Nantel et al.: Phys. Rev. Lett. **80**, 4442 (1998).

11. S. Pfalzner and P. Gibbon: Phys. Rev. E **57**, 4698 (1998).
12. G. Zwicknagel, C. Toepffer, and P.-G. Reinhard: Hyperfine Interactions **99**, 285 (1996); Alfred Müller and Andreas Wolf: Hyperfine Interactions **109**, 233 (1997).
13. J. Barnes and P. Hut: Nature **324**, 446 (1986).
14. J.K. Salmon and M.S. Warren: J. Comp. Phys. **111**, 136 (1994).
15. M.Ya. Amusia: *Atomic Photoeffect*, Plenum Press, New York (1990).
16. W. Lotz: Z. Phys. **216**, 241 (1968).
17. private communication Thomas Möller (2004).
18. V.P. Krainov: J. Phys. B **33**, 1585 (2000).
19. D.G. Yakovlev, O.Y. Gnedin, and A.Y. Potekhin: Contrib. Plasma Phys. **41**, 227 (2001).
20. I. Zahed: J. Phys. G **30**, S1267 (2004).

7 Resonance- and Chaos-Assisted Tunneling

Peter Schlagheck[1], Christopher Eltschka[1], and Denis Ullmo[2,3]

[1] Institut für Theoretische Physik, Universität Regensburg, 93040 Regensburg, Germany
[2] Department of Physics, Duke University, Durham NC, USA
[3] CNRS, Université Paris–Sud, LPTMS, UMR 8626, 91405 Orsay Cedex, France

Summary. We consider dynamical tunneling between two symmetry-related regular islands that are separated in phase space by a chaotic sea. Such tunneling processes are dominantly governed by nonlinear resonances, which induce a coupling mechanism between "regular" quantum states within and "chaotic" states outside the islands. By means of a random matrix ansatz for the chaotic part of the Hamiltonian, one can show that the corresponding coupling matrix element directly determines the level splitting between the symmetric and the antisymmetric eigenstates of the pair of islands. We show in detail how this matrix element can be expressed in terms of elementary classical quantities that are associated with the resonance. The validity of this theory is demonstrated with the kicked Harper model.

7.1 Introduction

Since the early days of quantum mechanics, tunneling has been recognized as one of the hallmarks of the wave character of microscopic physics. The possibility of a quantum particle to penetrate an energetic barrier represents certainly one of the most spectacular implications of quantum theory and has lead to various applications in atomic and molecular physics as well as in mesoscopic science. Typical scenarios in which tunneling manifests are the escape of a quantum particle from a quasi-bound region, the transition between two or more symmetry-related, but classically disconnected wells (which we shall focus on in the following), as well as scattering or transport through potential barriers. The spectrum of scenarios becomes even richer when the concept of tunneling is generalized to any kind of classically forbidden transitions in phase space, i.e. to transitions that are not necessarily inhibited by static potential barriers but by some other constraints of the underlying classical dynamics (such as integrals of motion). Such "dynamical tunneling" processes arise frequently in molecular systems [1] and were recently realized with cold atoms propagating in periodically modulated optical lattices [2,3].

Despite its genuinely quantal nature, tunneling is strongly influenced, if not entirely governed, by the structure of the underlying classical phase space (see [4] for a review). This is best illustrated within the textbook example of a one-dimensional symmetric double-well potential. In this simple case, the eigenvalue problem can be straightforwardly solved with the standard Wentzel–Kramers–Brillouin (WKB) ansatz [5]. The eigenstates of this system are, below the barrier height, obtained by the symmetric and antisymmetric linear combination of the local "quasi-modes" (i.e., of the wave functions that are semiclassically constructed on the quantized orbits within each well, without taking into account the classically forbidden coupling between the wells), and the splitting of their energies is given by an expression of the form

$$\Delta E = \frac{\hbar\omega}{\pi} \exp\left[-\frac{1}{\hbar} \int \sqrt{2m(V(x) - E)} \mathrm{d}x \right] . \tag{7.1}$$

Here E is the mean energy of the doublet, $V(x)$ represents the double well potential, m is the mass of the particle, ω denotes the oscillation frequency within each well, and the integral in the exponent is performed over the whole classically forbidden domain, i.e. between the inner turning points of the orbits in the two wells. Preparing the initial state as one of the quasi-modes (i.e., as the even or odd superposition of the symmetric and the antisymmetric eigenstate), the system will undergo Rabi oscillations between the wells with the frequency $\Delta E/\hbar$. The "tunneling rate" of this system is therefore given by the splitting (7.1) and decreases, keeping all classical parameters fixed, exponentially with $1/\hbar$, what gives rise to the statement that tunneling "vanishes" in the classical limit.

The above expression for the splitting can also be derived in a geometric way which is independent of the particular representation of the phase space. For this purpose, it is necessary to realize that the two symmetric wells are connected in the *complexified* classical phase space. This is most conveniently expressed in terms of the local action-angle variables (I, θ) of the well: describing the quantized torus with the action variable $I = I_n$ in the left well by $(p_L(I_n, \theta), q_L(I_n, \theta))$ and its counterpart in the right well by $(p_R(I_n, \theta), q_R(I_n, \theta))$ (p and q are the position and momentum variables), one can show that the analytic continuations of the two manifolds $(p_{L/R}(I_n, \theta), q_{L/R}(I_n, \theta))$ coincide when θ is permitted to assume complex values [6]. Generalizing standard semiclassical theory [7] to this complex Lagrangian manifold allows one to reproduce (7.1), where the exponent now contains the imaginary part of the action integral $\int p\mathrm{d}q$ along a path that connects the two tori in complex phase space [8].

This approach can be generalized to multidimensional, even non separable systems, as long as their classical dynamics is still integrable [8]. It breaks down, however, as soon as a non integrable perturbation is added to the system (e.g. if the one-dimensional double-well potential is exposed to a periodically time-dependent driving). In that case, invariant tori may, for weak perturbations, still exist due to the Kolmogorov–Arnol'd–Moser (KAM) theorem. It can be shown, however, that their analytic continuation to complex (multidimensional) angle domain generally encounters a natural boundary in form of weak singularities that arise at a given value of the imaginary part of the angle variable [9]. Only in very exceptional cases, the complex manifolds that originate from the two symmetry-related tori happen to meet in form of an intersection [10]. In such a situation, one can apply a semiclassical method introduced by Wilkinson, which leads to an expression of the type (7.1) for the splitting with a different $\hbar$-dependence in the prefactor [11].

Indeed, numerical calculations of model systems in the early nineties clearly indicated that tunneling in non integrable systems is qualitatively different from the above one-dimensional case. If the perturbation introduces an appreciable chaotic layer around the separatrix between the two wells (which in the Poincaré surface of section might still be located far away from the quantized tori), the tunnel splittings generally become strongly enhanced compared to the integrable limit. Moreover, they do no longer follow a smooth exponential scaling with $1/\hbar$ as expressed by (7.1), but display huge, quasi-erratic fluctuations at variations of $\hbar$ or any other parameter of the system [12,13].

These phenomena are traced back to the specific role that *chaotic* states play in such systems [14–17]. In contrast to the integrable case, the tunnel doublets of the localized quasi-modes are, in a mixed regular-chaotic system, no longer isolated in the spectrum, but resonantly interact with states that are associated with the chaotic part of phase space. Due to their delocalized nature, such chaotic states typically exhibit a significant overlap with the boundary regions of both regular wells. They may therefore provide an efficient coupling mechanism between the quasi-modes – which becomes particularly effective whenever one of the chaotic levels is shifted exactly on resonance with the tunnel doublet. This coupling mechanism generally enhances the tunneling rate, but may accidentally also lead to a complete *suppression* thereof, arising at very specific values of $\hbar$ or other parameters [18,19].

The validity of this "chaos-assisted" tunneling picture was essentially confirmed by a simple statistical ansatz in which the quantum dynamics within the chaotic part of the phase space was represented by a random matrix from the Gaussian orthogonal ensemble (GOE) [14, 15, 20]. In presence of small coupling coefficients between the regular states and the chaotic domain, this random matrix ansatz yields a truncated Cauchy distribution for the probability density to obtain a level splitting of the size ΔE. Such a distribution is indeed encountered in the exact quantum splittings, which was demonstrated for the two-dimensional quartic oscillator [20] as well as, later on, for the driven pendulum Hamiltonian that describes the tunneling process of cold atoms in periodically modulated optical lattices [21,22]. The random matrix ansatz can be straightforwardly generalized to the tunneling-induced decay of quasi-bound states in open systems, which is relevant for the ionization process of non dispersive electronic wave packets in resonantly driven hydrogen atoms [23]. Chaos-assisted tunneling is, furthermore, not restricted to quantum mechanics, but arises also in the electromagnetic context, as was shown by experiments on optical cavities [24] and microwave billiards [25,26].

A quantitative prediction of the *average* tunneling rate was not possible in the above-mentioned theoretical works. As we shall see later on, this average tunneling rate is directly connected to the coupling matrix element between the regular and the chaotic states, and the strength of this matrix element was unknown and introduced in an ad-hoc way. A natural way to tackle this problem would be to base the semiclassical description of the chaos-assisted tunneling process on *complex* trajectories [27], to be obtained, e.g., by solving Hamilton's equations of motion along complex time paths. Such trajectories would possibly involve a classically forbidden escape out of (and re-entrance into) the regular islands as well as classically allowed propagation within the chaotic sea.

For the case of time-dependent propagation processes, such as the evolution of a wave packet that was initially confined to a regular region, this ambitious semiclassical program can be carried out in a comparatively straightforward way, which is nevertheless hard to implement in practice [28, 29]. Indeed, Shudo, Ikeda and coworkers showed in this context that the selection of complex paths that contribute to the semiclassical propagator requires a careful consideration of the Stokes phenomenon, in order to avoid "forbidden" trajectories that would lead to an exponential increase (instead of decrease) of the tunneling amplitude [28,29]. This semiclassical method can be generalized to scattering problems in presence of non integrable barriers [30–32] and allows one to interpret structures in the tunneling tail of the wave function (such as plateaus and "cliffs") in terms of chaos in the complex classical domain.

A crucial step towards the semiclassical treatment of "time-independent" tunneling problems, such as the determination of the level splitting between nearly degenerate states in classically disconnected wells, was undertaken by Creagh and Whelan. They showed that the splitting-weighted density of states $\sum_n \Delta E_n \Delta(E - E_n)$ (where E_n are the mean energies and ΔE_n the splittings of the doublets) can be expressed as a Gutzwiller-like trace formula involving complex periodic orbits that connect the two wells [33, 34]. Such orbits also permit to determine the individual splittings ΔE_n, provided the wave functions of the associated quasimodes are known [35]. In practice, this approach can be successfully applied to fully chaotic wells that are separated by an energetic barrier [33,34,36], since in such systems the semiclassical tunneling process is typically dominated by a single instanton-type orbit. A generalization to chaos-assisted tunneling, where many different orbits would be expected to contribute, does not seem straightforward.

In time-dependent propagation problems as well as the in approaches based on the Gutzwiller trace formula [37], the Lagrangian manifolds that need to be constructed generally have simple analytical structures. This is the case for time-dependent problems because the initial wave function is usually chosen as a Gaussian wave packet or a plane wave, which are semiclassically associated with simple manifolds. In Gutzwiller-like approaches, this due to the inherent structure of the theory, but at the cost of involving long orbits when individual eigenlevels need to be described. For the problem of determining individual level splittings in chaos-assisted tunneling, however, the natural objects with which one would have to start with are the *invariant tori* of the system, the analytical structure of which is extremely involved in case of a mixed regular-chaotic dynamics. As was mentioned above, these tori cannot be analytically continued very far in the complex phase space as a natural boundary is usually encountered. As a consequence, it does not seem that a description in terms of complex dynamics could be easily developed for chaos-assisted tunneling.

The main purpose of our contribution is to show that a relatively comprehensive description of the tunneling between two regular island in a mixed phase space can nevertheless be obtained. This description involves a number of approximations which might be improved to reach better accuracy. It however leads to quantitative predictions for the tunneling rates which are in sufficiently good agreement with the exact quantum data to ensure that the *mechanism* that underlies this process is correctly accounted for. We shall, in this context, particularly focus on the classically forbidden transition from the regular island into the chaotic sea. As already pointed out above, the associated coupling matrix element determines the average tunnel splittings between the islands, which means that a simple semiclassical access to this matrix element would open the possibility to quantitatively estimate tunneling rates in systems with mixed dynamics.

An important first step in this direction was undertaken by Podolskiy and Narimanov: By assuming a perfectly clean, harmonic-oscillator like dynamics within the regular island and a structureless chaotic sea outside the outermost invariant torus of the island, a semiclassical expression of the form

$$\Delta E \simeq \gamma\hbar\frac{\Gamma(\nu,2\nu)}{\Gamma(\nu+1,0)} \overset{\nu\gg 1}{\simeq} \frac{\gamma\hbar}{\sqrt{2\pi\nu^3}}\mathrm{e}^{-(1-\ln 2)\nu} \quad \text{with} \quad \nu = A/(\pi\hbar) \tag{7.2}$$

was derived for the average eigenphase splitting [38]. Here A is the phase space area covered by the regular island, and $\Gamma(a,x)$ denotes the incomplete Gamma function [39]. The prefactor γ is system specific, but does not depend on $\hbar$, which permits a prediction of the general decay behavior of the splittings with $1/\hbar$. Good agreement was indeed found in a comparison with the exact splittings between near-degenerate optical modes that are associated with a pair of symmetric regular islands in a non integrable micro-cavity [38] (see also [40]).

The theory was furthermore applied to the dynamical tunneling process in periodically modulated optical lattices, for which the splittings between the left- and the right-moving stable eigenmodes were calculated in [21]. Those splittings seem to be very well described by (7.2) for low and moderate values of $1/\hbar$, but display significant deviations from this expression deeper in the semiclassical regime [38]. Indeed, the critical value of $1/\hbar$ beyond which this disagreement occurs coincides with the edge of an intermediate *plateau* in the splittings, which extends over the rather large, "macroscopic" range $10 < 1/\hbar < 30$. As we shall see in Sect. 7.3, such prominent plateau structures appear quite commonly in chaos-assisted tunneling processes (see also [41]), and a smooth, monotonously decreasing expression of the type (7.2) cannot account for them.

To understand the origin of such plateaus, it is instructive to step back to the conceptually simpler case of *nearly integrable* dynamics, where the perturbation from the integrable Hamiltonian is sufficiently small such that macroscopically large chaotic layers are not yet developed in the Poincaré surface of section. In such systems, the main effect of the perturbation consists in the manifestation of chain-like substructures in the phase space, which arise at *nonlinear resonances* between the eigenmodes of the unperturbed Hamiltonian, or, in periodically driven systems, between the external driving and the unperturbed oscillation within the well. In a similar way as for the quantum pendulum Hamiltonian, such resonances induce additional tunneling paths in the phase space, which lead to couplings between states that are located in the *same* well [42,43].

The relevance of this effect for the near-integrable tunneling process between two symmetry-related wells was first pointed out by Bonci et al. [44] who argued that such resonances may lead to a strong enhancement of the tunneling rate, due to couplings between lowly and highly excited states within the well which are permitted by near-degeneracies in the spectrum. In [45, 46], a quantitative semiclassical theory of near-integrable tunneling was formulated on the basis of this principal mechanism. This theory allows one to reproduce the exact quantum splittings on the basis of purely classical quantities that can be extracted from the phase space, and takes into account high-order effects such as the coupling via a sequence of different resonance chains [45, 46]. Recent studies by Keshavamurthy on classically forbidden coupling processes in model Hamiltonians that mimic the dynamics of simple molecules confirm that the "resonance-assisted" tunneling scenario prevails not only in one-dimensional systems that are subject to a periodic driving (such as the "kicked Harper" model which was studied in [45,46]), but also in autonomous systems with two and even three degrees of freedom [47,48].

Our main focus in this review is that such nonlinear resonances play an equally important role also in the mixed-regular chaotic case. Indeed, they have recently been shown to be primarily responsible for the coupling between the regular island and the chaotic sea in the semiclassical regime [49]. In combination with the above random matrix ansatz for the chaotic states, a simple analytical expression for the average tunneling rate is obtained in this way, which provides a straightforward interpretation of the plateau structures in the splittings in terms of multi-step coupling processes induced by the resonances.

To explain this issue in more detail, we start, in Sect. 7.2, with a description of the effect of nonlinear resonances on the quantum dynamics within a regular region in phase space. We then discuss how the presence of such resonances leads to a modification of the tunnel coupling in the near-integrable as well as in the mixed regular-chaotic regime, and present a simple semiclassical scheme that allows one to reproduce the associated tunneling rates. Applications to tunneling processes in the kicked Harper model are studied in Sect. 7.3.

7.2 Theory of Resonance-Assisted Tunneling

We restrict our study to systems with one degree of freedom that evolve under a periodically time-dependent Hamiltonian $H(p,q,t) = H(p,q,t+\tau)$. We suppose that, for a suitable choice of internal parameters, the classical phase space of H is mixed regular-chaotic and exhibits two symmetry-related regular islands that are embedded into the chaotic sea. This phase space structure is most conveniently visualized by a stroboscopic Poincaré section, where p and q are plotted at the times $t = n\tau (n \in \mathbb{Z})$. Such a Poincaré section typically reveals the presence of chain-like substructures within the regular islands, which arise due to nonlinear resonances between the external driving and the internal oscillation around the island's center. We shall assume now that the two islands exhibit a prominent $r{:}s$ resonance – i.e., where s internal oscillation periods match r driving periods, and r sub-islands are visible in the stroboscopic section.

The classical motion in the vicinity of the $r{:}s$ resonance is approximately integrated by secular perturbation theory [50] (see also [46]). For this purpose, we formally introduce a time-independent Hamiltonian $H_0(p,q)$ that approximately reproduces the regular motion in the islands and preserves the discrete symmetry of H. The phase space generated by this integrable Hamiltonian consequently exhibits two symmetric wells that are separated by an energetic barrier and "embed" the two islands of H. In terms of the action-angle variables (I,θ) describing the dynamics within each of the wells, the total Hamiltonian can be written as

$$H(I,\theta,t) = H_0(I) + V(I,\theta,t) \tag{7.3}$$

where V would represent a weak perturbation in the center of the island.

The nonlinear $r{:}s$ resonance occurs at the action variable $I_{r:s}$ that satisfies the condition

$$r\Omega_{r:s} = s\frac{2\pi}{\tau} \quad \text{with} \quad \Omega_{r:s} \equiv \left.\frac{\mathrm{d}H_0}{\mathrm{d}I}\right|_{I=I_{r:s}} . \tag{7.4}$$

We now perform a canonical transformation to the frame that corotates with this resonance. This is done by leaving I invariant and modifying θ according to

$$\theta \mapsto \vartheta = \theta - \Omega_{r:s} t \,. \tag{7.5}$$

This time-dependent shift is accompanied by the transformation $H \mapsto \mathcal{H} = H - \Omega_{r:s} I$ in order to ensure that the new corotating angle variable ϑ is conjugate to I. The motion of I and ϑ is therefore described by the new Hamiltonian

$$\mathcal{H}(I, \vartheta, t) = \mathcal{H}_0(I) + \mathcal{V}(I, \vartheta, t) \tag{7.6}$$

with

$$\mathcal{H}_0(I) = H_0(I) - \Omega_{r:s} I \,, \tag{7.7}$$

$$\mathcal{V}(I, \vartheta, t) = V(I, \vartheta + \Omega_{r:s} t, t) \,. \tag{7.8}$$

The expansion of $\mathcal{H}_0$ in powers of $I - I_{r:s}$ yields

$$\mathcal{H}_0(I) \simeq \mathcal{H}_0^{(0)} + \frac{(I - I_{r:s})^2}{2 m_{r:s}} + \mathcal{O}\left((I - I_{r:s})^3\right) \tag{7.9}$$

with a constant $\mathcal{H}_0^{(0)}$ and a quadratic term that is characterized by the effective "mass" parameter $m_{r:s}$. Hence, $\mathrm{d}\mathcal{H}_0/\mathrm{d}I$ is comparatively small for $I \simeq I_{r:s}$, which implies that the corotating angle ϑ varies slowly in time near the resonance. This justifies the application of adiabatic perturbation theory [50], which effectively amounts, in first order, to replacing $\mathcal{V}(I, \vartheta, t)$ by its time average over r periods of the driving (using the fact that $\mathcal{V}$ is periodic in t with the period $r\tau$) [51]. By making a Fourier series expansion for $V(I, \theta, t)$ in both θ and t, one can show that the resulting time-independent perturbation term is $(2\pi/r)$-periodic in ϑ. We therefore obtain, after this transformation, the time-independent Hamiltonian $\mathcal{H}_0(I) + V_{\mathrm{av}}(I, \vartheta)$ where V_{av} can be written as the Fourier series

$$V_{\mathrm{av}}(I, \vartheta) \equiv \frac{1}{r\tau} \int_0^{r\tau} \mathcal{V}(I, \vartheta, t) = \sum_{k=0}^{\infty} V_k(I) \cos(k r \vartheta + \phi_k) \,. \tag{7.10}$$

This effective Hamiltonian is further simplified by neglecting the action dependence of the Fourier coefficients of V_{av} – i.e., we use $V_k \equiv V_k(I = I_{r:s})$ in (7.10) – and by employing the quadratic approximation (7.9) of $H_0(I)$ around $I = I_{r:s}$. Leaving out constant terms, we finally obtain the effective integrable Hamiltonian

$$H_{\mathrm{eff}}(I, \vartheta) = \frac{(I - I_{r:s})^2}{2 m_{r:s}} + \sum_{k=1}^{\infty} V_k \cos(k r \vartheta + \phi_k) \,. \tag{7.11}$$

The structure of this Hamiltonian exhibits all ingredients that are necessary to understand the enhancement of tunneling. This can be most explicitly seen by applying quantum perturbation theory to the semiclassical quantization of H_{eff}, treating the "kinetic" term $\propto (\hat{I} - I_{r:s})^2$ as unperturbed part ($\hat{I} \equiv -\mathrm{i}\hbar\partial/\partial\vartheta$) and the "potential" term $\sum_{k=1}^{\infty} V_k \cos(kr\hat{\vartheta} + \phi_k)$ as perturbation: Within the unperturbed eigenbasis $\langle\vartheta|n\rangle \sim \exp(\mathrm{i}n\vartheta)$ (i.e., the eigenbasis of H_0 in action-angle variables), couplings are introduced between the states $|n\rangle$ and $|n + kr\rangle$ according to

$$\langle n + kr|H_{\text{eff}}|n\rangle = \frac{1}{2} V_k \mathrm{e}^{\mathrm{i}\phi_k} \,. \tag{7.12}$$

As a consequence, the "true" eigenstates $|\psi_n\rangle$ of H_{eff} contain admixtures from unperturbed modes $|n'\rangle$ that satisfy the selection rule $|n' - n| = kr$ with integer k. They are given by the perturbative expression

$$\begin{aligned} |\psi_n\rangle = |n\rangle + \sum_{k\neq 0} \frac{V_k \mathrm{e}^{\mathrm{i}\phi_k}/2}{E_n - E_{n+kr}} |n + kr\rangle + \\ + \sum_{k,k'\neq 0} \frac{V_k \mathrm{e}^{\mathrm{i}\phi_k}/2}{E_n - E_{n+kr}} \frac{V_{k'} \mathrm{e}^{\mathrm{i}\phi_{k'}}/2}{E_n - E_{n+kr+k'r}} |n + kr + k'r\rangle + \ldots \end{aligned} \tag{7.13}$$

where E_n denote the unperturbed eigenenergies of H_{eff}, i.e., the eigenvalues of $H_0(I) - \Omega_{r:s} I$.

Within the quadratic approximation of $H_0(I)$ around $I_{r:s}$, we obtain from (7.11)

$$E_n = \frac{(I_n - I_{r:s})^2}{2m_{r:s}} \tag{7.14}$$

where the quantized actions are given by

$$I_n = \hbar(n + 1/2) \tag{7.15}$$

(taking into account the generic Maslov index $\mu = 2$ for regular islands). This results in the energy differences

$$E_n - E_{n'} = \frac{1}{2m_{r:s}} (I_n - I_{n'})(I_n + I_{n'} - 2I_{r:s}) \,. \tag{7.16}$$

From this expression, we see that the admixture between $|n\rangle$ and $|n'\rangle$ becomes particularly strong if the r:s resonance is symmetrically located between the two tori that are associated with the actions I_n and $I_{n'}$ – i.e., if $I_n + I_{n'} \simeq 2I_{r:s}$. The presence of a significant nonlinear resonance within a region of regular motion provides therefore an efficient mechanism to couple the local "ground state" – i.e, the state that is semiclassically localized in the center of that region (with action variable $I_0 < I_{r:s}$) – to a highly excited state (with action variable $I_{kr} > I_{r:s}$).

It is instructive to realize that the Fourier coefficients V_k of the perturbation operator decrease rather rapidly with increasing k. Indeed, one can derive under quite general circumstances the asymptotic scaling law

$$V_k \sim (kr)^\gamma V_0 \exp[-kr\Omega_{r:s} t_{\mathrm{im}}(I_{r:s})] \tag{7.17}$$

for large k, which is based on the presence of singularities of the complexified tori of the integrable approximation $H_0(I)$ [46]. Here $t_{\mathrm{im}}(I)$ denotes the imaginary time that elapses from the (real) torus with action I to the nearest singularity in complex phase space, γ corresponds to the degree of the singularity, and V_0 contains information about the corresponding residue near the singularity as well as the strength of the perturbation. The expression (7.17) is of little practical relevance as far as the concrete determination of the coefficients V_k is concerned. It permits, however, to estimate the relative importance of different perturbative pathways connecting the states $|n\rangle$ and $|n+kr\rangle$ in (7.13). Comparing e.g. the amplitude $\mathcal{A}_2$ associated with a single step from $|n\rangle$ to $|n+2r\rangle$ via V_2 and the amplitude $\mathcal{A}_1$ associated with two steps from $|n\rangle$ to $|n+2r\rangle$ via V_1, we obtain from (7.16) and (7.17) the ratio

$$\mathcal{A}_2/\mathcal{A}_1 \simeq \frac{2^\gamma r^{2-\gamma}\hbar^2}{m_{r:s}V_0} e^{\mathrm{i}(\phi_2 - 2\phi_1)} \tag{7.18}$$

under the assumption that the resonance is symmetrically located in between the corresponding two tori. Since V_0 can be assumed to be finite in mixed regular-chaotic systems, we infer that the second-order process via the stronger coefficient V_1 will more dominantly contribute to the coupling between $|n\rangle$ and $|n+2r\rangle$ in the semiclassical limit $\hbar \to 0$.

A similar result is obtained from a comparison of the one-step process via V_k with the k-step process via V_1, where we again find that the latter more dominantly contributes to the coupling between $|n\rangle$ and $|n+kr\rangle$ in the limit $\hbar \to 0$. We therefore conclude that in mixed regular-chaotic systems the semiclassical tunneling process can be adequately described by an effective pendulum-like Hamiltonian in which the Fourier components V_k with $k > 1$ are completely neglected:

$$H_{\mathrm{eff}}(I, \vartheta) = \frac{(I - I_{r:s})^2}{2m_{r:s}} + 2V_{r:s}\cos r\vartheta \tag{7.19}$$

with $2V_{r:s} \equiv V_1$ [49] (we assume $\phi_1 = 0$ without loss of generality). This simple form of the effective Hamiltonian allows us to determine the parameters $I_{r:s}$, $m_{r:s}$ and $V_{r:s}$ from the Poincaré map of the classical dynamics, without explicitly using the transformation to the action-angle variables of H_0. To this end, we numerically calculate the monodromy matrix $M_{r:s} \equiv \partial(p_f, q_f)/\partial(p_i, q_i)$ of a stable periodic point of the resonance (which involves r iterations of the stroboscopic map) as well as the phase space areas $S^+_{r:s}$ and $S^-_{r:s}$ that are enclosed by the outer and inner separatrices of the resonance, respectively (see also Fig. 7.1). Using the fact that the trace of $M_{r:s}$

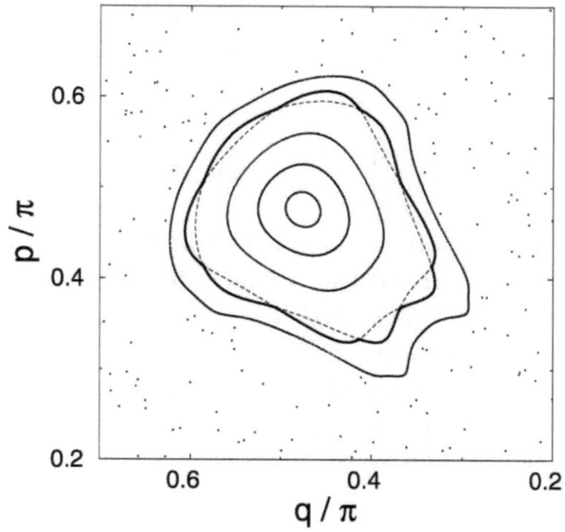

Fig. 7.1. Classical phase space of the kicked Harper Hamiltonian (at $\tau = 3$), showing a regular island with an embedded 9:8 resonance. The *thick solid line* represents the "outer" and the *thin dashed line* the "inner" separatrix of the resonance.

as well as the phase space areas $S^{\pm}_{r:s}$ remain invariant under the canonical transformation to (I, ϑ), we infer

$$I_{r:s} = \frac{1}{4\pi}(S^{+}_{r:s} + S^{-}_{r:s}) \, , \tag{7.20}$$

$$\sqrt{2m_{r:s}V_{r:s}} = \frac{1}{16}(S^{+}_{r:s} - S^{-}_{r:s}) \, , \tag{7.21}$$

$$\sqrt{\frac{2V_{r:s}}{m_{r:s}}} = \frac{1}{r^2\tau} \arccos(\mathrm{tr}\, M_{r:s}/2) \tag{7.22}$$

from the integration of the dynamics generated by H_{eff}.

Quantum mechanically, the tunneling process between the nth excited quantized torus and its counterpart in the symmetry-related island manifests itself in a small level splitting between the associated symmetric and antisymmetric eigenstates. In our case of a periodically driven system with one degree of freedom, these eigenstates arise from a diagonalization of the unitary time evolution operator U over one period τ of the driving, and the splitting is defined by the difference

$$\Delta\varphi_n = |\varphi_n^+ - \varphi_n^-| \tag{7.23}$$

between the corresponding eigenphases $\varphi_n^{\pm}$ of the symmetric and antisymmetric state. In the integrable limit, these eigenphase splittings are trivially related to the energy splittings $\Delta E_n^{(0)}$ of the unperturbed Hamiltonian H_0 via

$$\Delta\varphi_n^{(0)} = \tau \Delta E_n^{(0)}/\hbar \, . \tag{7.24}$$

The latter can be semiclassically calculated by the analytic continuation of the tori to the complex domain [8], and are given by

$$\Delta E_n^{(0)} = \frac{\hbar \Omega_n}{\pi} \exp(-\sigma_n/\hbar) \tag{7.25}$$

where Ω_n is the oscillation frequency of the nth quantized torus and σ_n denotes the imaginary part of the action integral along the complex path that joins the two symmetry-related tori.

In the near-integrable case [45, 46], the presence of a prominent $r{:}s$ resonance provides an efficient coupling between the ground state and highly excited states within the regular region. Within perturbation theory, we obtain for the modified ground state splitting

$$\Delta\varphi_0 = \Delta\varphi_0^{(0)} + \sum_k |\mathcal{A}_{kr}^{(r:s)}|^2 \Delta\varphi_{kr}^{(0)} \tag{7.26}$$

where $\mathcal{A}_{kr}^{(r:s)} = \langle kr|\psi_0\rangle$ denotes the admixture of the (kr)th excited component to the perturbed ground state according to (7.13). The rapid decrease of the amplitudes $\mathcal{A}_{kr}^{(r:s)}$ with k is compensated by an exponential increase of the unperturbed splittings $\Delta\varphi_{kr}^{(0)}$, arising from the fact that the tunnel action σ_n in (7.25) generally decreases with increasing n. The maximal contribution to the modified ground state splitting is generally provided by the state $|kr\rangle$ for which $I_{kr} + I_0 \simeq 2I_{r:s}$ – i.e., which in phase space is most closely located to the torus that lies symmetrically on the opposite side of the resonance chain. This contribution is particularly enhanced by a small energy denominator (see (7.16)) and typically dominates the sum in (7.26).

In the mixed regular-chaotic case, invariant tori exist only up to a maximum action variable I_c corresponding to the outermost boundary of the regular island in phase space. Beyond this outermost invariant torus, multiple overlapping resonances provide various couplings and pathways such that unperturbed states in this regime can be assumed to be strongly connected to each other. Under such circumstances, the classically forbidden coupling between the two symmetric islands does not require any "direct" tunneling process of the type (7.25); it can be achieved by a resonance-induced transition from the ground state to a state within the chaotic domain.

The structure of the effective Hamiltonian that describes this coupling process is depicted in Fig. 7.2. We assume here that the perturbation induced by the nonlinear $r{:}s$ resonance is adequately described by the simplified pendulum Hamiltonian (7.19). Separating the Hilbert space into an "even" and "odd" subspace with respect to the discrete symmetry of H and eliminating

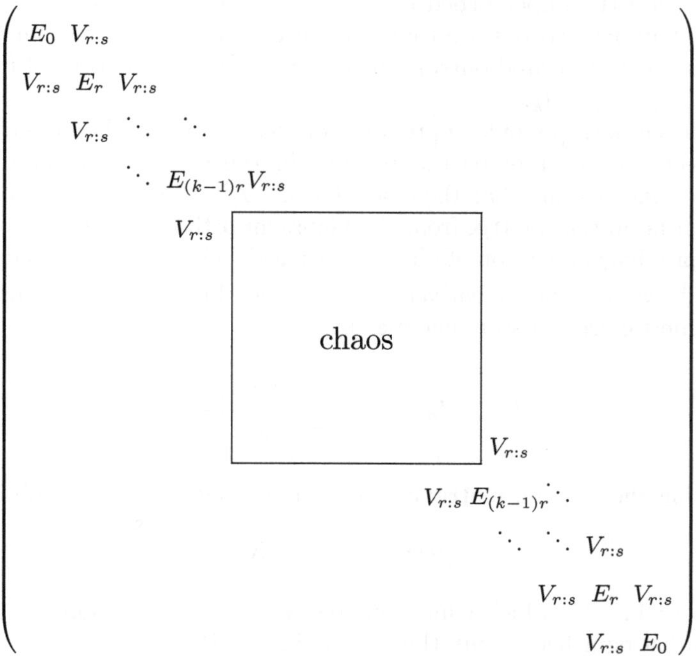

Fig. 7.2. Sketch of the effective Hamiltonian matrix that describes tunneling between the symmetric quasi-modes in the two separate regular islands. The regular parts (*upper left* and *lower right* band) includes only components that are coupled to the island's ground state by the $r{:}s$ resonance. The chaotic part (*central square*) consists of a full sub-block with equally strong couplings between all basis states with actions beyond the outermost invariant torus of the islands.

intermediate states within the regular island leads to an effective Hamiltonian matrix of the form

$$H_{\text{eff}}^{\pm} = \begin{pmatrix} E_0 & V_{\text{eff}} & 0 & \cdots & 0 \\ V_{\text{eff}} & H_{11}^{\pm} & \cdots & \cdots & H_{1N}^{\pm} \\ 0 & \vdots & & & \vdots \\ \vdots & \vdots & & & \vdots \\ 0 & H_{N1}^{\pm} & \cdots & \cdots & H_{NN}^{\pm} \end{pmatrix} . \tag{7.27}$$

for each symmetry class. The effective coupling matrix element between the ground state and the chaos block ($H_{ij}^{\pm}$) is given by

$$V_{\text{eff}} = V_{r:s} \prod_{l=1}^{k-1} \frac{V_{r:s}}{E_0 - E_{lr}} \tag{7.28}$$

where E_n are the unperturbed energies (7.14) of H_{eff}. Here $|kr\rangle$ represents the lowest unperturbed state that is connected by the $r{:}s$ resonance to the ground state and located outside the outermost invariant torus of the island (i.e., $I_{(k-1)r} < I_c < I_{kr}$).

In the simplest possible approximation, which follows the lines of [15,20], we neglect the effect of partial barriers in the chaotic part of the phase space [14] and assume that the chaos block $(H_{ij}^{\pm})$ is adequately modeled by a random hermitian matrix from the Gaussian orthogonal ensemble (GOE). After a pre-diagonalization of $(H_{ij}^{\pm})$, yielding the eigenstates $\phi_j^{\pm}$ and eigenenergies $\mathcal{E}_j^{\pm}$, we can perturbatively express the shifts of the symmetric and antisymmetric ground state energies by

$$E_0^{\pm} = E_0 + V_{\mathrm{eff}}^2 \sum_{j=1}^{N} \frac{|\langle kr|\phi_j^{\pm}\rangle|^2}{E_0 - \mathcal{E}_j^{\pm}} \, . \tag{7.29}$$

Performing the random matrix average for the eigenvectors, we obtain

$$|\langle kr|\phi_j^{\pm}\rangle|^2 \simeq 1/N \tag{7.30}$$

for all $j = 1 \ldots N$, which simply expresses the fact that none of the basis states is distinguished within the chaotic block (H_{ij}).

As was shown in [20], the random matrix average over the eigenvalues $\mathcal{E}_j^{\pm}$ gives rise to a Cauchy distribution for the shifts of the ground state energies, and consequently also for the splittings

$$\Delta E_0 = |E_0^+ - E_0^-| \tag{7.31}$$

between the symmetric and the antisymmetric ground state energy. For the latter, we specifically obtain the probability distribution

$$P(\Delta E_0) = \frac{2}{\pi} \frac{\overline{\Delta E_0}}{(\Delta E_0)^2 + (\overline{\Delta E_0})^2} \tag{7.32}$$

with

$$\overline{\Delta E_0} = \frac{2\pi V_{\mathrm{eff}}^2}{N \Delta_c} \tag{7.33}$$

where Δ_c denotes the mean level spacing in the chaos at energy E_0. This distribution is, strictly speaking, valid only for $\Delta E_0 \ll V_{\mathrm{eff}}$ and exhibits a cutoff at $\Delta E_0 \sim 2V_{\mathrm{eff}}$, which ensures that the statistical expectation value $\langle \Delta E_0 \rangle = \int_0^\infty x P(x) \mathrm{d}x$ does not diverge.

Since tunneling rates and their parametric variations are typically studied on a logarithmic scale (i.e., $\log(\Delta E_0)$ rather than ΔE_0 is plotted vs. $1/\hbar$, see Figs. 7.3–7.5 below), the relevant quantity to be calculated from (7.32) and compared to quantum data is not the mean value $\langle \Delta E_0 \rangle$, but rather the average of the *logarithm* of ΔE_0. We therefore define our "average" level splitting $\langle \Delta E_0 \rangle_g$ as the *geometric* mean of ΔE_0, i.e.

$$\langle \Delta E_0 \rangle_g \equiv \exp\left[\langle \ln(\Delta E_0) \rangle\right] \tag{7.34}$$

and obtain as result the scale defined in (7.33),

$$\langle \Delta E_0 \rangle_g = \overline{\Delta E_0} \, . \tag{7.35}$$

This expression further simplifies for our specific case of periodically driven systems, where the time evolution operator U is modeled by the dynamics under the effective Hamiltonian (7.27) over one period τ. In this case, the chaotic eigenphases $\varphi_j^{\pm} \equiv \mathcal{E}_j^{\pm}\tau/\hbar$ are uniformly distributed in the interval $0 < \varphi_j^{\pm} < 2\pi$. We therefore obtain

$$\Delta_c = \frac{2\pi\hbar}{N\tau} \tag{7.36}$$

for the mean level spacing near E_0. This yields

$$\langle \Delta\varphi_0 \rangle_g \equiv \frac{\tau}{\hbar} \langle \Delta E_0 \rangle_g = \left(\frac{\tau V_{\text{eff}}}{\hbar} \right)^2 \tag{7.37}$$

for the geometric mean of the ground state's eigenphase splitting. Note that this final result does not depend on how many of the chaotic states do actually participate in the sub-block $(H_{ij}^{\pm})$; as long as this number is sufficiently large to justify the validity of the Cauchy distribution (7.32) (see [20]), the geometric mean of the eigenphase splitting is essentially given by the square of the coupling V_{eff} from the ground state to the chaos.

The distribution (7.32) also permits the calculation of the logarithmic variance of the eigenphase splitting: we obtain

$$\left\langle \left[\ln(\Delta\varphi_0) - \langle \ln(\Delta\varphi_0) \rangle\right]^2 \right\rangle = \frac{\pi^2}{4} \, . \tag{7.38}$$

This universal result predicts that the actual splittings may be enhanced or reduced compared to $\langle \Delta\varphi_0 \rangle_g$ by factors of the order of $\exp(\pi/2) \simeq 4.8$, independently of the values of $\hbar$ and external parameters. Indeed, we shall show in the following section that short-range fluctuations of the splittings, arising at small variations of $\hbar$, are well characterized by the standard deviation that is associated with (7.38).

7.3 Application to the Kicked Harper Model

To demonstrate the validity of our approach, we apply it to the "kicked Harper" model [52]

$$H(p, q, t) = \cos p + \sum_{n=-\infty}^{\infty} \tau\delta(t - n\tau) \cos q \, . \tag{7.39}$$

This model Hamiltonian is characterized by the parameter $\tau > 0$ which corresponds to the period of the driving as well as to the strength of the perturbation from integrability. The classical dynamics of this system is described by the map $(p, q) \mapsto (p', q') \equiv \mathcal{T}(p, q)$ with

$$p' = p + \tau \sin q \tag{7.40}$$

$$q' = q - \tau \sin p' \tag{7.41}$$

that generates the stroboscopic Poincaré section at times immediately before the kick. The phase space of the kicked Harper is 2π periodic in position and momentum, and exhibits, for not too large perturbation strengths τ, a region of bounded regular motion centered around $(p, q) = (0, 0)$ (see the upper panel of Fig. 7.3).

The associated time evolution operator of the quantum kicked Harper is given by

$$U = \exp\left(-\frac{\mathrm{i}\tau}{\hbar}\cos\hat{p}\right)\exp\left(-\frac{\mathrm{i}\tau}{\hbar}\cos\hat{q}\right) \tag{7.42}$$

where $\hat{p}$ and $\hat{q}$ denote the position and momentum operator, respectively. The quantum eigenvalue problem drastically simplifies for $\hbar = 2\pi/N$ with integer $N > 0$, since the two phase-space translation operators $T_1 = \exp(2\pi\mathrm{i}\hat{p}/\hbar)$ and $T_2 = \exp(-2\pi\mathrm{i}\hat{q}/\hbar)$ mutually commute with U and with each other in that case [52]. This allows us to make a simultaneous Bloch ansatz in both position and momentum – i.e., to choose eigenstates with the properties

$$\psi(q + 2\pi) = \psi(q)\exp(\mathrm{i}\xi_q) \tag{7.43}$$

$$\hat{\psi}(p + 2\pi) = \hat{\psi}(p)\exp(\mathrm{i}\xi_p) \tag{7.44}$$

where $\hat{\psi}$ denotes the Fourier transform of ψ. Since the subspace of wave functions satisfying (7.43), (7.44) at fixed Bloch phases ξ_q and ξ_p has finite dimension N, finite matrices need to be diagonalized to obtain the eigenstates of U.

Quantum tunneling can take place between the central regular region around $(0, 0)$ and its periodically shifted counterparts. The spectral manifestation of this classically forbidden coupling process is a finite bandwidth of the eigenphases $\varphi_n \equiv \varphi_n^{(\xi_q,\xi_p)}$ of U that are associated with the nth excited quantized torus within this region. We shall not discuss this bandwidth in the following (the calculation of which would require diagonalizations for many different values of ξ_q and ξ_p), but consider a simpler, related quantity, namely the difference

$$\Delta\varphi_n = \left|\varphi_n^{(0,0)} - \varphi_n^{(\pi,0)}\right| \tag{7.45}$$

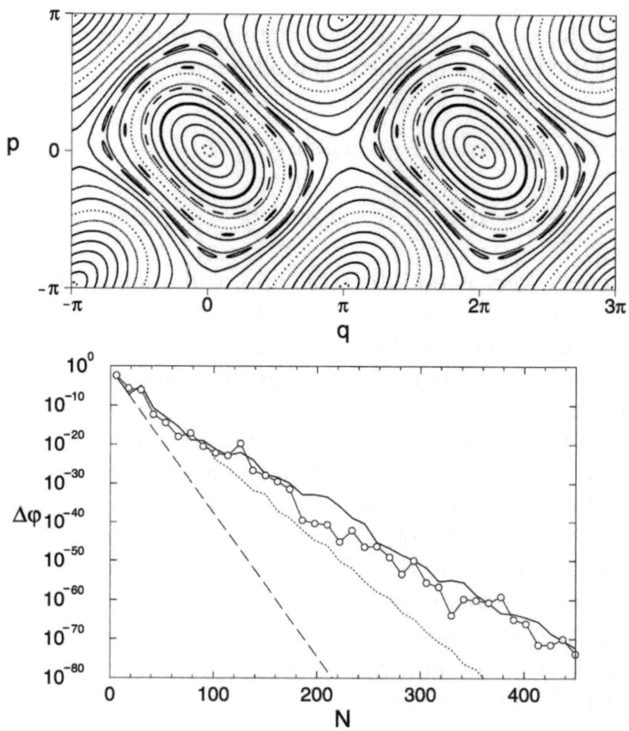

Fig. 7.3. Classical phase space (*upper panel*) and quantum eigenphase splittings (*lower panel*) of the kicked Harper at $\tau = 1$. The latter are calculated for the nth excited states at $N \equiv 2\pi/\hbar = 6(2n+1)$ – i.e., for all states that are semiclassically localized on the torus with action variable $\pi/6$. The decay of the exact quantum splittings (*circles*) is quite well reproduced by the semiclassical prediction (*thick solid line*) that is based on the 16:2, the 10:1, and the 14:1 resonance (highlighted in the *upper panel*, together with the above torus). The *dotted* and dashed lines show, respectively, the semiclassical splittings that are obtained by taking into account only the 10:1 resonance, and the "unperturbed" splittings $\Delta\varphi_n^{(0)}$ that would result from the integrable approximation (7.46).

between the eigenphases of the periodic ($\xi_q = 0$) and the anti-periodic ($\xi_q = \pi$) state in position, at fixed Bloch phase $\xi_p = 0$ in momentum. In this way, we effectively map the tunneling problem to a double well configuration, with the two symmetric wells given e.g. by the regions around $(0, 0)$ and $(2\pi, 0)$.

Figure 7.3 shows the eigenphase splittings of the kicked Harper in the near-integrable regime at $\tau = 1$. The splittings were calculated for the nth excited states at $N = 6(2n + 1)$, i.e. for all possible states that are, in phase space, localized on the same classical torus with action variable $\pi/6$. These states were identified by comparing the overlap matrix elements of the eigenstates of U with the nth excited eigenstate (as counted from the center of the region) of the time-independent Hamiltonian

$$H_0(p,q) = \cos p + \cos q - \frac{\tau}{2} \sin p \sin q \\ - \frac{\tau^2}{12}\left(\cos p \sin^2 q + \cos q \sin^2 p\right) - \frac{\tau^3}{48} \sin(2p)\sin(2q) \,, \qquad (7.46)$$

which represents a very good integrable approximation of H at small τ [46]. Multiple precision arithmetics, based on the GMP library [53], was used to compute eigenphase splittings below the ordinary machine precision limit.

The semiclassical calculation of the eigenphase splittings is based on three prominent nonlinear resonances that are located between the torus with action variable $\pi/6$ and the separatrix: the 16:2 resonance [54], the 10:1 resonance, and the 14:1 resonance. Below $N \simeq 100$, we find that the tunneling process is entirely induced by the 10:1 resonance (i.e., the most dominant one, according to the criterion that r and s be minimal). In this regime, the eigenphase splittings are reproduced by an expression of the form (7.26) (with the resonance parameters $I_{r:s}$, $m_{r:s}$, $V_{r:s}$ extracted from the classical phase space), where the unperturbed splittings $\Delta\varphi_n^{(0)}$ are derived, via (7.25), from the imaginary action integrals σ_n along complex orbits between the two symmetry-related regular regions.

The other two resonances come into play above $N \simeq 100$, where they "assist" at the transition across the 10:1 resonance. In that regime, a recursive application of the resonance-assisted coupling scheme, taking into account all possible perturbative pathways that involve those resonances (see [46]), is applied to calculate the semiclassical tunnel splittings. We see that the result systematically overestimates the exact quantum splittings above $N \simeq 200$, and does not properly describe their fluctuations. We believe that this mismatch might be due to incorrect energy denominators and coupling matrix elements that result from the simplified form (7.19) of the effective pendulum Hamiltonian. The average exponential decay of the quantum splittings, however, is well reproduced by the semiclassical theory. A comparison with the unperturbed splittings $\Delta\varphi_n^{(0)}$ calculated from the integrable approximation (7.46) (dashed line in Fig. 7.3) clearly demonstrates the validity of the resonance-assisted tunneling mechanism.

At $\tau = 2$, the phase space of the kicked Harper becomes mixed regular-chaotic, and the regular region around $(0,0)$ turns into an island that is embedded into the chaotic sea. Figure 7.4 shows the phase space together with the corresponding eigenphase splittings. The latter were calculated here for the ground state of the island, with the dimension N ranging from 4 to 300 in integer steps. We clearly see that the fluctuations of the splittings are much more pronounced than in the near-integrable regime.

The semiclassical calculation of the eigenphase splittings is based on a prominent 8:2 resonance within the island (visible in the upper panel of

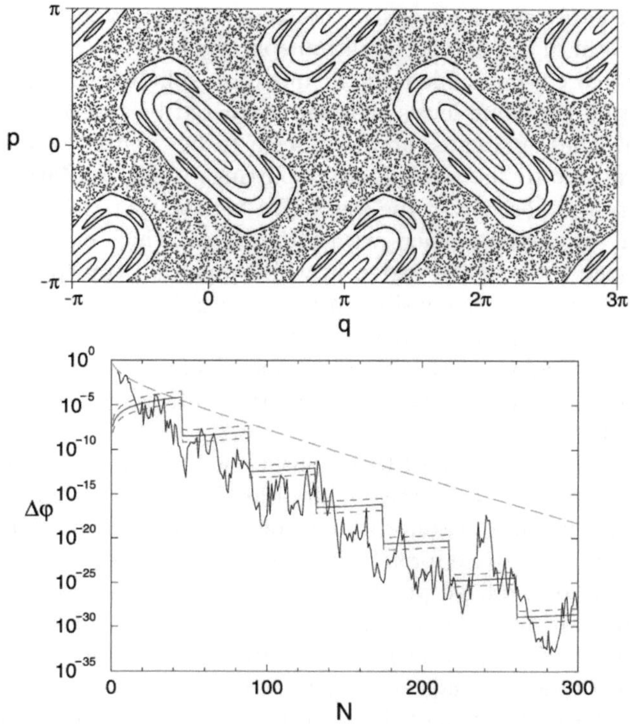

Fig. 7.4. Classical phase space (*upper panel*) and quantum eigenphase splittings (*lower panel*) of the kicked Harper at $\tau = 2$. The splittings are calculated for the ground state within the regular island, as a function of $N \equiv 2\pi/\hbar$. The *thick solid line* in the *lower panel* represents the semiclassical prediction of the eigenphase splittings, which is based on the 8:2 resonance displayed in the *upper panel.* The size of the logarithmic standard deviation according to (7.38) is indicated by the *two dashed lines* that accompany the semiclassical splittings; these *dashed lines* are defined by $\langle \Delta\varphi_0 \rangle_g \times \exp(\pm\pi/2)$. The *long-dashed curve* represents the prediction of the eigenphase splittings according to the theory of [38] (see (7.47)).

Fig. 7.4) and evaluated according to (7.37). The sharp steps of $\langle \Delta\varphi_0 \rangle_g$ arise from the artificial separation between perfect regularity inside and perfect chaos outside the island: when $\hbar$ drops below the value at which the (kr)th excited state of the island is exactly localized on the outermost invariant torus, $(k+1)$ instead of k perturbative steps are required to connect the ground state to the chaotic domain, and the corresponding coupling matrix element $V_{\rm eff}$ (7.28) acquires an additional factor. In reality, the transition to the chaos is "blurred" by the presence of high-order nonlinear resonances and "Cantori" [55,56] in the vicinity of the island. The latter provide efficient barriers to the quantum flow [57–59] and therefore change the structure of the block $(H^{\pm}_{ij})$ in the matrix (7.27). Hence, except for the case of a very "clean", structureless chaotic sea, the steps are not expected to appear in this sharp form in the actual, quantum splittings.

It is therefore remarkable that the quantum splittings exhibit plateaus at approximately the same levels that are predicted by the resonance-assisted mechanism. These plateaus, however, seem to be "shifted" to the left-hand side with respect to the semiclassical splittings – i.e., the latter apparently overestimate the position of the steps. We believe that this is due to a rich substructure of partial barriers within the chaos, which effectively increases the phase space area of the region in which quantum transport is inhibited. Such a substructure is not visible in the Poincaré surface of section, but its existence becomes indeed apparent when individual trajectories are propagated in the vicinity of the island. A more refined approach, taking into account such partial barriers in the chaos, would probably be required to obtain a better agreement.

The central fixed point $(0, 0)$ becomes unstable at $\tau = 2$, when the 2:1 resonance emerges in the center of the island. This 2:1 resonance develops into a symmetric pair of regular islands located along the $p = -q$ axis, which dominate the phase space for $\tau > 2.5$ together with their counterparts from the bifurcation of the fixed point at (π, π). The stable periodic points of the 2:1 resonance are fixed points of the twicely executed kicked Harper map $\mathcal{T}^2$ (i.e., where (7.40), (7.41) is applied twice). Therefore, quantum states associated with those islands appear, at fixed Bloch phases ξ_q, ξ_p, as doublets in the eigenphase spectrum of the corresponding time evolution operator U^2. We shall show now that the splitting between the levels of such a doublet is again described by the resonance- and chaos-assisted tunneling scenario in the semiclassical regime.

The doublets associated with the ground state of the 2:1 resonance islands are calculated by diagonalizing U^2 and by identifying those eigenstates that most sharply localized around the centers of the two islands (we restrict ourselves to the pair that is located on the $q = -p$ axis). Figure 7.5 shows the corresponding eigenphase splittings as a function of $N = \hbar/2\pi$, computed for periodic boundary conditions $\xi_q = \xi_p = 0$ at $\tau = 2.8$ (left column) and $\tau = 3$ (right column). The semiclassical calculation of the eigenphase splittings is performed in the same way as for $\tau = 2$, with the little difference that 2τ rather than τ is used as period in (7.37). For the coupling to the chaos, we identify a relatively large 5:4 sub-resonance within the 2:1 resonance islands at $\tau = 2.8$, and a smaller 9:8 sub-resonance at $\tau = 3$.

It is instructive to notice that a rather small variation $\Delta\tau = 0.2$ of the perturbation parameter can lead to qualitatively different features of the tunneling rates (compare the lower panels of Fig. 7.5), and that these features are indeed reproduced by the resonance-assisted tunneling scheme. At $\tau = 2.8$, the 5:4 sub-resonance is located sufficiently closely to the center of the island that it induces *near-resonant internal transitions* within the island, i.e. couplings from the ground state to the $(5l)$th excited states that are strongly

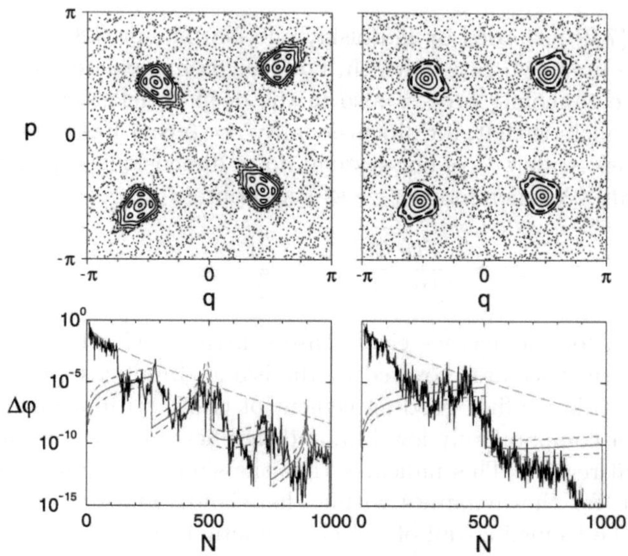

Fig. 7.5. Chaos-assisted tunneling in the kicked Harper at $\tau = 2.8$ (*left column*) and $\tau = 3$ (*right column*). Calculated are the eigenphase splittings between the symmetric and the antisymmetric quasimodes that are localized on the two bifurcated islands (i.e., the two islands facing each other on the $q = -p$ axis), for periodic boundary conditions $\xi_q = \xi_p = 0$. The semiclassical calculation (*thick lines* in the *lower panels*) is based on a 5:4 resonance at $\tau = 2.8$ and on a 9:8 resonance at $\tau = 3$. The huge peaks in the *lower left panel* are induced by internal near-resonant transitions within the island, which are permitted since the 5:4 resonance lies rather closely to the center of the island. As in Fig. 7.4, the *two dashed lines* accompanying the semiclassical splittings indicate the size of the logarithmic standard deviation according to (7.38), and the *long-dashed curve* represents prediction according to the theory of [38].

enhanced in the expression (7.28) due to nearly vanishing energy denominators $E_0 - E_{5l}$. As a consequence, pronounced peaks in the tunneling rate are obtained in the vicinity of such internal near-degeneracies [60]. Such peaks are indeed displayed by the exact quantum splittings as well, though not always at exactly the same position and with the same height as predicted by semiclassics. A completely different scenario is encountered at $\tau = 3$, where the coupling to the chaos is mediated by a tiny 9:8 sub-resonance that is closely located to the outermost invariant torus of the island. This high-order resonance induces rather larges plateaus in the chaos-assisted tunneling rates, which are clearly manifest in the quantum splittings, and which indicate that the coupling to the chaos is essentially governed by the same matrix element over a wide range of $\hbar$.

In addition to the resonance-assisted eigenphase splittings (7.37), we also plot in Figs. 7.4 and 7.5 the prediction that is based on the semiclassical expression (7.2) proposed by Podolskiy and Narimanov [38]. This expression involves the unknown rate γ which, however, is independent of $\hbar$ and characterizes the tunneling process also in the deep anticlassical regime $\hbar \sim 1$. Hence, we conclude that, for dimensionality reasons, γ has to be of the order of the intrinsic scale τ^{-1} of our system, up to dimensionless prefactors of the order of unity. We therefore set $\gamma\tau \equiv 1$ and obtain

$$\Delta\varphi \simeq \frac{\Gamma(\nu, 2\nu)}{\Gamma(\nu+1, 0)} \simeq \frac{1}{\sqrt{2\pi\nu^3}} \mathrm{e}^{-(1-\ln 2)\nu} \tag{7.47}$$

as prediction for the average eigenphase splittings, with $\nu = A/(\pi\hbar)$ where A is the phase space area covered by the island. In all cases that were studied in this work, we find good agreement of (7.47) with the exact quantum splittings for comparatively low $1/\hbar$, and significant deviations deeper in the semiclassical regime. This indicates that the simple harmonic-oscillator approximation for the dynamics within the island (which is needed in order to predict the tunneling tail of the eigenfunction) is correct for large $\hbar$, but becomes invalid as soon as nonlinear resonances come into play.

While the exact quantum splittings may, depending on the size of $\hbar$, considerably deviate from both semiclassical predictions (7.37) and (7.47) of the average, the short-range fluctuations of the splittings seem to be well characterized by the universal expression (7.38) for the logarithmic variance: As is shown by the dashed lines accompanying $\langle \Delta\varphi_0 \rangle_g$ in Fig. 7.4, the amplitudes of those fluctuations are more or less contained within the range that is defined by the standard deviation associated with (7.38). This confirms the general validity of the chaos-assisted tunneling scenario.

7.4 Conclusion

In summary, we have presented a straightforward semiclassical scheme to reproduce tunneling rates between symmetry-related regular islands in mixed systems. Our approach is based on the presence of a prominent nonlinear resonance which induces a coupling mechanism between regular states within and chaotic states outside the islands. The associated coupling matrix element can be directly extracted from classical quantities that are associated with the resonance. Assuming the presence of a structureless chaotic sea, a random matrix ansatz can be made for the chaotic part of the Hamiltonian, which results in a simple expression for the average tunneling rate in terms of the above matrix element.

Application to the kicked Harper model shows good overall agreement and confirms that plateau structures in the quantum splittings originate indeed from the influence of nonlinear resonances. A significant overestimation of the quantum splittings is generally found for "weakly" chaotic systems where a large part of the phase space is covered by regular islands. This is tentatively attributed to the presence of a rich substructure of partial barriers (such as Cantori or island chains) in the chaotic sea. Such partial barriers generically manifest in the immediate vicinity of regular islands and inhibit, for not too small $\hbar$, the quantum transport in a similar way as invariant tori [57, 58]. Semiclassical studies in the annular billiard [16, 17] do indeed indicate that quantum states localized in this hierarchical region [59] play an important role in the dynamical tunneling process [61].

In the "anticlassical" regime of large $\hbar$, our semiclassical theory systematically underestimates the exact quantum splittings. This might be due to the approximations that are involved in the present implementation of the resonance-assisted tunneling scheme (we neglect, e.g., the action dependence of the Fourier coefficients of the effective potential in (7.10), which could play an important role in this regime). It is also possible that the coupling to the chaos is induced by a *different* mechanism at large $\hbar$, which effectively amounts to extracting the associated matrix element from the overlap of the tunneling tail of the local semiclassical wave function with the chaotic phase space. Indeed, we find that the simple semiclassical expression (7.2) introduced by Podolskiy and Narimanov [38], which is essentially based on that scheme, reproduces the tunneling rates quite well in this regime.

The validity of the resonance-assisted tunneling mechanism was confirmed not only for the kicked Harper, but also for the kicked rotor [49] as well as for the driven pendulum Hamiltonian that describes dynamical tunneling of cold atoms in periodically modulated optical lattices [21]. The theory can be furthermore generalized to describe chaos-assisted decay processes of quasi-bound states in open systems, such as the ionization of non dispersive wave packets in microwave-driven hydrogen [62]. Ongoing studies on dynamical tunneling in autonomous model Hamiltonians with two and three degrees of freedom [47, 48] clearly reveal that nonlinear resonances play an equally important role in more complicated systems as well. The mechanism presented here might therefore provide a feasible scheme to predict, understand, and possibly also control dynamical tunneling in a variety of physical systems.

Acknowledgement

It is a pleasure to thank E. Bogomolny, O. Bohigas, O. Brodier, A. Buchleitner, D. Delande, S. Fishman, S. Keshavamurthy, P. Leboeuf, A.M. Ozorio de Almeida, and S. Tomsovic for fruitful and inspiring discussions.

References

1. M.J. Davis and E.J. Heller, J. Chem. Phys. **75**, 246 (1981).
2. W.K. Hensinger *et al.*, Nature **412**, 52 (2001).
3. D.A. Steck, W.H. Oskay, and M.G. Raizen, Science **293**, 274 (2001).

4. S. Creagh, in *Tunneling in Complex Systems*, edited by S. Tomsovic (World Scientific, Singapore, 1998), p. 1.
5. L.D. Landau and E.M. Lifshitz, *Quantum Mechanics: Non-Relativistic Theory* (Pergamon, Oxford, 1958).
6. This assumes, of course, that the double well potential V is analytic.
7. V.P. Maslov and M.V. Fedoriuk, *Semiclassical Approximations in Quantum Mechanics* (Reidel, Dordrecht, 1981).
8. S.C. Creagh, J. Phys. A **27**, 4969 (1994).
9. J.M. Greene and I.C. Percival, Physica **3D**, 530 (1981).
10. S.C. Creagh and M.D. Finn, J. Phys. A **34**, 3791 (2001).
11. M. Wilkinson, Physica **21D**, 341 (1986).
12. W.A. Lin and L.E. Ballentine, Phys. Rev. Lett. **65**, 2927 (1990).
13. O. Bohigas, D. Boosé, R. Egydio de Carvalho, and V. Marvulle, Nucl. Phys. A **560**, 197 (1993).
14. O. Bohigas, S. Tomsovic, and D. Ullmo, Phys. Rep. **223**, 43 (1993).
15. S. Tomsovic and D. Ullmo, Phys. Rev. E **50**, 145 (1994).
16. E. Doron and S.D. Frischat, Phys. Rev. Lett. **75**, 3661 (1995).
17. S.D. Frischat and E. Doron, Phys. Rev. E **57**, 1421 (1998).
18. F. Grossmann, T. Dittrich, P. Jung, and P. Hänggi, Phys. Rev. Lett. **67**, 516 (1991).
19. V. Averbukh, S. Osovski, and N. Moiseyev, Phys. Rev. Lett. 89, 253201 (2002 **89**, 253201 (2002).
20. F. Leyvraz and D. Ullmo, J. Phys. A **29**, 2529 (1996).
21. A. Mouchet *et al.*, Phys. Rev. E **64**, 016221 (2001).
22. A. Mouchet and D. Delande, Phys. Rev. E **67**, 046216 (2003).
23. J. Zakrzewski, D. Delande, and A. Buchleitner, Phys. Rev. E **57**, 1458 (1998).
24. J.U. Nöckel and A.D. Stone, Nature **385**, 45 (1997).
25. C. Dembowski *et al.*, Phys. Rev. Lett. **84**, 867 (2000).
26. R. Hofferbert *et al.*, Phys. Rev. E **71**, 046201 (2005).
27. W.F. Miller and T.F. George, J. Chem. Phys. **56**, 5668 (1972).
28. A. Shudo and K.S. Ikeda, Phys. Rev. Lett. **74**, 682 (1995).
29. A. Shudo and K.S. Ikeda, Phys. Rev. Lett. **76**, 4151 (1996).
30. T. Onishi, A. Shudo, K.S. Ikeda, and K. Takahashi, Phys. Rev. E **64**, 025201 (2001).
31. K. Takahashi and K.S. Ikeda, J. Phys. A **36**, 7953 (2003).
32. K. Takahashi and K.S. Ikeda, Europhys. Lett. **71**, 193 (2005).
33. S.C. Creagh and N.D. Whelan, Phys. Rev. Lett. **77**, 4975 (1996).
34. S.C. Creagh and N.D. Whelan, Phys. Rev. Lett. **82**, 5237 (1999).
35. S.C. Creagh and N.D. Whelan, Ann. Phys. **272**, 196 (1999).
36. S.C. Creagh and N.D. Whelan, Phys. Rev. Lett. **84**, 4084 (2000).
37. M.C. Gutzwiller, *Chaos in Classical and Quantum Mechanics* (Springer, New York, 1990).
38. V.A. Podolskiy and E.E. Narimanov, Phys. Rev. Lett. **91**, 263601 (2003).
39. *Handbook of Mathematical Functions*, edited by M. Abramowitz and I. Stegun (Dover, New York, 1972).
40. V.A. Podolskiy and E.E. Narimanov, Opt. Lett. **30**, 474 (2005).
41. R. Roncaglia *et al.*, Phys. Rev. Lett. **73**, 802 (1994).
42. A.M. Ozorio de Almeida, J. Phys. Chem. **88**, 6139 (1984).
43. T. Uzer, D.W. Noid, and R.A. Marcus, J. Chem. Phys. **79**, 4412 (1983).

44. L. Bonci, A. Farusi, P. Grigolini, and R. Roncaglia, Phys. Rev. E **58**, 5689 (1998).
45. O. Brodier, P. Schlagheck, and D. Ullmo, Phys. Rev. Lett. **87**, 064101 (2001).
46. O. Brodier, P. Schlagheck, and D. Ullmo, Ann. Phys. **300**, 88 (2002).
47. S. Keshavamurthy, J. Chem. Phys **122**, 114109 (2005).
48. S. Keshavamurthy, nlin.CD/0505020 (2005).
49. C. Eltschka and P. Schlagheck, Phys. Rev. Lett. **94**, 014101 (2005).
50. A.J. Lichtenberg and M.A. Lieberman, *Regular and Stochastic Motion* (Springer-Verlag, New York, 1983).
51. This step involves, strictly speaking, another time-dependent canonical transformation $(I, \vartheta) \mapsto (\widetilde{I}, \widetilde{\vartheta})$ which slightly modifies I and ϑ (see also [46]).
52. P. Leboeuf, J. Kurchan, M. Feingold, and D.P. Arovas, Phys. Rev. Lett. **65**, 3076 (1990).
53. `http://www.swox.com/gmp/`.
54. Due to the internal symmetries of the kicked Harper, all $r{:}s$ resonances with relative prime r, s and r being a multiple of 4 have $2r$ instead of r islands. To be consistent with (7.19), we name them $2r{:}2s$ resonances.
55. R.S. MacKay, J.D. Meiss, and I.C. Percival, Phys. Rev. Lett. **52**, 697 (1984).
56. J.D. Meiss and E. Ott, Phys. Rev. Lett. **55**, 2742 (1985).
57. T. Geisel, G. Radons, and J. Rubner, Phys. Rev. Lett. **57**, 2883 (1986).
58. N.T. Maitra and E.J. Heller, Phys. Rev. E **61**, 3620 (2000).
59. R. Ketzmerick, L. Hufnagel, F. Steinbach, and M. Weiss, Phys. Rev. Lett. **85**, 1214 (2000).
60. In the numerical evaluation of the expression (7.28), we took care that the admixture of the (rl)th excited state to the ground state may not exceed the upper limit 1, for all $l = 1 \ldots (k-1)$. The prominent peak at $N \simeq 500$ in the lower panel of Fig. 7.5 is therefore rounded.
61. It should be noted that the annular billiard is exceptionally nongeneric insofar as it exhibits a coexistence of an *exactly integrable* dynamics in the regular islands with a mixed dynamics in the chaotic sea. Nonlinear classical resonances do therefore not at all manifest within the islands, which means that the central mechanism leading to dynamical tunneling between the islands might be completely different from the generic case.
62. S. Wimberger, P. Schlagheck, C. Eltschka, and A. Buchleitner, in preparation.

8 Effects of Carrier-Envelope Phase of Few-Cycle Pulses on High-Order Harmonic Generation

Mauro Nisoli[1], Sandro De Silvestri[1], Giuseppe Sansone[1], Luca Poletto[2], Paolo Villoresi[2], Salvatore Stagira[1], Caterina Vozzi[1], and Orazio Svelto[1]

[1] National Laboratory for Ultrafast and Ultraintense Optical Science – CNR-INFM
Department of Physics, Politecnico, Milan, Italy
mauro.nisoli@fisi.polimi.it

[2] Laboratory for Ultraviolet and X-Ray Optical Research – CNR-INFM
D.E.I. – Università di Padova, Padova, Italy

8.1 Introduction

The development of sub-10-fs laser sources allows the traditional field of time resolved spectroscopy to be extended to extreme performances. In particular, the generation of high-peak-power light pulses in the 5-fs regime by the hollow fiber compression technique [1], has opened up new frontiers for experimental physics. One of these, especially appealing for future investigation of molecular or solid state physics, is extreme nonlinear optics, i.e., the wealth of phenomena taking place when ultrashort pulses are focused to unprecedented peak intensities so that the electric field of the pulse, rather than the intensity profile, is relevant [2]. In this context, a key parameter of the electric field of the light pulse, which significantly influences the strong-field interaction, is the phase of the carrier frequency with respect to the envelope (the so-called carrier-envelope phase, CEP) [3].

The investigation of the CEP role of few-cycle pulses is particularly important, both from a fundamental and a technical point of view, for the production of extreme-ultraviolet (XUV) pulses by high-order harmonic generation (HHG) in noble gases. The physical processes leading to the generation of XUV or soft-X-ray radiation by high-order harmonic generation can be understood using the so-called three step model [4]. In the framework of this semi-classical model, an electron exposed to an intense, linearly polarized electromagnetic field is emitted from the atom by tunnel ionization. The freed electron may be driven back towards its parent ion by the external field and, with small probability, it recombines to the ground state, thus emitting a photon with an energy equal to the sum of the ionization potential and the electron kinetic energy gained in the laser field. Since this process is periodically repeated every half-cycle of the optical radiation, it gives rise to the generation of the odd harmonics of the fundamental fields.

As theoretically suggested in 1992 by Farkas and Toth [5], the HHG process can be used for the generation of attosecond pulses. This has been experimentally demonstrated in the last few years. In the case of multi-cycle driving pulses, trains of attosecond pulses, with a discrete spectrum containing the odd harmonics of the fundamental radiation, are produced [6, 7]. Using few-cycle driving pulses, one can confine the emission process to one half-oscillation period of the fundamental radiation, so that isolated X-ray sub-femtosecond pulses can be generated [8]. It has been demonstrated that the use of phase-stabilized sub-6-fs light pulses allows one to generate isolated attosecond pulses [9]. In this contest, the role of CEP is crucial.

In this chapter we will review the main experimental evidences of the influence of the carrier-envelope phase on strong-field photoionization and high-order harmonic generation. The chapter is organized as follows: in Sect. 8.2 we briefly discuss the hollow fiber compression technique. Section 8.3 reports on the first experimental demonstration of the influence of CEP of few-cycle pulses in strong field photoionization. The fundamental role of CEP on high-order harmonic generation is analyzed in Sect. 8.4. The experimental results are interpreted in the framework of the strong-field approximation using the nonadiabatic saddle-point method, briefly outlined in Sect. 8.5. In Sects. 8.6 and 8.7 the role of CEP on HHG by multiple-cycle pulses is considered.

8.2 Hollow-Fiber Compression Technique

So far the hollow fiber compression technique is the only proven way for the generation of high-peak power light pulses in the sub-5-fs regime. Wave propagation along hollow fibers can be thought of as occurring by grazing incidence reflections at the dielectric inner surface. Since the losses caused by these reflections greatly discriminate against higher order modes, only the fundamental mode, with large and scalable size, will be transmitted through a sufficiently long fiber [10]. Furthermore, by proper mode matching the incident radiation can be dominantly coupled into the fundamental EH_{11} hybrid mode, whose intensity profile as a function of the radial coordinate r can be written as $I_0(r) \propto J_0^2(2.405\ r/a)$, where a is the fiber inner radius and J_0 is the zero-order Bessel function. Moreover, the hollow fiber preserves the polarization of the input radiation and, since it acts as a distributed spatial filter, suppressing high-frequency spatial components possibly present, the output beam is expected to be diffraction limited. Profile measurements at different distances from the output of the fiber show indeed that the beam is virtually diffraction limited.

In 1996 the hollow-fiber compression technique was first demonstrated using 140-fs, 0.66-mJ pulses from a Ti:sapphire laser system [1]. Spectral broadening was obtained in a 70-cm long, 140-μm diameter hollow fiber filled with krypton at a pressure of 2 bar. Pulse compression was achieved using Brewster-cut prism pairs. Pulses as short as 10 fs, with an energy up to 0.24 mJ were generated. A second set of experiments was performed using 20-fs input pulses and a 60-cm-long, 160-μm diameter hollow fiber filled with krypton. Using a dispersive delay line based on chirped mirrors and two pairs of fused silica prisms of small apex angle, we obtained pulses as short as 4.5 fs [11]. Scaling the input energy to higher values, the generation of sub-10-fs pulses at the sub-terawatt peak power level was also demonstrated. Sub-5-fs pulses with hundreds of microjoules energy have been generated using a tapered hollow fiber (input diameter 0.5 mm, output diameter 0.3 mm) and a set of ultrabroadband chirped mirrors [12].

The extension of the hollow-fiber technique to the sub-4-fs regime requires the extension of the induced spectral broadening, preserving single-transverse-mode operation and the development of ultrabroadband dispersive delay lines for dispersion compensation. In 2002 we introduced a novel spectral broadening technique based on hollow fiber cascading, which allows the generation of a supercontinuum extending over a bandwidth exceeding 510 THz with excellent spatial beam quality [13]. 25-fs pulses were coupled into an argon-filled (gas pressure 0.2 bar) hollow fiber, (0.25-mm radius). Gas pressure was chosen in order to obtain pulses with duration of about 10 fs, after compression with completely negligible wings using broadband chirped mirrors. Such pulses were then injected into a second argon-filled hollow fiber (0.15-mm radius). The output beam presents excellent spatial characteristics (single-mode operation) and it is diffraction limited. The pulse spectrum at the output of the second fiber extends from ~400 nm to >1000 nm. The possibility to take advantage of such ultrabroadband spectrum is strictly related to the development of dispersive delay lines capable of controlling the frequency-dependent group delay over large bandwidths. In 2003, ultrabroadband dispersion compensation has been achieved using a liquid-crystal spatial light modulator (SLM). The beam at the output of this hollow-fiber cascading was collimated and sent into a pulse shaper consisting of a 640-pixel liquid-crystal SLM, two 300-line/mm grating, and two 300-mm focal-length spherical mirrors (4-f setup). Pulse characterization was performed using the spectral phase interferometry for direct electric field reconstruction (SPIDER) technique, optimized for sub-10-fs pulses. The measured spectral phase was used to compress the pulse iteratively: compression was started with an initially flat phase written on the liquid-crystal mask. Then, the measured spectral phase was inverted and added to the phase applied to the SLM. Typically five iterations were necessary to yield the shortest pulse. Pulses as short as 3.8 fs have been generated with this technique [14].

Therefore, by using the hollow fiber compression technique, it is possible to generate high-peak power light pulses consisting of only two or even less than two optical cycles in full-width at half maximum. As a consequence, the envelope of the electric field of the light pulse changes almost as fast as the field itself, and its temporal evolution is strongly influenced by the phase of the carrier frequency with respect to the peak envelope. Since strong field interactions in this temporal regime are governed by the electric field of the driving pulses, rather than by their envelope, they are significantly affected by the pulse CEP. In the following sections we will report on experimental observations of the role of CEP of few-cycle pulses in strong field photoionization and high-order harmonic generation.

8.3 Role of Carrier-Envelope Phase in Strong-Field Photoionization

The first experimental evidence of the CEP role of few-cycle pulses has been obtained in strong-field photoionization, using 6-fs pulses with random CEP [15]. For few-cycle pulses, depending on the CEP, the generation of photoelectrons violates inversion symmetry [16]. The CEP is expected to cause an anticorrelation in the number of electrons escaping in opposite sides with respect to the propagation direction of the laser beam. In the experiment two electron detectors were placed in opposite directions with respect to the laser focus. For each laser pulse, the number of electrons detected with both detectors was recorded. Accordingly, each laser pulse is characterized by two numbers: that is, the number of electrons detected in the left-hand and that in the right-hand arm of the experimental apparatus. We then interpret these pairs of numbers as x and y coordinates and accumulate them, for each laser shot, in a map (contingency map). As the signature of an effect from the CEP

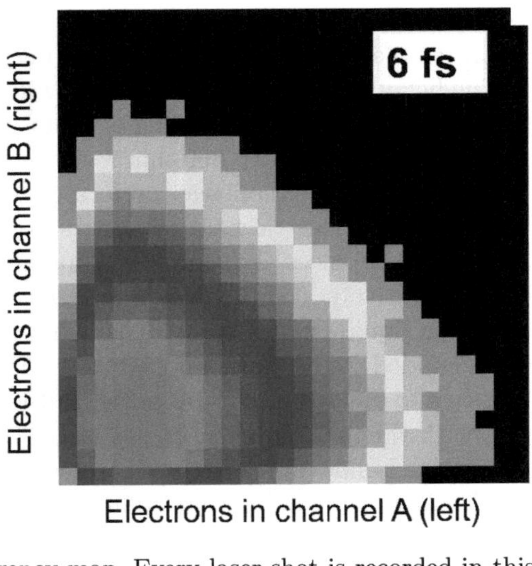

Fig. 8.1. Contingency map. Every laser shot is recorded in this plot according to the number of electrons measured in the *left* (channel A) and the *right* (channel B) arm of the electron spectrometer. In the color code, the pixel colors represent the number of laser shots with electron numbers given by the coordinates of the pixel

is an anticorrelation, one would then expect a structure inclined at $-45°$ in the contingency map. Figure 8.1 shows a contingency map obtained for electron emission from krypton at an intensity of $5 \times 10^{13}\,\mathrm{W/cm^2}$. 6-fs driving pulses have been used with circular polarization. The signature of anticorrelation, indicated by the features inclined at approximately $-45°$, is clearly visible. It is worth mentioning that CEP effects in the reported experimental measurements completely disappear for light pulse duration exceeding 8 fs. Much stronger CEP effects have been measured in HHG using few- and multiple-optical cycle light pulses, as we will show in the next sections.

8.4 CEP Effects in High-Order Harmonic Generation: Few-Cycle Regime

A crucial aspect for the application of high-peak power few-cycle pulses is the stabilization of their CEP. The self-referencing phase stabilization technique [17–19], first introduced for frequency metrology, has revolutionized the field of attosecond physics.

In our experiments, the carrier-envelope phase of the light pulses has been stabilized using an experimental setup similar to that described by Baltuška et al. [20]. The CEP of the Ti:sapphire oscillator has been stabilized using an active feedback loop based on a f-to-$2f$ nonlinear interferometer, which drives an acousto-optic modulator controlling the power of the transmitted pump laser beam. The CEP is reproduced every four round trips in the Ti:Sa oscillator. The phase-stabilized pulses have been amplified in a multipass Ti:Sa amplifier operating at a 1-kHz repetition rate. The amplification process and the beam pointing fluctuation introduce a slow CEP drift. Such drift has been monitored using a second f-to-$2f$ nonlinear interferometer. To this purpose a small fraction of the amplified beam was focused into a 2-mm-thick sapphire plate for the generation of an octave-spanning spectrum by self-phase modulation. Frequency doubling of the infrared portion of the broadened spectrum is accomplished using a BBO crystal. The interference between the high-frequency portion of the broadened spectrum and the frequency-doubled light has been measured using an optical multichannel analyzer. The position of the interference fringes depends on the pulse CEP and can be used to extract the CEP drift, using a Fourier-transform-based algorithm. The residual slow CEP drift has been corrected using a second feedback loop operating on the acousto-optic modulator on the oscillator pump beam. Using the two feedback loops a good stability of the pulse CEP has been achieved. The pulses used for the experiments present an overall CEP residual fluctuation of ~90 mrad (rms). The CEP can be adjusted by introducing in the beam path, before the hollow fiber, a glass plate with variable thickness. At 800 nm a 2π CEP variation is induced by the addition of 52 μm of fused silica, without affecting the pulse duration. To this purpose we have used a pair of wedges mounted on a stepper motor. A change δz of the glass thickness in the beam path determines a change of the CEP of the driving pulse, $\Delta\psi = (\pi/26)\delta z[\mu m]$.

The harmonic radiation, generated in a gas jet with intense linearly polarized few- and multiple-cycle laser pulses, is sent to the spectrometer by a grazing-incidence toroidal mirror. The spectrometer is composed of one toroidal mirror mounted at grazing incidence for stigmatic imaging with almost unitary magnification, and one variable-line-spaced grating, mounted after the mirror [21]. The spectral aberrations are minimized by a proper choice of the distribution of the line spacing on the grating surface. The resulting spectrum is stigmatic and almost flat even in a wide spectral region. The spectrometer spans the 5–75-nm region with a 600 grooves/mm grating, and is operated without an entrance slit because of the limited size of the emitting source. The detector is a multichannel-plate intensifier with phosphor screen optically coupled to a low-noise fast-readout CCD camera, which is moved by a linear drive to acquire different portions of the spectrum.

Figure 8.2 shows two XUV spectra generated in neon by sub-5-fs pulses, measured for two different amounts δz of glass in the beam path, which correspond to different CEP values of the driving pulses, ψ_0 and $\psi_0 + \pi/2$. The gas jet was placed ~ 1 mm after the laser focus in order to select the short quantum paths. As already pointed out by the authors [22] in the lower plateau region (photon energy up to ~50 eV) odd harmonics are generated, which are not affected by the pulse CEP. The upper plateau region shows a com-

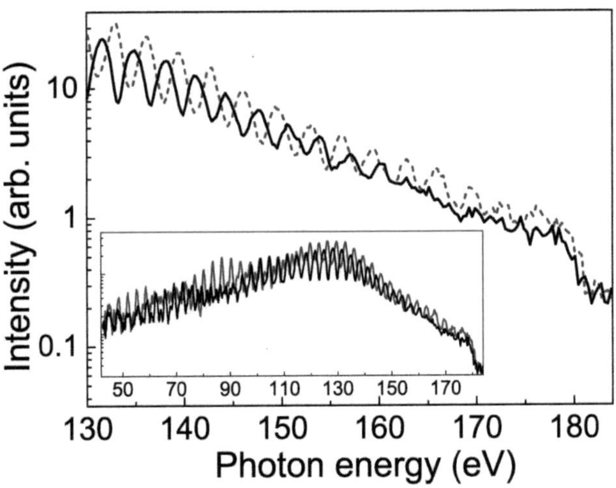

Fig. 8.2. Cutoff region of harmonic spectra generated in neon by phase stabilized sub-5-fs pulses for two different variations δz of the glass thickness in the laser beam path ($\delta z = 3.42\,\mu m$ for the *black curve*, $\delta z = 17.1\,\mu m$ for the *red dashed curve*). The *inset* shows the complete harmonic spectra. Pulse peak intensity on the gas jet 8.8×10^{14} W/cm^2

plex structure with low contrasted CEP-dependent spectral peaks. At higher photon energy (> 90 eV), the spectral position of the peaks is significantly affected by the pulse CEP. Moreover, whilst the cutoff region of the XUV spectrum corresponding to the red dashed curve is modulated, in the case of the black solid curve a broad featureless continuum is evident at photon energies larger than ~ 160 eV.

We point out that the generation of a continuous spectrum in the cutoff region is possible only by using 5-fs or shorter driving pulses, with the proper CEP. Upon increasing the pulse duration, the transition between a modulated and a continuous spectrum as a function of the CEP cannot be obtained. Indeed, Fig. 8.3 shows two harmonic spectra generated in neon by 6-fs pulses with different CEPs (in this case the spectra have been acquired on a single-shot basis). In the cutoff region the spectra are always characterized by a deep CEP-dependent modulation, for any phase value. This observation is particularly important for the generation of attosecond pulses: as demonstrated by Kienberger et al. [9] isolated attosecond pulses can be generated by using 5-fs (or shorter) driving pulses, while using 7-fs pulses a satellite pulse is unavoidable for any value of the CEP.

These experimental results can be analyzed in the framework of the strong-field approximation (SFA). In particular the use of the quantum paths concept, which is the peculiar characteristic of the saddle-point method, allows one to obtain a simple and intuitive picture of the process. The quantum paths are the complex trajectories followed by the electron from the ionization instant to the recombination with the parent ion. The theoretical background at the basis of the nonadiabatic saddle-point method is reviewed in the following section.

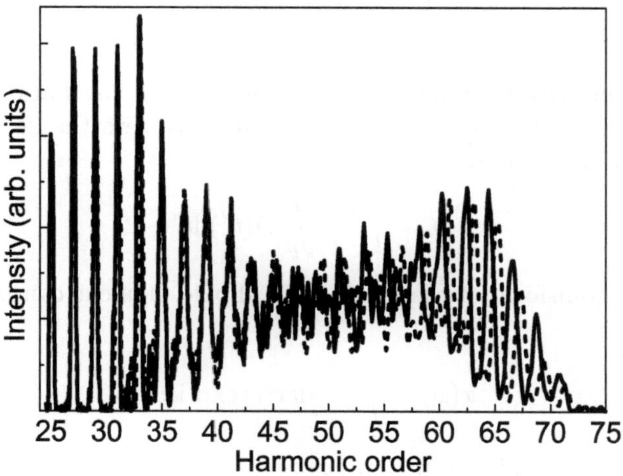

Fig. 8.3. Single-shot harmonic spectra generated in neon by 6-fs pulses. Gas jet located 2 mm after the laser focus; pulse intensity on the gas jet ~ 6.2×10^{14} W/cm^2

8.5 Nonadiabatic Saddle-Point Method

Using the SFA, we give the nonlinear dipole moment by the following expression (in atomic units) [23]

$$x(t) = \mathrm{i} \int_{-\infty}^{t} \mathrm{d}t' \int d^3p E(t') d_x^*[p - A(t)] d_x[p - A(t')] \exp[-\mathrm{i}S(p,t,t')] + \mathrm{c.c.} \tag{8.1}$$

where: $E(t)$ is the electric field of the light pulse, linearly polarized along the x-direction; $A(t)$ is the associated vector potential; p is the canonical momentum; d_x is the dipole matrix element for bound-free transitions; $S(p,t,t')$ is the quasiclassical action, given by

$$S(p,t,t') = \int_{t'}^{t} \mathrm{d}t'' \left(\frac{[p - A(t'')]^2}{2} + I_p \right) \tag{8.2}$$

and I_p is the ionization potential. The linearly polarized electric field used in the calculation has the following expression

$$E(t) = E_0 \cos^2(t/\tau) \cos(\omega_0 t + \psi) \, , \tag{8.3}$$

where: ω_0 is the fundamental angular frequency; ψ is the CEP; τ is related to the driving pulse duration T (full-width at half maximum) by the following expression: $T = 2\tau \arccos(2^{-1/4})$. As shown in [23], the integration over the momentum space in (8.1) can be performed using the saddle-point approximation (SPA), giving

$$\begin{aligned} x(t) = \mathrm{i} \int_{-\infty}^{t} \mathrm{d}t' \left[\frac{\pi}{\epsilon + i(t - t')/2} \right]^{3/2} & E(t') d_x^*[p_\mathrm{s} - A(t)] \\ d_x[p_\mathrm{s} - A(t')] \exp[-\mathrm{i}S(p_\mathrm{s}, t, t')] + \mathrm{c.c.} & \end{aligned} \tag{8.4}$$

where ϵ is a positive regularization constant and p_s is the stationary value of the momentum, which is obtained by equating to zero the derivative, with respect to p, of the action $S(p,t,t')$:

$$p_\mathrm{s} = \frac{1}{t - t'} \int_{t'}^{t} A(t'') \mathrm{d}t'' \tag{8.5}$$

The Fourier transform of the single atom dipole moment can be calculated as

$$x(\omega) = \int_{-\infty}^{+\infty} \mathrm{d}t x(t) \exp(\mathrm{i}\omega t) \tag{8.6}$$

and the harmonic emission rate $W(\omega)$ is

$$W(\omega) \propto \omega^3 |x(\omega)|^2 \, . \tag{8.7}$$

Taking into account (8.4), we have applied the SPA to the integral (8.6) over dt and dt' [24]. This method requires the solution of the saddle-point equations, obtained by equating to zero the derivatives of the Legendre transformed action $\Theta(p_s, t, t') = \omega t - S(p_s, t, t')$, with respect to t and t':

$$\left.\frac{\partial \Theta}{\partial t}\right|_{t_s} = \omega - \frac{[p_s - A(t_s)]^2}{2} - I_p = 0 \tag{8.8}$$

$$\left.\frac{\partial \Theta}{\partial t'}\right|_{t'_s} = \frac{[p_s - A(t'_s)]^2}{2} + I_p = 0 \tag{8.9}$$

Using the SPA the Fourier transform of the dipole moment, $x(\omega)$, can be written as a coherent superposition of the contributions from the complex electron quantum paths corresponding to the complex saddle-point solutions (p_s, t_s, t'_s). Note that the full electric field (8.3) of the laser pulse is used for the calculation, so that the model is generalized to account for nonadiabatic effects. Therefore $x(\omega)$ can be written as

$$x(\omega) = \sum_s x_s(\omega) \exp[\mathrm{i}\Phi_s(\omega)] = \sum_s \frac{\mathrm{i}2\pi}{\sqrt{\det(S'')}} \left[\frac{\pi}{\epsilon + \mathrm{i}(t_s - t'_s)/2}\right]^{3/2} E(t'_s) \times$$
$$\times\, d_x^* \left[p_s - A(t_s)\right] d_x \left[p_s - A\left(t'_s\right)\right] \exp\left[-\mathrm{i}S(p_s, t_s, t'_s) + \mathrm{i}\omega t_s\right] \tag{8.10}$$

where: $\Phi_s(\omega)$ is the phase of the complex function $x_s(\omega)$; $\det(S'')$ is the determinant of the 2×2 matrix of the second derivatives of Θ with respect to t and t', evaluated in correspondence of the saddle-point solutions:

$$\det(S'') = \det(\Theta'') = \left[\frac{\partial^2 S}{\partial t \partial t'}\right]^2 - \frac{\partial^2 S}{\partial t^2}\frac{\partial^2 S}{\partial t'^2} \tag{8.11}$$

where:

$$\frac{\partial^2 S}{\partial t \partial t'} = \frac{[p_s - A(t_s)][p_s - A(t'_s)]}{t_s - t'_s} \tag{8.12}$$

$$\frac{\partial^2 S}{\partial t^2} = -\frac{2(\omega - I_p)}{t_s - t'_s} + E(t_s)[p_s - A(t_s)] \tag{8.13}$$

$$\frac{\partial^2 S}{\partial t'^2} = \frac{2I_p}{t_s - t'_s} - E(t'_s)[p_s - A(t'_s)] \tag{8.14}$$

For a given value of the photon energy, $\hbar\omega$, a series of complex saddle-point solutions (p_s, t_s, t'_s) is obtained, which can be ordered depending on the time $\tau_s = t_s - t'_s$ spent by the electron in the continuum. The most relevant electron trajectories are characterized by a travel time $\mathrm{Re}(\tau)$ shorter than one or two periods (T_0) of the electric field of the driving pulse. In particular, two classes of quantum paths give the most relevant contribution: the so-called short and long paths, with return times of the order of half and one optical period, respectively. In the following section we will consider the contribution of the short quantum paths.

8.5.1 CEP Effects on the Short Electron Quantum Paths

Equation (8.10) can be written as follows:

$$x(\omega) = \sum_{\text{short}} x_s(\omega)\exp[\mathrm{i}\Phi_s(\omega)] + \sum_{\text{long}} x_s(\omega)\exp[\mathrm{i}\Phi_s(\omega)] \,. \tag{8.15}$$

The first sum in (8.15) corresponds to the coherent superposition of the short quantum paths, while the second sum is the coherent superposition of the long paths. We point out that, from an experimental point of view, efficient selection of the short quantum paths can be achieved when the gas jet is placed after the laser focus; whereas, when the gas jet is located before the laser focus, the contribution of the long quantum paths increases [25, 26].

Figure 8.4 shows two XUV spectra generated by a 6-fs light pulse calculated as coherent superposition of the contributions of the short quantum paths, assuming two different CEP values. In good agreement with the experimental results (see Fig. 8.3), three distinct regions can be identified in the final spectrum. In the lower-plateau region the spectra show well resolved peaks, corresponding to the odd harmonics of the fundamental radiation. In the upper-plateau region irregular peaks are present. Finally, in the cutoff region the spectra are characterized by well resolved peaks, which do not necessarily correspond to harmonics of the fundamental beam, whose spectral position is related to the CEP of the driving pulses.

The use of the nonadiabatic SPA allowed us to analyze and explain the general characteristics of the generated harmonic spectra. The single-atom emission rate, $F(\omega)$, generated by the contribution of the relevant short quantum paths has been calculated as

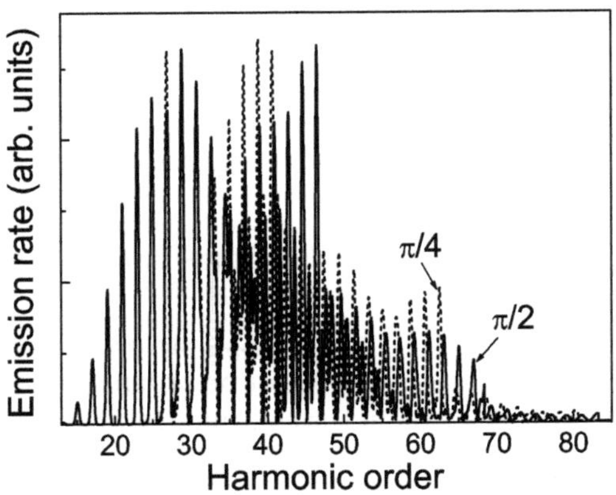

Fig. 8.4. Harmonic emission rate in neon calculated using the nonadiabatic SPA as coherent superposition of the relevant short quantum paths, for two CEPs of a 6-fs driving pulse. Same parameters of Fig. 8.3

$$F(\omega) \propto \omega^3 \left| \sum_{j,\text{short}} x_j(\omega) \exp[\mathrm{i}\Phi_j(\omega)] \right|^2 \tag{8.16}$$

The position of the corresponding spectral peaks is determined by the condition of constructive interference

$$\Delta\Phi_{j+2,j}(\omega) = \Phi_{j+2}(\omega) - \Phi_j(\omega) = 2m\pi \qquad (m = 0, \pm 1, \pm 2, ...) \tag{8.17}$$

where j and $j+2$ indicate two consecutive short paths. As shown in (8.10), the phase $\Phi_j(\omega)$, associated to the quantum path j, is given by three different contributions

$$\Phi_j(\omega) = \omega t_j(\omega) - S_j(\omega) + \chi_j(\omega) \tag{8.18}$$

where $\chi_j(\omega)$ takes into account the total phase of the various terms preceding the exponential function in (8.10). In the adiabatic (monochromatic)

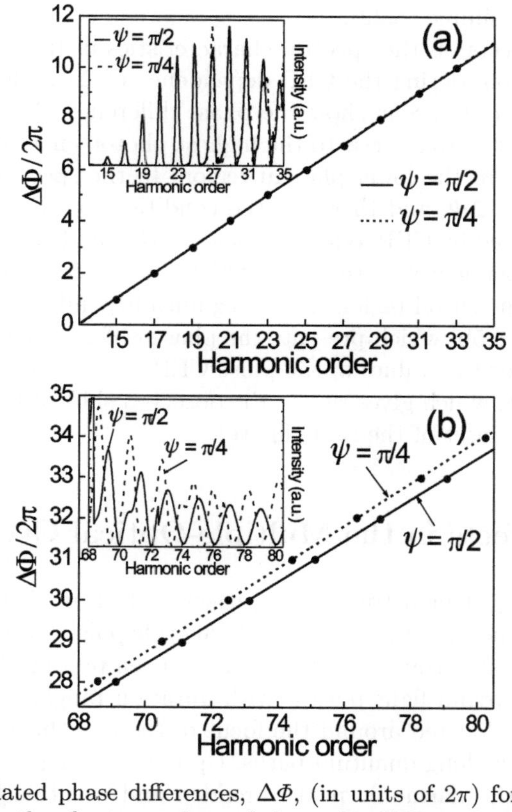

Fig. 8.5. Calculated phase differences, $\Delta\Phi$, (in units of 2π) for two consecutive short quantum paths, for two values of the pulse CEP. **a** Lower-plateau and **b** cutoff region. The *dots* mark the spectral positions corresponding to the condition of constructive interference. The *insets* show the corresponding total emission rates

approach, the variations of these three phase terms between two consecutive paths are

$$\omega t_{j+2}(\omega) - \omega t_j(\omega) = q\pi \quad (8.19)$$

$$S_{j+2}(\omega) - S_j(\omega) = 0 \quad (8.20)$$

$$\chi_{j+2}(\omega) - \chi_j(\omega) = \pi \quad (8.21)$$

where $\omega = q\omega_0$ (q is the harmonic order), and (8.21) takes into account the reversal of the electric field in two successive semi-cycles. Therefore in the adiabatic case we have

$$\Delta\Phi_{\text{ad}} = \Delta\Phi_{j+2,j} = (q+1)\pi \quad (8.22)$$

Equations (8.17) and (8.22) lead to the obvious conclusion that, in the adiabatic case, the condition for constructive interference is satisfied in the case of odd harmonics. In the nonadiabatic case (8.19)–(8.21) do not hold anymore; therefore the phase difference, $\Delta\Phi$, between two consecutive paths is not given by the simple (8.22).

The CEP effects on the spectral characteristics of the emission rate can be understood considering the CEP dependence of $\Delta\Phi$. The two coincident curves reported in Fig. 8.5a show the phase difference, $\Delta\Phi$, of the two consecutive short paths giving rise to the highest photon energies, calculated for two CEP values, in the lower-plateau region. In the spectral region considered in Fig. 8.5a, $\Delta\Phi$, and therefore the condition for constructive interference, is not affected by CEP: odd harmonics of the fundamental radiation are generated, independently of the pulse CEP. On the contrary, in the upper-plateau and in the cutoff regions $\Delta\Phi$ is significantly influenced by CEP. This is shown in Fig. 8.5b, which presents the phase differences for the same pair of short paths, for two values of the pulse CEP. The two curve are displaced by a phase-offset, which gives rise to the observed shift of the spectral peaks (as shown in the inset of the same figure).

8.6 CEP Effects in the Multiple-Optical Cycle Regime

We have recently demonstrated the presence of clear CEP effects in harmonic spectra generated by multiple-optical cycle pulses (with duration up to more than 40 fs) [27]. Harmonic emission has been produced focusing phase-stabilized femtosecond light pulses (with duration ranging from 20 to 40 fs) into an argon jet located around the focus of the laser beam to enhance the contribution of the long quantum paths. Upon increasing the intensity of the driving pulses, the harmonic peaks broaden and eventually overlap in the spectral region between consecutive odd harmonics, where distinct spectral peaks, whose position is CEP dependent, are formed, as shown in Fig. 8.6. The experimental results can be nicely reproduced using the nonadiabatic

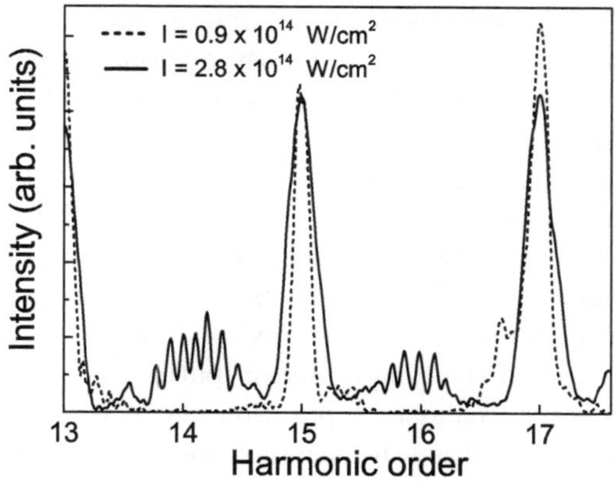

Fig. 8.6. Normalized harmonic spectra generated in argon by 35-fs pulses, with stable CEP, at two peak intensities

saddle-point method [27, 28]. We have demonstrated that the peaks corresponding to odd harmonics of the fundamental radiation shown in Fig. 8.6 mainly originate from the contributions of the short quantum paths. The peaks located between adjacent odd harmonics are due to contributions from the long paths, which are sensitive to the CEP. We have demonstrated that such spectral peaks are generated by the interference between consecutive harmonics and that such interference pattern can be used to measure the phase difference between adjacent harmonics, as it will be reviewed in the following section.

8.7 Measurement of the Phase Difference Between Harmonics

It is well known that, in the case of multiple-cycle pulses, the harmonic generation process gives rise to the production of trains of attosecond pulses. The main issue in order to measure the temporal structure of the XUV radiation is the determination of the relative phase of consecutive harmonics. This can be obtained by the technique called reconstruction of attosecond beating by interference of two-photon transition (RABITT) [6, 29]. RABITT is based on the measurement of photoelectron spectra generated in noble gases ionized by a superposition of odd harmonics in the presence of the fundamental infrared (IR) light of the driving beam. The phase difference between consecutive harmonics can be determined by measuring the magnitude of the generated sidebands peaks as a function of the delay between the IR pulse and the harmonic field. Recently it has been demonstrated that this method can be considered as a particular case of a more general technique called frequency-resolved optical gating for complete reconstruction of attosecond bursts (FROG CRAB), suitable for the characterization of isolated and trains of attosecond pulses [30]. We have recently demonstrated that it is possible to directly measure the phase difference between consecutive harmonics employing a self-referencing technique [31].

As already mentioned (see Fig. 8.6), upon increasing the intensity of the driving pulses the harmonics broaden and eventually overlap in the spectral region between consecutive odd harmonics. This is due to the fact that the temporal variation of the driving pulse intensity produces a harmonic chirp, which broadens the spectrum of each individual harmonic. This effect is more significant in the case of the long quantum paths [32]. If the spectral broadening is larger than the frequency separation between consecutive harmonics, the high-frequency side of the qth-harmonic spectrum, generated on the leading edge of the IR pulse, overlaps the low-frequency side of the spectrum of $(q+2)$th harmonic, generated on the pulse trailing edge, thus giving rise to the observed interference effect.

In order to analyze the CEP dependence of the interference pattern we have then varied the CEP, ψ, of the driving pulses. Figure 8.7 shows the portion between 13th and 15th harmonics of nine spectra for different amounts δz of glass in the beam path, corresponding to different CEPs in the range $\psi_0 < \psi < \psi_0 + \pi$. Each horizontal line represents a spectrum measured at a fixed CEP. The positions of the interference peaks continuously shift by changing the pulse CEP. The beat pattern periodically changes for a CEP variation $\Delta\psi = \pi$. The same behavior was observed for all the pairs of consecutive harmonics. Using the algorithm of Fourier transform spectral interferometry [33], one can retrieve, from the interference pattern, the phase difference, $\Delta\phi_q(\omega) = \phi_{q+2}(\omega) - \phi_q(\omega)$, between consecutive harmonics in the overlapping region.

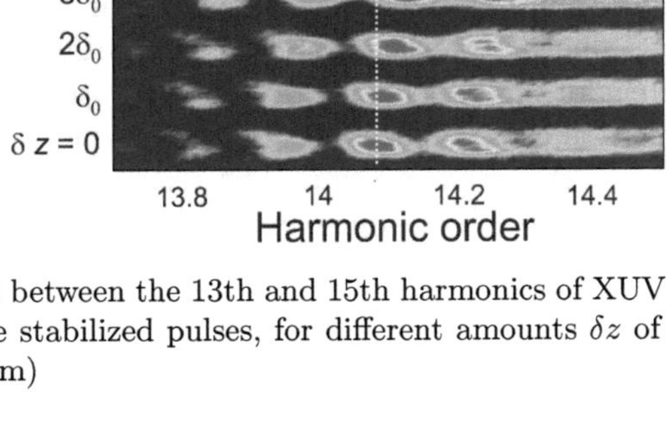

Fig. 8.7. Portion between the 13th and 15th harmonics of XUV spectra generated in argon by phase stabilized pulses, for different amounts δz of glass in the beam path ($\delta_0 = 3.42\,\mu\text{m}$)

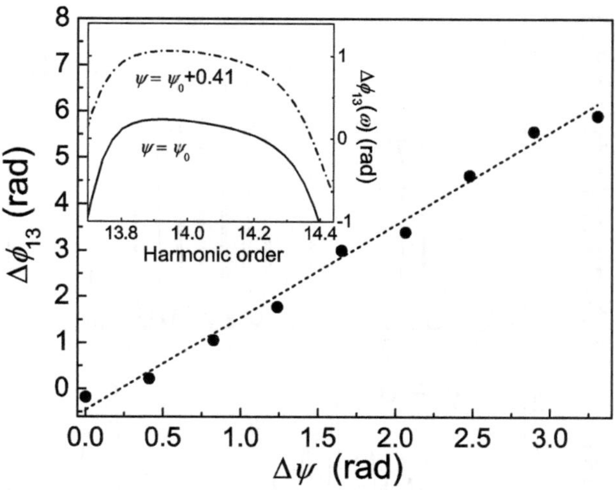

Fig. 8.8. $\Delta\phi_{13}(\omega = 14\omega_0)$ as a function of the CEP variation $\Delta\psi$ calculated from the experimental spectra reported in Fig. 8.7; the *dashed line* is a linear fit $\Delta\phi_{13} = 2\Delta\psi - 0.46$. *Inset*: retrieved $\Delta\phi_{13}(\omega)$ for two CEP values (consecutive spectra of Fig. 8.7)

The inset of Fig. 8.8 shows the retrieved $\Delta\phi_{13}(\omega)$ for two CEP values. A CEP variation $\Delta\psi$ determines a 2 $\Delta\psi$ variation of $\Delta\phi_{13}(\omega)$. This is clearly shown in Fig. 8.8, which displays $\Delta\phi_{13}(\omega = 14\omega_0)$ as a function of $\Delta\psi$: the experimental points can be well fitted by a line (dashed line in Fig. 8.8) with slope 2. The same results are valid for all the other pairs of consecutive harmonics. From the experimental results we can conclude that the harmonic phase, $\phi_q(\omega)$, is related to the CEP of the driving pulses by the following expression: $\phi_q(\omega) = \theta_q(\omega) + q\psi$, where $\theta_q(\omega)$ does not depend on ψ. This conclusion is confirmed by the results of numerical simulations based on the use of the nonadiabatic saddle-point method. Moreover such calculations demonstrate that the validity of the previous conclusion can be extended also to the contributions of the short paths.

Then we have considered the temporal structure of the emitted XUV radiation using the nonadiabatic saddle point method. Since the measured beat pattern is due to the long quantum paths, we have calculated the spectrum produced by the coherent superposition of the long path contributions. By taking the inverse Fourier transform it is possible to calculate the corresponding temporal structure. A train of attosecond pulses, separated by one-half the optical period is obtained, as shown by the dashed curve in Fig. 8.9a. In the calculated spectrum we have then selected the interference pattern, by proper spectral windows centered in the regions between consecutive odd harmonics. The associated temporal structure (solid curve in Fig. 8.9a) corresponds to two trains of attosecond pulses, located one on the leading and the other on the trailing edge of the complete train, separated in time by about one-half

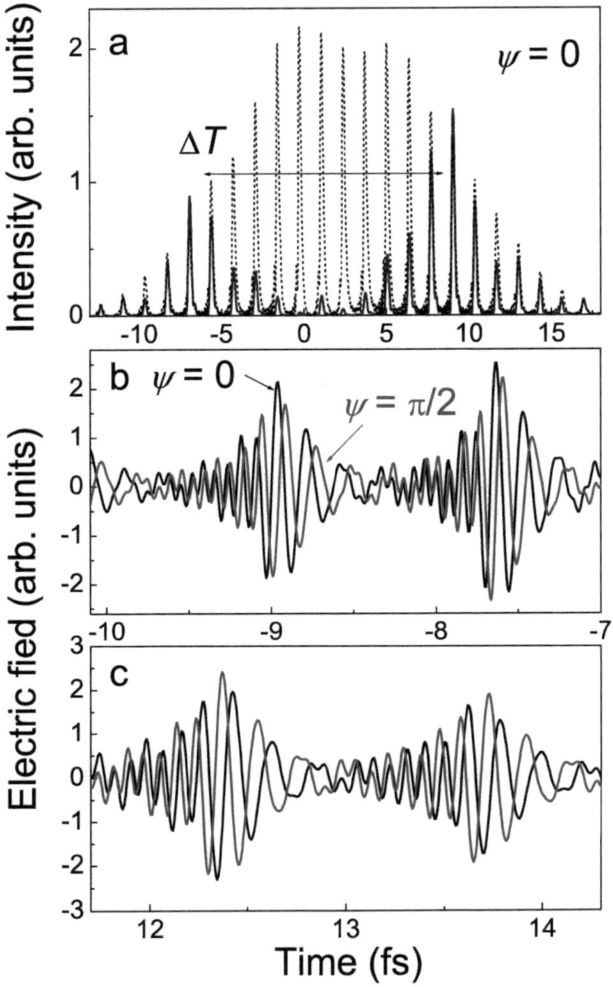

Fig. 8.9. a *Dashed curve*: intensity profile of the attosecond pulse train generated by the long paths, calculated as described in the text; *solid blue curve*: calculated temporal structure associated to the interference pattern generated by the long paths. Calculated electric field of the attosecond pulses generated on the leading **b** and on the trailing edge **c** of the driving pulse, for two CEP values of the driving pulse. Driving pulse duration 30 fs; peak intensity $I = 2.6 \times 10^{14}\,\mathrm{W/cm^2}$

the duration of the driving pulse. This is the picture in the temporal domain of the interference effect previously described in the frequency domain: the spectral structures between the odd harmonics originate from the interference between the electric fields of such two groups of attosecond pulses. In addition, the temporal analysis offers a simple explanation of the physical origin of the CEP dependence of the interference pattern. We found that, in agreement with our experimental results, the temporal delay, ΔT, between the interfering fields is not significantly affected by the CEP of the driving pulse. Therefore, we can conclude that the CEP dependence of the beat pattern is related to the temporal evolution of the electric field of the interfering attosecond pulses, which turns out to be directly influenced by ψ. Indeed, Fig. 8.9b and 8.9c show two different portions of the attosecond pulse train calculated for two values of ψ (the two trains have been temporally shifted to overlap the intensity envelopes of the attosecond pulses; the same shift determines the overlap of the electric fields of the corresponding IR pulses). The temporal evolution of the electric field of the attosecond pulses generated on the leading and trailing edges of the IR pulse is significantly influenced by ψ, while around the peak of the driving pulse the attosecond pulses (not show in Fig. 8.9) are not influenced by ψ. We have then calculated the temporal evolution of the electric field of the attosecond pulses generated by the long quantum path at low laser intensity and by the short paths; in these cases the electric field of the attosecond pulses is not significantly influenced by the CEP of the driving pulses.

8.8 Conclusions

Strong-field processes in the few-optical cycle regime are significantly affected by the carrier-envelope phase of the driving pulses and the control of such parameter is essential for many application, particularly for the generation of isolated and reproducible attosecond pulses. In this chapter we have reviewed, both from an experimental and a theoretical point of view, the influence of CEP on high-order harmonic generation. Moreover the role of CEP on the phase of the individual harmonics has been determined, using a self-referencing technique based on the interference of attosecond light pulses.

Acknowledgements

This work was partially supported by European Community under project MRTN-CT-2003-505138 (XTRA)

References

1. M. Nisoli, S. De Silvestri and O. Svelto: Appl. Phys. Lett. **68**, 2793 (1996)
2. T. Brabec and F. Krausz: Rev. Mod. Phys. **72**, 545 (2000)
3. S.T. Cundiff: J. Phys. D **35**, R43 (2002)
4. P.B. Corkum: Phys. Rev. Lett. **71**, 1994 (1993)
5. G. Farkas and C. Toth: Phys. Lett. A **168** 447 (1992)
6. P.M. Paul, E.S. Toma, P. Breger et al: Science **292**, 1689 (2001)

7. P. Tzallas, D. Charalambidis, N.A. Papadogiannis et al: Nature **426**, 267 (2003)
8. I.P. Christov, M.M. Murnane and H. Kapteyn: Phys. Rev. Lett. **78**, 1251 (1997)
9. R. Kienberger, E. Goulielmakis, M. Uiberacker et al: Nature **427**, 817 (2004)
10. S. De Silvestri, M. Nisoli, G. Sansone et al: Few-cycle pulses by external compression. In: *Few-Cycle Laser Pulse Generation and its Applications*, vol 95, ed by F. Kärtner (Springer, Berlin Heidelberg 2004) pp 137–177
11. M. Nisoli, S. De Silvestri, O. Svelto et al: Opt. Lett. **22**, 522 (1997)
12. J-P. Caumes, G. Sansone, E. Benedetti et al: J. Mod. Optics, **53**, 97 (2006)
13. M. Nisoli, G. Sansone, S. Stagira et al: Appl. Phys. B **75**, 601 (2002)
14. B. Schenkel, J. Biegert, U. Keller et al: Opt. Lett. **28**, 1987 (2003)
15. G.G. Paulus, F. Grasbon, H. Walther et al: Nature **414**, 182 (2001)
16. P. Dietrich, F. Krausz and P.B. Corkum: Opt. Lett. **25**, 16 (2000)
17. J. Reichert, R. Holzwarth, Th. Udem et al: Opt. Commun **172**, 59 (1999)
18. H.T. Telle, G. Steynmeyer, A.E. Dunlop et al: Appl. Phys. B **69**, 327 (1999)
19. D.J. Jones, S.A. Diddams, J.K. Ranka et al: Science **288**, 635 (2000)
20. A. Baltuška, Th. Udem, M. Uiberacker et al: Nature **421**, 611 (2003)
21. L. Poletto, S. Bonora, M. Pascolini et al: Rev. Sci. Instr. **75**, 4413 (2004)
22. M. Nisoli, G. Sansone, S. Stagira et al: Phys. Rev. Lett. **91**, 213905 (2003)
23. M. Lewenstein, Ph. Balcou, M.Y. Ivanov et al: Phys. Rev. A **49**, 2117 (1994)
24. G. Sansone, C. Vozzi, S. Stagira and M. Nisoli: Phys. Rev. A **70**, 013411 (2004)
25. P. Salières, B. Carré, L. Le Déroff et al: Science **292**, 902 (2001)
26. M. Bellini, C. Lyngå, A. Tozzi et al: Phys. Rev. Lett. **81**, 297 (1998)
27. G. Sansone, C. Vozzi, S. Stagira et al: Phys. Rev. Lett. **92**, 113904 (2004)
28. J-P. Caumes, M. Anscombe, G. Sansone et al: Laser Physics **15**, 205 (2005)
29. P. Agostini and L.F. DiMauro: Rep. Prog. Phys. **67**, 813 (2004)
30. Y. Mairesse and F. Quéré: Phys. Rev. A **71**, 011401 (2005)
31. G. Sansone, E. Benedetti, J-P. Caumes et al: Phys. Rev. Lett. **94**, 193903 (2005)
32. M.B. Gaarde, F. Salin, E. Constant et al: Phys Rev. A **59**, 1367 (1999)
33. M. Takeda, H. Ina and S. Kobayashi: J. Opt. Soc. Amer. **72**, 156 (1982)

9 Short-Pulse Laser-Produced Plasmas

Jean-Claude Gauthier

Centre Lasers Intenses et Applications (CELIA)
UMR 5107 CNRS, CEA, Université Bordeaux 1, 33405 Talence (France)
gauthier@celia.u-bordeaux1.fr

Summary. In this review, the physics of short-pulse laser-produced plasmas at moderate intensities is described, together with applications to X-ray sources and material processing.

9.1 Introduction

In the 1980s, advances in laser science and optical technologies opened new possibilities in the study of laser-produced plasmas. One of these advances is the implementation of the Chirped Pulse Amplification (CPA) technique [1] which has created new opportunities in the domain of ultra-intense and ultrafast laser physics. The interactions of ultra high-intensity laser pulses with matter have opened the field of optics in relativistic plasmas, a new topic of high-field science presented in comprehensive reviews [2, 3]. Indeed, intense lasers have been used to accelerate beams of electrons [4] and protons [5] to energies of several megaelectronvolts in distances of only microns. Recent improvements in particle energy spread [6] may allow compact laser-based radiation sources to be useful someday for cancer hadrontherapy [7] and as injectors into conventional accelerators [8], which are critical tools for X-ray and nuclear physics research. They might also be used for "fast ignition" [9] of inertial fusion targets. The ultrashort pulse duration of these particle bursts and the X-rays they can produce, hold great promise as well to resolve chemical, biological or physical reactions on ultrafast time scales and on the spatial scale of atoms [10]. Indeed, the time duration of these pulses being less than 100-fs, this is shorter than the time-scale of significant hydrodynamic motion of ions or solid target surfaces. Consequently, solid-density matter may be heated from room temperature to several hundreds of electronvolts without the usual change in density that accompanies long-pulse irradiation [11]. Ultrafast plasmas have important applications in material processing [12], thin film growth using ultrafast pulsed-laser deposition [13], and ultrashort pulse X-ray sources [14].

In this short review, we concentrate on "low" (non-relativistic) laser intensities, i.e. $I\lambda^2$ below a few $10^{18}\,\mathrm{W\,cm^{-2}\,\mu m^2}$, where I and λ are the laser intensity and wavelength, respectively. We give credit to pioneering experimental works with femtosecond lasers aimed at producing dense plasmas in Sect. 9.2. In Sect. 9.3 we describe optical techniques to study ultra-short pulse plasma phenomena, including pump-probe techniques and Frequency-Domain Interferometry (FDI), a very powerful optical technique with many applications in ultrafast science. Section 9.4 presents the most significant results in X-ray spectroscopy of ultrafast plasmas and describes recent advances in X-ray sources from short-pulse laser-produced plasmas. Section 9.5 reviews ultrafast plasma modeling and Sect. 9.6 details some practical applications of femtosecond laser-matter interaction.

9.2 Pioneering Works on Ultrafast Plasmas

In the "long" (nanosecond) pulse regime, laser-plasma interaction physics has been studied extensively to explore the efficiency of collisional absorption by inverse bremsstrahlung, parametric instability growth rates [15], filamentation, X-ray conversion efficiency and hydrodynamic instabilities [16]. In the short pulse regime of laser interaction with solid targets, a key issue is the understanding of the initial processes leading to the production of hot dense plasmas [17, 18]. It is crucial to obtain a time- and space-resolved picture of the very steep gradient around the critical density region to study the laser absorption mechanisms. Indeed, most of the energy transfer between the laser and the solid target occurs [19–21] in plasma regions where $0.5n_c < n_e < 10n_c$ where n_c is the critical density. With lasers of $\approx$100-fs duration, hydrodynamic simulations (see Sect. 9.5) show that the gradient scale length L can be very much smaller than the laser wavelength λ and that typical expansion velocities are in the 0.1-nm/fs range.

Absorption occurs within a layer thickness of a skin depth ($\approx$10-nm) [23–26]. At normal incidence in the Fresnel limit ($\lambda \ll L$) and a metallic target, laser absorption is quite small and the coupling of laser energy to target electrons is very inefficient (see Fig. 9.1). In the WKB limit ($\lambda \geq L$), absorption can be calculated simply [15] using "long" laser pulse formulas and in P-polarization, a peculiar absorption mechanism dubbed "resonance absorption" occur when the component of the laser field perpendicular to the target surface drives electron plasma waves at or below critical density (depending on the incidence angle). In the intermediate regime, numerical solutions of the Helmholtz equations must be derived. The basics of laser absorption mechanisms have been described in detail in [11].

Driven by X-ray sources applications and to circumvent the low absorption registered in the short pulse regime [27,28], there had been a great deal of interest in methods that could enhance the X-ray yield. The influence of various laser and target conditions has been the subject of many recent studies. Preplasma formation has been investigated in detail as one of the prominent ways of improving the X-ray yields [29]. While significant enhancement in the emission is noticed, it has also been shown that the X-ray pulse duration tends to become longer in such cases [30, 31]. The role of modulation/roughness of

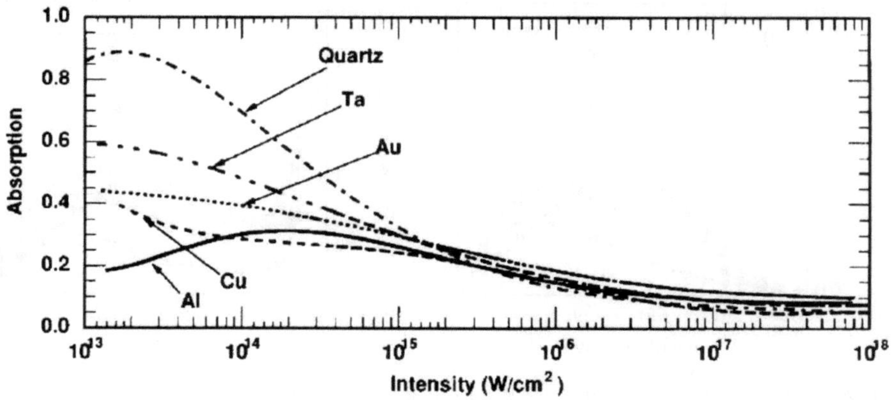

Fig. 9.1. Absorption coefficient of a ≈100-fs duration laser in different materials as a function of intensity (From [22])

the surface in increasing the coupling of the input light into the plasma, which results in an enhancement in the X-ray yield, has been studied in detail. Several authors [32, 33] have shown laser light absorption of over 90% into the plasma formed on grating targets as well as those coated with metal clusters. More recently, impressive enhancements of X-ray flux have been achieved in nanohole alumina, porous silicon targets [34] and nickel velvet targets [35].

Figure 9.2 shows the absorption coefficient as a function of the incidence angle for different plasma scale length L calculated with a model of laser-plasma interaction with a grating target [33]. We find a good qualitative agreement, both in incidence angle position and absorption fraction (see Fig. 9.2a), between the crude model and more involved calculations [20,21] with a flat target. In particular, the fact that the absorption maximum is displaced towards grazing angles for steep electron density gradients is in accord with more elaborate calculations of resonance absorption [15].

In Fig. 9.2b, we have plotted the absorption coefficient calculated for a 100-nm depth grating. The groove spacing d of the grating has to be correctly chosen to realize the "phase matching" condition ($k_g = 2\pi/d$; $k_y = k_g + k_0 \cos\theta$) where k_0 and k_y are the radiation wavevectors in vacuum and perpendicular to the grooves, respectively. A sharp peak is observed close to the resonance angle for the 10-nm gradient scale length. A larger absorption is observed below 20° in the case of the grating compared to the case of the flat target. This can be easily explained because, even near normal incidence, there is always a component of the incident electric field parallel to the local electron density gradient. Efficient X-ray production above 1-keV using femtosecond has been demonstrated [36] with laser-produced plasmas on silicon gratings of 1600-nm period and 250-nm groove depth. For long pulse irradiations (1.5-ps), the spectral shape of shifted-Kα transitions is similar to the one obtained previously [37] with flat targets. For short pulses (120-fs) and

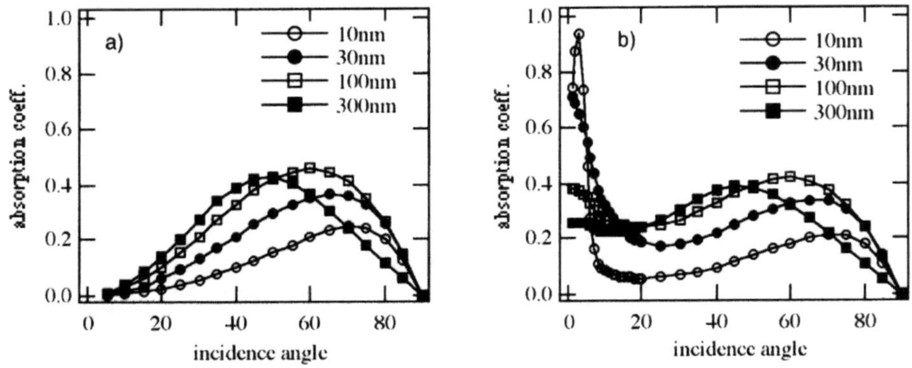

Fig. 9.2. Calculated absorption fraction as a function of the angle of incidence on **a** flat target and **b** grating, as a function of the gradient scale length in nm. λ=1000-nm, groove spacing 890-nm, groove depth 100-nm

good laser intensity contrast (10^{-6}), the spectral width of these transitions is surprisingly large, indicating electron densities well in excess of $10^{23}\,\mathrm{cm}^{-3}$. Two-dimensional particle-in-cell simulations indicate that the grating corrugation shape remains imprinted in the plasma during several tens of femtoseconds, increasing the period of time during which laser absorption is resonantly enhanced and favoring radiation emission at high electron densities.

9.3 Optical Characterization and Pump-Probe Techniques

Different techniques have been used to study the early evolution of ultrafast plasmas on a picosecond or subpicosecond time scale. Most of them rely on pump/probe techniques in which a short duration probe pulse interrogates the plasma surface at different time delays after the pump, plasma-producing, pulse. For example, subpicosecond time-resolved Schlieren measurements have been used to locate the critical density layer of a plasma [38]. However, diffraction effects limit the spatial resolution along the target normal to a value of the order of a few laser wavelengths. Analysis of the spectrum of the specular backscattered laser have been used to monitor the critical surface expansion [39–41]. Similar experiments have been performed in gases [42–44] and clusters [45]. A more efficient method to probe the plasma expansion relies on the spectral analysis of a reflected auxiliary beam at different delays. In some experiments, [46–48] expansion velocities have been inferred from Doppler shifts. However, the large spectral width $\Delta\omega \times \Delta t \approx 2\pi$ of Fourier-transform-limited short pulses makes it difficult to estimate frequency shifts much less than $\Delta\omega$. The measurement of the plasma reflectivity difference for S- and P-polarized light as a function of the angle of incidence has also been used to extract information on the gradient scale length [49]. All these methods rely on intensity or spectroscopic measurements which are very sensitive to the detrimental shot-to-shot frequency and spatial fluctuations of the lasers.

Optical methods giving access to the phase shift of a probe beam in transmission or reflection have shown their usefulness in the study of surfaces, interfaces, and thin films [50]. In the realm of laser-produced plasmas, Frequency-Domain Interferometry (FDI) is a new technique [51] that enables to measure both the amplitude and the phase of the complex reflection coefficient of a plasma with simultaneously high spatial and temporal resolution. In recent years, this technique has been used in ultrafast laser-matter interaction experiments including molecular spectroscopy [52], laser-induced dielectric damage [53, 54], laser wakefield particle acceleration [55], and the study of the propagation and breakout of femtosecond shock waves [56–58]. In laboratory plasmas, this technique has been successfully exploited to measure the electron density gradient scale-length [59], the onset of optical breakdown in dielectric materials [60], the collisionality of ultrafast plasmas [61], and its applicability has been extended to single-shot measurements [62, 63]. FDI has also been used at short wavelengths to probe plasmas with high order harmonic generation in gases [64].

To summarize, FDI probes the plasma surface with two successive twin femtosecond pulses separated in time by an external Michelson interferometer. The power spectrum of a double pulse sequence presents a fringe pattern with an envelope which is the Fourier transform of the pulse duration, modulated with a period inversely proportional to the pulse separation. Any phase shift due to the plasma generated between the two pulses can therefore be detected directly in the reflected spectrum as a fringe shift. This technique allows to achieve very accurate measurements of the phase with an incertitude of ± 0.01 radian, with a time resolution of the order of the duration of the probe pulse, and a spatial resolution along the focal diameter of ≈ 3-μm [55]. Recent applications of phase measurement techniques involve direct observation of the ponderomotive force exerted on a plasma [65].

Figure 9.3 shows the phase shift of a S-polarized probe beam when two laser pulses (laser intensities of 10^{15} and 5×10^{17} W/cm^2) separated by 6-ps, the prepulse and the main pulse, are interacting with the target. Only the region corresponding to the center of the main pulse focal spot is shown. We clearly see that the interaction with the main pulse almost stops the plasma expansion. Ponderomotive force effects on plasmas have been used to increase the electron density in spectroscopic studies [66].

9.4 Ultrafast X-ray Spectroscopy and X-ray Sources

Once the low-absorption problem of ultrafast laser interaction has been solved, laser-plasmas have a number of characteristics [30, 67] that make

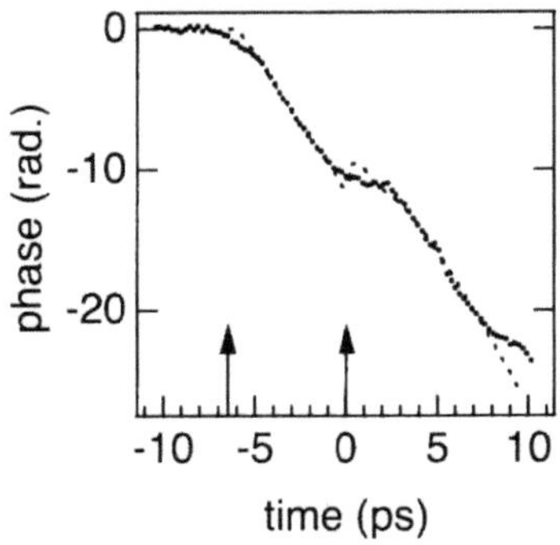

Fig. 9.3. Phase shift of S-polarized probe beam as a function of time following the interaction of a pure carbon target with a 400-fs duration laser pulse and prepulse (*arrows*). Experiment (*dots*) and FILM-fs simulations (*solid line*), see Sect. 9.5, are compared for a prepulse delay of 6-ps. The peak of the main pulse is at time zero

them valuable as X-ray sources in time-resolved X-ray measurements: i) a wide range of pulse durations from a few hundred femtoseconds to tens of picoseconds [68, 69], ii) very bright X-ray sources can be generated with source sizes as small as a few microns, and iii) laser-plasma X-ray sources can be accurately synchronized with other events that can be driven, triggered or stimulated by the same laser light. Thermal X-ray emission from laser-produced plasmas in the subpicosecond regime occurs with pulse durations of the order of a few picoseconds [70–72], measured by a streak camera. Conversion efficiencies (i.e. the ratio of X-ray energy to laser energy) for radiation above one keV are of the order of 10^{-5} in a single line, but conversions efficiencies around 200-eV are higher, about 0.1% for a 20% bandwidth [73].

Aluminum and silicon have been used as benchmark materials to spectroscopically study ultrafast plasmas. The conversion efficiency into narrow spectral lines and the duration of Al and Si thermal lines have been studied for a variety of experimental situations.

Helium-like Lines and Satellites

Primarily, He-like ion and satellite emissions from lower-charge state ions have been studied extensively [37, 74–76]. H-like ions have been seldom seen in experiments because their emergence temperature is much higher at the laser intensities considered here. Stark broadening of the spectral lines and the strong influence of satellite lines originating from doubly-excited levels are the main results of these studies. Figure 9.4 shows the striking difference between short pulse (Fig. 9.4a) and long pulse (Fig. 9.4b) laser irradiation of a silicon target around the He-like resonance line and its Li-like and Be-like dielectronic satellites. Time duration of these kilovolt emissions have been studied with a sub-ps resolution streak camera [77, 78],

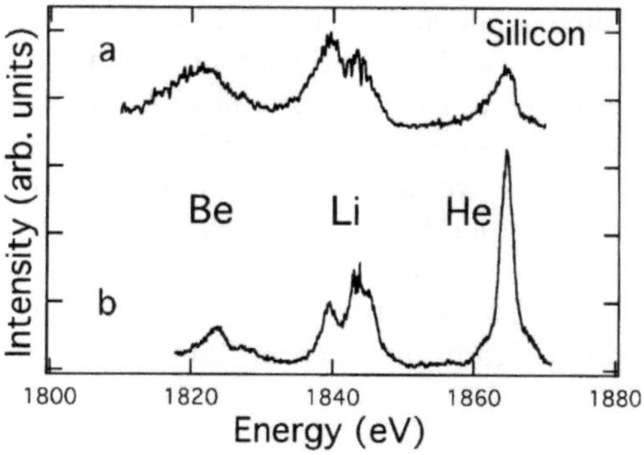

Fig. 9.4. Experimental Silicon spectrum associated to the Be-, Li-, and He-like ions, **a** short-pulse (0.12-ps), **b** long-pulse (2.5-ps)

and conversion efficiencies have been determined for more energetic X-rays [79]. Spectroscopy of ultrafast plasmas has been performed with contrast-clean laser pulses obtained through frequency doubling [80, 81] and tamped Al-tracer layer-targets [82]. Isochoric heating of solid aluminum at temperatures around 300-eV has been achieved. This corresponds to a pressure or energy density of ≈0.4-Gbar in the heated dense aluminum. Free standing foils, 25-nm thin, have also been used to reach very high electron densities in contrast-clean experiments [83, 84]. The small foil thickness in these experiments minimized thermal gradients in the longitudinal direction and allowed a simple physical interpretation of the spectral data.

Kα Lines

It is well known [85] that ultra-intense laser pulse interaction with solid targets produce copious amounts of hot electrons. Collective absorption mechanisms transfer part of the laser energy into hot electrons which are accelerated to multi-kilovolt energies and penetrate into the "cold" solid behind the plasma where they generate X-rays via K-shell ionization and bremsstrahlung [18]. Resonance absorption is one of these collective mechanism that explains the sensitivity of hot electron conversion efficiency to laser polarization. Indeed, the conversion efficiency from laser light into K-shell line radiation in P-polarization is 100-times higher than the one measured in S-polarization. Suprathermal electrons produced by non-collisional absorption mechanisms have proved to be a convenient way of generating X-rays in the photon energy range above one keV [86]. In the 1- to 10-keV energy range, efficient production of K_α radiation in aluminum, calcium, and iron has been demonstrated. The X-ray throughput was controlled by varying the energy contrast ratio between the main ultrashort pulse and its nanosecond pedestal. The main characteristics of this type of source are a source diameter of about 10-μm, repetition rate is 10-Hz and a total number of photons per shot of about 3×10^8 to 10^{10}, depending on laser intensities. Kα lines produced from ultrafast plasmas have been shown to be of a time duration similar to the laser pulse [87–89]. Supra-thermal electrons have also been used to investigate isochoric heating of solid matter. Temperatures of 500-eV and electron densities up to $5\times10^{23}\,\mathrm{cm}^{-3}$ have been found. The heated depth is consistent with the range of 20-keV electrons, and the energy deposited in the heated layer was estimated to be a significant fraction (20 to 25%) of the incident laser energy [90].

Broadband Sources

Thermal emission from ultrafast plasmas is composed of discrete lines or very narrow bands of lines when targets of low atomic number Z are used. The effect of an increase of Z on the emission is that it changes the character of the thermal radiation from line to bands. In order to realize absorption spectroscopy one needs a backlight flash (multi-keV range) that is quasi continuous in some region of interest. M-shell emission spectra, obtained with ultrafast lasers interacting with high-Z material, have been measured recently [91, 92]. These spectra emitted by dense plasmas display broadband emissions related to Unresolved Transitions Arrays [93]. In Fig. 9.5 is shown broadband emission spectra of the 3d–4f transitions obtained for two different values of the energy contained in the hot electrons (equivalent to the laser intensity) when thick Ta targets are irradiated at $5\times10^{18}\,\mathrm{W/cm^2}$ with 400-fs

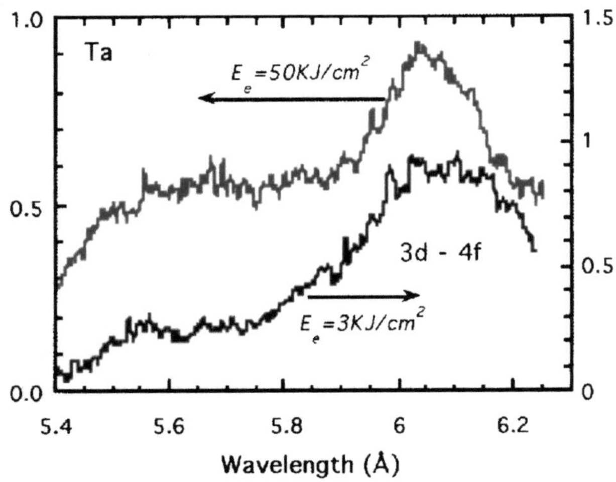

Fig. 9.5. Time-integrated spectra of the Ta 3d–4f transitions obtained for two different values of the hot electron energy density on target. Laser pulse is 400-fs long and intensity is $5\times10^{18}\,\mathrm{W/cm^2}$. Adapted from [94]

pulses. The observed ionization states seem to be slightly higher in presence of hot electrons (mostly Ga-like), compared to the result obtained with less hot electrons (mostly As-like) [94].

Laser-produced X-ray plasma sources can be distinguished by their high *peak* brightness. Increasing the laser repetition rate to boost their *average* brightness is not always practical [95, 96]. Conventional laser-produced plasma targets produce debris which may destroy or coat sensitive X-ray components such as multilayer optics, Bragg crystals, or even the sample under study. The interaction of small rare gas clusters with short pulse high intensity lasers may be applied, in the future, to the production of short pulse X-rays without the complications of laser solid-target interaction [97].

Harmonic Generation on Solids

High-harmonic generation with ultrashort laser sources has recently attracted great interest [98] as a convenient technique for the production of coherent EUV and XUV radiation. Until now, most high-harmonic-generation processes were studied in inert-gas and cluster targets [99]. High-harmonic generation from an overdense plasma surface operates at higher intensities [100]. The generation of harmonics from a ponderomotively driven oscillating plasma surface was studied theoretically by Lichters et al. [101] who used fully relativistic one-dimensional particle-in-cell simulations. Recent experiments show the importance of the laser intensity contrast ratio on harmonic generation [102, 103]. A plasma mirror was designed and used to obtain 60-fs 10-TW laser pulses with a temporal contrast of 10^8 on a nanosecond time scale and 10^6 on a picosecond time scale. With these high-contrast pulses high harmonics were generated by nonlinear reflection on a plasma with a steep electronic density gradient. Well-collimated harmonics up to 20th order were observed for a laser intensity of 3×10^{17} W/cm^2, whereas no harmonics are obtained without the plasma mirror [102]. Harmonics from the rear side of laser-irradiated thin foils are useful to probe plasmas. The highest harmonic generated in this way is near the plasma frequency of the dense foil. Above this frequency a cut off occurs, which is observed in experiments [104], and is of interest for diagnostics to determine the maximum density in the foil during the interaction.

9.5 Ultrafast Plasma Modeling

The various hydrodynamics codes which have been developed for laser-induced fusion [105] have all been rapidly adapted to treat the case of ultrashort pulse interaction. Different flavors of hydrocodes like FILM-fs [106],

MULTI-fs [107] and UBC [108] are now available. The set of fluid equations that all these codes solve is:

$$\frac{\partial \rho}{\partial t} + \rho^2 \frac{\partial u}{\partial m} = 0 \tag{9.1}$$

$$\frac{\partial u}{\partial t} + \frac{\partial (p_e + p_i)}{\partial m} = 0 \tag{9.2}$$

$$\frac{\partial \epsilon_e}{\partial t} + p_e \frac{\partial u}{\partial m} = -\frac{\partial S_e}{\partial m} - \sum_k \frac{\partial S_k}{\partial m} + \frac{\partial S_L}{\partial m} - \gamma(T_e - T_i) \tag{9.3}$$

$$\frac{\partial \epsilon_i}{\partial t} + p_i \frac{\partial u}{\partial m} = \gamma(T_e - T_i) \tag{9.4}$$

The system is closed by the equation of state (EOS) of the target material which relates the electron ϵ_e, p_e and ion ϵ_i, p_i energies and pressures to the density ρ and temperatures T_e and T_i. S_k is the radiative flow of photon group k with energies $h\nu_k \leq h\nu \leq h\nu_{k+1}$. S_e is the electron heat flow. S_L is the energy deposited by the laser. In the ultrashort pulse regime where energy is impinging on a steep density gradient, S_L is calculated by self-consistent solution of the Helmholtz equations using the Drude approximation to the dielectric constant. FILM-fs [106] incorporates the effects of the ponderomotive pressure (see above). These codes have not been developed to treat the relativistic regime $\mathrm{I} \geq 10^{18}\,\mathrm{W/cm^2}$. Within the framework of the approximations made, such as the Drude model for electron conduction, there are a few adjustable parameters (within a restricted range governed by comparison to various experiments) to match the isochoric solid/warm plasma transition. These parameters are: the momentum transfer collision frequency which enters the dielectric constant formula, the electron–ion collision frequency for energy transfer entering the parameter γ in the hydrodynamic equations above, and the correct expressions of the electron and thermal conductivities.

LTE is seldom found in short pulse laser-produced plasmas because of time-dependent effects and steep spatial gradients. To simulate plasma line and continuum emissions, one has to rely on collisional-radiative models, such as TRANSPEC as a postprocessor [110]. In this type of code, one solves the full master equations for the ion level populations $\{N\}$:

$$\frac{\mathrm{d}\{N\}}{\mathrm{d}t} = \{A\}\{N\} \tag{9.5}$$

where $\{A\}$ is the matrix of collisional and radiative excitation and de-excitation rate constants. Rates can be calculated in various ways, from detailed term accounting to the formalism of super-configuration arrays [111]. Experimental results and numerical simulations of K-shell emission from Be-like, Li-like and He-like ions from a 100-fs laser-produced aluminium plasma at intensities in excess of $10^{16}\,\mathrm{W/cm^2}$ have been obtained [75] that show (see Fig. 9.6) that time-dependent modelling explains the much shorter than the laser pulse X-ray pulse duration of the Be-like and Li-like emissions [75].

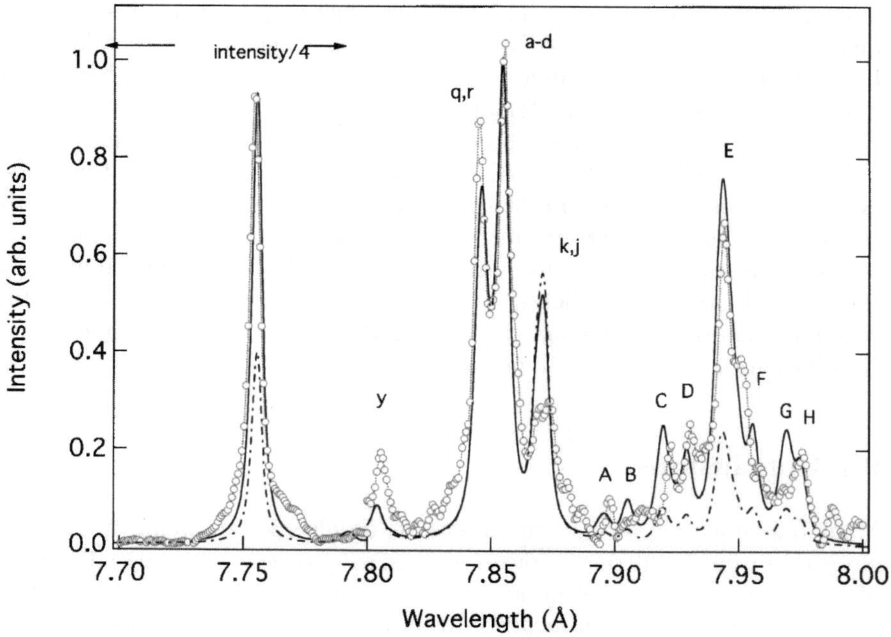

Fig. 9.6. Al spectrum of a 200-nm thin film irradiated at 2×10^{16} W/cm^2 compared to the result of time-dependent atomic physics and hydrodynamic calculations at 10^{16} W/cm^2 (*full*) and 3×10^{16} W/cm^2 (*chain*). Heights are normalized to the experimental **a**–**d** peaks (following the notation of Gabriel [109]). The intensity of the He-like resonance line below 0.78-nm has been divided by four

9.6 Applications of Ultrafast Plasmas at Low Pulse Energies

Absorption of ultra-short low-energy laser pulses takes place on a timescale much shorter than energy transfer of excited electrons to the lattice atoms (typically a few ps). The result is the creation of a highly ionized surface layer (plasma) at near solid density. Surface melting can virtually be eliminated along with plasma shielding, which takes place in typical nanosecond laser ablation [112]. Hence, femtosecond laser ablation can yield precise materials processing resulting from efficient energy deposition while simultaneously minimizing heat conduction and thermal damage to surrounding material. Focussed intensities $I > 10^{13}$ W/cm^2 are easily obtained with micro-joule pulses and the processing of, for example, normally transparent dielectrics can be achieved through multi-photon absorption [113]. Metals can be ablated quite easily [114] and the absence of a heat affected zone (HAZ), like in ordinary ns-pulse material processing, makes femtosecond pulse micro-machining a fast developing technology. For the foreseeable future, femtosecond-based machining will be restricted mainly to applications that cannot be addressed with other machining technologies [115, 116]. One such application is glass micromachining, a much more subtle endeavor than machining of metals [112]. In summary, the hydrodynamics of material ejection into vacuum have been modeled to consider particle formation following energetic femtosecond laser ablation. The ejected material is quenched from 1 to 3 orders of magnitude more rapidly than is material on the bulk surface and therefore stands the best chance of trapping potentially unique material states associated with rapid quenching of an extreme material state. The mean particle diameter is estimated to range from 1- to 10-nm, and it should be possible to exercise some control over nanoparticle size by varying the laser fluence. Experimental determination of both nanoparticle size and associated electronic and structural properties remains an important goal [117].

9.7 Conclusions

Many fundamental processes in physics, materials science, chemistry, and biology occur on the ultrafast (picosecond or subpicosecond) time scale. Some of these processes can be initiated by transient optical excitation, and followed in their time evolution by ultrafast infrared, visible and ultraviolet spectroscopy. Pump-probe optical techniques are sensitive to electronic excitations, whereas extension to the sub-nanometer wavelength range should make possible the direct monitoring of atomic positions. Ultrafast laser-based plasma studies will play a pivotal role in the development of experimental dense plasma physics in the coming years. Ultrafast plasmas will also have important applications in material processing, thin film growth using ultrafast pulsed-laser deposition, and ultrashort pulse X-ray sources. Moreover, in the future, femtosecond X-ray pulses will have tremendous applications in broad areas of science, including condensed matter physics, chemistry, biology, and engineering because the ability to time-resolve atomic motions by X-ray diffraction (XRD) could open entirely new fields of scientific research [118]. The potentially most rewarding, but also most demanding application of femtosecond XRD will be the characterization of ultrafast structural processes in complex biological molecules.

Acknowledgments

It is a pleasure to acknowledge the strong support of my colleagues Jean-Paul Geindre, Patrick Audebert, Claude Chenais-Popovics from Laboratoire d'Utilisation des Lasers Intenses, and Antoine Rousse from Laboratoire d'Optique Appliquée.

References

1. P. Maine, D. Strickland, M. Pessot, J. Squier, P. Bado, G. Mourou, and D. Harter, *Chirped Pulse Amplification: Present and Future* in *Ultrafast Phenomena VI*, ed by T. Yajima, K. Yoshihara, C.B. Harris, and S. Shionoya (Springer-Verlag, New York, 1988), p. 205, and references therein.
2. G.A. Mourou, C.P. Barty, and M.D. Perry, Physics Today, January 98, 22 (1998).
3. D. Umstadter, Phys. Plasmas **8**, 1774 (2001).
4. V. Malka *et al.*, Science **298**, 1596 (2002).
5. M. Roth *et al.*, Phys. Rev. STAB **5**, 061031 (2002) and references therein.
6. S.P.D. Mangles *et al.*, Nature **431**, 535 (2004); C.G.R. Geddes *et al.*, ibid. 535 (2004); J. Faure *et al.*, ibid. 541 (2004).
7. S.V. Bulanov, T. Zh. Esirkepov, V.S. Khoroshkov, A.V. Kuznetsov and F. Pegoraro, Physics Letters A **299**, 247 (2002).
8. T.E. Cowan *et al.*, Phys. Rev. Lett. **92**, 204801 (2004).
9. M. Tabak, Physics of Plasmas **12**, 057305 (2005).
10. C. Rose-Petruck, R. Jimenez, T. Guo, A. Cavalleri, C.W. Siders, F. Ràski, J.A. Squier, B.C. Walker, K. Wilson and C.P.J. Barty, Nature **398**, 310 (1999).
11. J.C. Gauthier, *Dense Ultrafast Plasmas*, in Atoms, Solids, and Plasmas in Super-Intense Fields, ed D. Batani *et al.*(Kluwer Academic/Plenum Publishers, 2001) pp. 193–231.
12. X. Liu, D. Du, and G. Mourou, IEEE J. Quantum Electron. **33**, 1706 (1997).
13. E.G. Gamaly, A.V. Rode, and B. Luther-Davies, J. Appl. Phys. **85**, 4213 (1999).
14. A. Rousse, P. Audebert, J.P. Geindre, F. Falliès, J.-C. Gauthier, A. Mysyrowicz, G. Grillon, and A. Antonetti, Phys. Rev. E **50**, 2200 (1994).
15. W.L. Kruer, *The Physics of Laser Plasma Interactions* (Addison-Wesley, Redwood-City, 1988).
16. E.M. Campbell, Phys. Fluids B **4**, 3781 (1992).
17. J.-C. Gauthier, *Short Pulse Laser Interaction With Solid Targets* in Laser Interaction with Matter, ed S. Rose (Institute of Physics Publishing, Bristol, 1995).
18. P. Gibbon and E. Förster, Plasma Phys. Control. Fusion **38**, 769 (1996).
19. J.-C. Kieffer, P. Audebert, M. Chaker, J.-P. Matte, H. Pépin, T.W. Johnston, P. Maine, D. Meyerhoffer, J. Delettrez, D. Strickland, P. Bado, and G. Mourou, Phys. Rev. Lett. **62**, 760 (1989).
20. H.M. Milchberg and R.R. Freeman, J.O.S.A. B **6**, 1351 (1989).
21. R. Fedosejevs, R. Ottmann, R. Sigel, G. Kuhnle, S. Szatmari, and F.P. Schäfer, Phys. Rev. Lett. **64**, 1250 (1990).
22. D.F. Price, R.M. More, R.S. Walling, G. Guethlein, R.L. Shepherd, R.E. Stewart, and W.E. White, Phys. Rev. Lett. **75**, 252 (1995).
23. F. Brunel, Phys. Rev. Lett. **59**, 152 (1987).
24. M.K. Grimes, Y.-S. Lee, A.R. Rundquist and M.C. Downer, Phys. Rev. Lett. **82**, 4010 (1999).
25. W. Rozmus and V.T. Tikhonchuk, Phys. Rev. A **42**, 7401 (1990); ibid. **46**, 7810 (1992).
26. A.A. Andreev, K. Yu. Platonov, and J.-C. Gauthier, Phys. Rev. E **58**, 2424 (1998).

27. M.M. Murnane, H.C. Kaypten, and R.W. Falcone, Phys. Rev. Lett. **62**, 155 (1989).
28. P. Audebert, J.-P. Geindre, J.-C. Gauthier, J.-P. Chambaret, A. Mysyrowicz, and A. Antonetti, Europhysics Letters **19**, 189 (1992).
29. S. Bastiani, A. Rousse, J.-P. Geindre, P. Audebert, C. Quoix, G. Hamoniaux, A. Antonetti, and J.-C. Gauthier, Phys. Rev. E **56**, 7179 (1997).
30. J.F. Pelletier, M. Chaker, and J.-C. Kieffer, J. Appl. Phys. **81**, 5980 (1997).
31. H. Nakano, T. Nishikawa, H. Ahn, and N. Uesugi, Appl. Phys. Lett. **69**, 2992 (1996).
32. M.M. Murnane, H.C. Kapteyn, S.P. Gordon, J. Bokor, E.N. Glytsis, and R.W. Falcone, Appl. Phys. Lett. **62**, 1068 (1993).
33. J.-C. Gauthier, S. Bastiani, P. Audebert, J.-P. Geindre, K. Neuman, T. Donnelly, M. Hoffer, R.W. Falcone, R. Shepherd, D. Price, and B. White, SPIE Proceedings **2523**, 242 (1995).
34. T. Nishikawa, H. Nakano, N. Uesugi, M. Nakao, and H. Masuda, Appl. Phys. Lett. **75**, 4079 (1999); ibid. **70**, 1653 (1997).
35. G. Kulcsár, D. AlMawlawi, F.W. Budnik, P.R. Herman, M. Moskovits, L. Zhao, and R.S. Marjoribanks, Phys. Rev. Lett. **84**, 5149 (2000).
36. J.-C. Gauthier, S. Bastiani, P. Audebert, J.P. Geindre, K. Neuman, T. Donnelly, M. Hoffer, R.W. Falcone, R. Sheperd, D. Price and B. White, Proc. SPIE **2523**, 242 (1995).
37. R.C. Mancini, P. Audebert, J.-P. Geindre, A. Rousse, F. Fallis, J.-C. Gauthier, A. Mysyrowicz, J.-P. Chambaret, andA. Antonetti, J. Phys. B: Atom. Optical Mol. Phys. **27**, 1671 (1994).
38. R. Benattar, J.P. Geindre, P. Audebert, J.-C. Gauthier, J.P. Chambaret, A. Mysyrowicz and A. Antonetti, Optics Commun. **88**, 376 (1992).
39. H.M. Michberg and R.R. Freeman, Phys. Rev. A **41**, 2211 (1990).
40. R. Sauerbrey, J. Fure, S.P. Le Blanc, B. van Wonterghem, U. Teubner and F.P. Schäfer, Phys. Plasmas **1**, 1635 (1994).
41. S.P. Le Blanc, R. Sauerbrey, J. Opt. Soc. Am. B **13**, 72 (1996).
42. Y.M. Li and R. Fedosejevs, Phys. Rev. E **54**, 2166 (1996).
43. W.M. Wood, C.W. Siders, and M.C. Downer, Phys. Rev. Lett. **67**, 3523 (1991).
44. C.W. Siders, N.C. Turner III, M.C. Downer, A. Babine, A. Stepanov, and A.M. Sergeev, J. Opt. Soc. Am. B **13**, 330 (1996).
45. K.Y. Kim *et al.*, Phys. Rev. A **71**, 011201 (2005).
46. T. Dewandre, J.R. Albritton, and E.A. Williams, Phys. Fluids **24**, 528 (1981).
47. O.L. Landen and W.E. Alley, Phys. Rev. A **48**, 5089 (1992).
48. X. Liu and D. Umstadter, Phys. Rev. Lett. **69**, 1935 (1992).
49. O.L. Landen, D.G. Stearns, and E.M. Campbell, Phys. Rev. Lett. **63**, 1475 (1989).
50. S. Bera, A.J. Sabbah, C.G. Durfee, and J.A.Squier, Optics Letters **30**, 373 (2005).
51. J.-P. Geindre *et al.*, Optics Letters**19**, 1997 (1994).
52. N. Scherer *et al.*, J. Chem. Phys. **95**, 1487 (1991).
53. P. Audebert *et al.*, Phys. Rev. Lett. **73**, 1990 (1994).
54. F. Quéré, S. Guizard, and P. Martin, Europhysics Letters **56**, 138 (2001).
55. J.-R. Marquès *et al.*, Phys. Rev. Lett. **76**, 3566 (1996).
56. R. Evans *et al.*, Phys. Rev. Lett. **77**, 3359 (1996).
57. K.T. Gahagan, D.S. Moore, D.J. Fink, R.L. Rabie, and S.J. Bulow, Phys. Rev. Lett. **85**, 3205 (2000).

58. D.J. Fink *et al.*, Phys. Rev. B **64**, 115114 (2001).
59. P. Blanc *et al.*, J.O.S.A. B **13**, 118 (1996).
60. C. Quoix *et al.*, Eur. Phys. J. AP **5**, 163 (1999).
61. C. Quoix *et al.*, J.Q.R.S.T. **65**, 455 (2000).
62. A. Benuzzi-Mounaix *et al.*, Phys. Rev. E **60**, R2488 (1999).
63. S. Rebibo *et al.*, Laser and Particle Beams **19**, 67 (2001).
64. D. Descamps *et al.*, Optics Letters **25**, 135 (2000).
65. P. Audebert, J.-P. Geindre, S. Rebibo, and J.-C.Gauthier, Phys. Rev. E **64**, 056412 (2001).
66. O. Peyrusse, M. Busquet, J.-C. Kieffer, Z. Jiang, and C.Y. Côté, Phys. Rev. Lett. **75**, 3862 (1995).
67. J.C. Kieffer and O. Peyrusse, Phys. Rev. E **56**, 992 (1997).
68. I.C. Turcu and J.B. Dance, *X-Rays From Laser Plasmas* (John Wiley and Sons, Chichester, 1999).
69. See several contributions in *New Ultrafast X-Ray Sources and Applications*, ed J.-C. Gauthier, C.R. Acad. Sci. Paris, Séries IV **1** (Elsevier, Paris, 2000).
70. X. Liu, U. Ellenberger, J.S. Coe, C.Y. Chien, and D. Umstadter, Phys. Rev. Lett. **75**, 2324 (1995).
71. J.C. Kieffer, Optics Letters **21**, 1040 (1996).
72. U. Ellenberger, J.S. Coe, X.-Y. Chien, and D. Umstadter, J.O.S.A. **13**, 125 (1996).
73. D. Alterbernd, U. Teubner, P. Gibbon, E. Förster, P. Audebert, J.P. Geindre, J.C. Gauthier, G. Grillon, and A. Antonetti, J. Phys. B: Atom. Mol. Opt. Phys. **30**, 3969 (1997).
74. G.A. Kyrala, R.D. Fulton, E.K. Wahlin, L.A. Jones, G.J. Schappert, and J.A. Cobble, Appl. Phys. Lett. **60**, 2195 (1992).
75. P. Audebert, J.-P. Geindre, A. Rousse, F. Fallis, J.-C. Gauthier, A. Mysyrowicz, G. Grillon, and A. Antonetti, J. Phys. B: Atom. Optical Mol. Phys. **27**, 3303 (1994).
76. R.C. Mancini, A.S. Shlyaptseva, P. Audebert, J.-P. Geindre, S. Bastiani, J.-C. Gauthier, G. Grillon, A. Mysyrowicz, and A. Antonetti, Phys. Rev. E **54**, 4147 (1996).
77. Z. Jiang, J.-C. Kieffer, J.P. Matte, M. Chaker, O. Peyrusse, D. Gilles, G. Korn, A. Maksimchuk, S. Coe, and G. Mourou, Phys. Plasmas **2**, 1702 (1995).
78. C.Y. Côté, J.-C. Kieffer, Z. Jiang, A. Ikhlef, and H. Pépin, J. Phys. B: At. Mol. Opt. Phys. **31**, L883 (1998).
79. J. Yu, Z. Jiang, J.-C. Kieffer, and A. Krol, Phys. Plasmas **6**, 1318 (1999).
80. K. Eidmann, A. Saemann, U. Andiel, I.E. Golovkin, R.C. Mancini, E. Anderson, and E. Förster, J. Quant. Spectr. Rad. Transfer **65**, 173 (2000).
81. U. Andiel, K. Eidmann, K. Witte, I. Uschmann, and E. Förster, Appl. Phys. Lett. **80**, 198 (2002).
82. A. Saemann, K. Eidmann, I.E. Golovkin, R.C. Mancini, E. Anderson, E. Förster, and K. Witte, Phys. Rev. Lett. **82**, 4843 (1999).
83. P. Audebert, R. Shepherd, K.B. Fournier, O. Peyrusse, D. Price, R.W. Lee, P. Springer, J.-C. Gauthier, and L. Klein, Phys. Rev. E **66**, 066412 (2002).
84. P. Audebert, R. Shepherd, K.B. Fournier, O. Peyrusse, D. Price, R. Lee, P. Springer, J.-C. Gauthier, and L. Klein, Phys. Rev. Lett. **89**, 265001 (2002).
85. J. Lindl, Phys. Plasmas **2**, 3933 (1995).
86. A. Rousse, P. Audebert, J.-P. Geindre, F. Falliès, J.-C. Gauthier, A. Mysyrowicz, G. Grillon, and A. Antonetti, Phys. Rev. E **50**, 2200 (1994).

87. C. Rischel *et al.*, Nature **390**, 497 (1997).
88. Ch. Reich, P. Gibbon, I. Uschmann, and E. Förster, Phys. Rev. Lett. **84**, 4846 (2000).
89. T. Feurer *et al.*, Phys. Rev. E **65**, 016412 (2001).
90. K. Eidmann *et al.*, in Atomic Processes in Plasmas, ed J. Cohen, S. Mazevet, and D. Kilcrease, AIP Conference Proceedings **730**, pp. 81–93 (2004).
91. C. Chenais-Popovics *et al.*, in Laser-generated and other laboratory X-ray and EUV sources, optics, and applications, San Diego, CA, SPIE proceedings vol. 5196, 2003. Bellingham, WA: SPIE-International Society for Optical Engineering; 2004. p. 20512.
92. M. Busquet *et al.*, Phys. Rev. E **61**, 801 (2000).
93. J. Bauche and C. Bauche-Arnoult, in Laser Interactions with Atoms, Solids and Plasmas, ed R.M. More (Plenum Press, New York, 1994).
94. J.C. Kieffer, C.Y. Chien, F. Dorchies, P. Forget, P. Gallant, Z. Jiang and H. Pépin, in New Ultrafast X-Ray Sources and Applications, ed J.-C. Gauthier, C.R. Acad. Sci. Paris, Séries IV, **1** (Elsevier, Paris, 2000), pp. 297–303.
95. Y. Jiang, T. Lee, and C.G. Rose-Petruck, J. Opt. Soc. Am. B **20**, 229 (2003).
96. A. Thoss *et al.*, J. Opt. Soc. Am. B **20**, 224 (2003).
97. F. Dorchies, T. Caillaud, F. Blasco, C. Bonté, H.Jouin, S. Micheau, B. Pons, and J. Stevefelt, Phys. Rev. E **71**, 066410 (2005).
98. P. Balcou, R. Haroutunian, S. Sebban *et al.*Appl. Phys. B **74**, 509 (2002).
99. T.D. Donnelly, T. Ditmire, K. Neuman, M.D. Perry, and R.W. Falcone, Phys. Rev. Letters **76**, 2472 (1996).
100. D. von der Linde, Appl. Phys. B **68**, 315 (1999).
101. R. Lichters, J. Meyer-ter-Vehn and A. Pukhov, Phys. Plasmas **3**, 3425 (1996).
102. P. Monot *et al.*, Optics Letters **29**, 893 (2004).
103. G. Doumy *et al.*, Physical Review E **69**, 026402 (2004).
104. U. Teubner *et al.*, Phys. Rev. Lett. **92**, 185001 (2004).
105. G.B. Zimmerman and W.L. Kruer, Comm. in Plasma Physics **11**, 51 (1975).
106. U. Teubner, P. Gibbon, E. Förster, F. Falliès, P. Audebert, J.-P. Geindre, and J.-C. Gauthier, Phys. Plasmas **3**, 2679 (1996).
107. K. Eidmann, J. Meyer-ter-Vehn, T. Schlegel, and S. Hüller, Phys.Rev. E **62**, 1202 (2000).
108. A. Forsman, A. Ng, G. Chiu, and R.M. More, Phys.Rev. E **58**, R1248 (1998).
109. A.H. Gabriel, Mon. Not. R. Astron. Soc. **160**, 99 (1972).
110. O. Peyrusse, Phys. Fluids **B4**, 2007 (1992); O. Peyrusse, J.Q.R.S.T. **51**, 281 (1994).
111. O. Peyrusse, Nucl. Fusion **44**, S202 (2004).
112. E.G. Gamaly, A.V. Rode, B. Luther-Davies, and V.T. Tikhonchuk, Phys. Plasmas **9**, 949 (2002).
113. B.C. Stuart *et al.*, Phys. Rev. B **53**, 1749 (1996).
114. S. Nolte *et al.*, J. Opt. Soc. Am. B **14**, 2716 (1997).
115. X. Liu, D. Du, and G. Mourou, IEEE J. Quantum Electron. **33**, 1706 (1997).
116. P.P. Pronko, P.A. Van Rompay, Z. Zhang, and J.A. Nees, Phys. Rev. Lett. **83**, 2596 (1999).
117. T.E. Glover, J. Opt. Soc. Am. B **20**, 131 (2003).
118. *Time-Resolved Diffraction*, ed by J.R. Helliwell and P.M. Rentzepis (Oxford Science Publications, Clarendon, Oxford, 1997).

10 Ultraintense Electromagnetic Radiation in Plasmas

Maurizio Lontano[1] and Matteo Passoni[1,2]

[1] Plasma Physics Institute "P. Caldirola", C.N.R., Milan, Italy
lontano@ifp.cnr.it
[2] Nuclear Engineering Dept., Polytechnic of Milan, Milan, Italy
matteo.passoni@polimi.it

10.1 Introduction

In the last two decades, the introduction of the Chirped Pulse Amplification (CPA) technique [1,2], together with the development of ultrashort pulse (below 1 picosecond) generators and of solid state (Nd:glass, Yb:glass) amplifiers, has made it possible to develop lasers capable of producing peak values of the specific intensity at the focal spot of the order of $I \approx 10^{20}$–$10^{21}\,\mathrm{W/cm^2}$, and correspondingly laser powers at the petawatt ($10^{15}\,\mathrm{W}$) level [3]. Without going into the technical details, which we leave to the specialized literature [1, 2], we wish to mention what are the main features of the CPA technique. First, in the amplification phase of the initial radiation pulse, in order to prevent the appearance of distortions due to the nonlinearity of the active medium, the wave-packet is stretched by a factor 10^3–10^4, at constant total energy, by means of a dispersive grating. Then the long pulse is amplified by a factor 10^{10} without being distorted. Finally, it is compressed back of the same factor 10^3–10^4, by means of the inverse dispersive process, thus obtaining exceptionally high intensities. The production of ultra-intense laser radiation goes along with the achievement of ultra-short laser pulses, from 100 down to 10 femtosecond ($1\,\mathrm{fs} = 10^{-15}\,\mathrm{s}$), corresponding to energies in the range of 0.1–10 Joule, depending on the specific experimental conditions. Different combinations of pulse intensity and duration make the CPA technique very flexible in developing laser systems of various size (roughly corresponding to the total pulse energy), from compact systems for university laboratories (10–100 mJ) [4] to powerful lasers for inertial confinement studies ($E > 1 - 10\,\mathrm{J}$) [5,6].

At increasing values of the intensity at the focal spot (which nowadays can be brought down to the diffraction limit) one can identify various physical processes, that can dominate the corresponding interaction regimes. Let's assume that a laser beam with angular frequency ω is injected in an H gas or plasma and define the specific intensity of the electromagnetic (EM) field $\boldsymbol{E}(\boldsymbol{r}, t)$ as the power per unit surface

$$I\left(\frac{\mathrm{W}}{\mathrm{cm}^2}\right) = c\frac{E^2}{8\pi} \approx 1.33 \times 10^{-3}|E|^2\left(\frac{\mathrm{V}}{\mathrm{cm}}\right) , \qquad (10.1)$$

where c is the speed of light. Depending on the value of the radiation electric field, the following interaction conditions can be envisaged:

a) In an initially uniform plasma, with electron density $n_e(\mathrm{cm}^{-3})$ and temperature $T_e(\mathrm{eV})$, if the electron kinetic energy density in the laser field, $\frac{1}{2}n_e m \langle v_e^2 \rangle$ becomes of the same order as the plasma kinetic pressure $\frac{3}{2}n_e T_e$, then the laser beam acts on the plasma electrons through the so-called average *ponderomotive force* or *Miller force* [7, 8] $\boldsymbol{F}_{pm} = -\nabla \Phi_{pm}$ where the *ponderomotive potential* $\Phi_{pm}(\boldsymbol{r}) = e^2E^2/2m\omega^2$, has been introduced, e and m being the electron charge and mass, respectively. Plasma electrons are then expelled from the spatial region where the electromagnetic energy is concentrated, leading to an increasing localization of the laser beam: this process is called *self-focusing* of the laser beam and it is one of the most striking effects of the non-linear wave-plasma interaction.

From the above pressure balance, the threshold laser field amplitude and the corresponding specific intensity above which ponderomotive effects are important can be found

$$E > E_{pm} = 7.8 \times 10^7 \frac{T_e^{1/2}(\mathrm{eV})}{\lambda(\mu\mathrm{m})} \frac{V}{\mathrm{cm}} \qquad I > I_{pm} = 8.04 \times 10^{12} \frac{T_e(\mathrm{eV})}{\lambda^2(\mu\mathrm{m})} \frac{\mathrm{W}}{\mathrm{cm}^2} \,. \tag{10.2}$$

In the case of a Ti:Sa ($\lambda = 0.8\,\mu$m) laser in a 10 eV plasma, $E_{pm} \approx 3 \times 10^8$ V/cm and $I_{pm} \approx 1.3 \times 10^{14}$ W/cm^2.

b) The intra-atomic electric field, and the corresponding intensity, can be estimated as the ratio of the unitary electric charge e and the squared Bohr radius $a_0 = \hbar/me^2$, that is

$$E_{at} = \frac{m^2e^5}{\hbar^4} \approx 5.15 \times 10^9 \frac{\mathrm{V}}{\mathrm{cm}} \qquad I_{at} \approx 3.5 \times 10^{16} \frac{\mathrm{W}}{\mathrm{cm}^2} \,. \tag{10.3}$$

For laser intensities lower than I_{at} the main ionization mechanism is *tunneling ionization*, and a high ionization degree of the gas target is achieved after many laser cycles. Multiphoton ionization processes become important at higher intensity leading, for example, to high-order harmonic generation [9]. Generally speaking, the ionization process tends to *defocus* the laser beam. Indeed, during its propagation in a neutral gas, the laser beams ionizes predominantly the gas along its axis of propagation, closer to the EM energy peak than at the wings of the spatial distribution. Then, the central part of the beam propagates in a relatively high density plasma, with a refractive index less than unity, and an initially flat phase front becomes convex in the forward direction, leading to the beam spreading.

However, with increasing intensity, new features appear. If the gas pressure is sufficiently high, the laser beam can undergo a strong side scattering due to the ionization itself [10]. On the contrary, at low working pressure, if a saturation mechanism acts in such a way as to prevent further electron density production along the laser beam axis, while electrons are continuosly produced at the beam wings, the radiation can be channeled for several Rayleigh lengths [11].

When a short laser pulse with intensity approaching I_{at} is considered, the complete ionization of the gas target is achieved in few field cycles, and one should consider that the front of the pulse propagates in the still neutral gas, while its rear part feels the plasma produced by the early cycles. Then the pulse is strongly distorted during its propagation from its initial shape.

c) When the laser intensity is large enough as to impart a relativistic momentum to an electron in the laser field, that is $p_e \approx mc$, so-called relativistic non linearities arise [12–14], which open the laser-plasma system up to mess of peculiar physical processes [15]. The relativistic regime of interaction is characterized by the threshold values of electric field and intensity given by

$$E_{\rm rel} = 3.2 \times 10^{10} \frac{1}{\lambda(\mu{\rm m})} \frac{\rm V}{\rm cm} \qquad I_{\rm rel} = 1.4 \times 10^{18} \frac{1}{\lambda^2(\mu{\rm m})} \frac{\rm W}{\rm cm^2} \,. \tag{10.4}$$

When $I > I_{\rm rel}$ the magnetic contribution to the Lorentz force acting on an electron, $-(e/c)\boldsymbol{v} \times \boldsymbol{B}$ becomes as important as the electric part, $-e\boldsymbol{E}$ and the electron trajectory strongly deviates from its shape for lower intensity (see Sect. 10.2.1). As a consequence at macroscopic level several new effects occur in the target, as for example relativistic harmonic generation [16], ultra-high quasi-stationary magnetic field production [17–20], excitation of large amplitude plasma waves, which can accelerate electrons to GeV energies over few mm [21–25], generation of relativistic EM solitons [26–32], effective proton and ion extraction from thin solid targets and their acceleration [33–36], nuclear fission and fusion reactions [37–39]. At the same time, the EM energy inside the pulse is redistributed due to relativistic self-focusing, self-compression and self-splitting [40]. Some of these peculiar physical processes will be dealt with in more detail in the next sections of this chapter. Presently, the largest laser intensity available is $\approx 10^{21}\,{\rm W/cm^2}$ in pulses of 500 fs [41]. The corresponding normalized laser field is $a = e\tilde{E}/m\omega_0 c \approx 27 \approx \gamma$ where γ is the relativistic factor for an electron under its action, and $\lambda = 1\,\mu{\rm m}$ has been considered. Therefore, on one side the laser is able to detach bound electrons with binding energies up to 4 keV, on the other side the motion of the freed electrons should be treated in a fully relativistic frame.

d) This astonishing survey of the physical processes which can be driven by existing petawatt class lasers does not exhaust the potentialities which are expected to become a reality once higher laser intensities will be available [42]. If the laser intensity is increased such that the work done on an electron by the electric field over a Compton length ($\lambda_{\rm C} = \hbar/mc \approx 3.86 \times 10^{-11}\,{\rm cm}$) is of the order of mc^2, copious production of electron-positron pairs is possible, that is QED effects can be experimentally investigated. The laser electric field and intensity corresponding to the request that $eE\lambda_C \approx mc^2$ can be written

$$E_{qed} = 1.33 \times 10^{16} \frac{\mathrm{V}}{\mathrm{cm}} \qquad I_{qed} = 2.35 \times 10^{29} \frac{\mathrm{W}}{\mathrm{cm}^2} \,. \tag{10.5}$$

Such extremely high values are far from being implemented in laboratory. However it is possible to observe appreciable laser-induced pair-production by exploiting the interaction of a "lower" intensity laser pulse, say $E = 10^{12}\,\mathrm{V/cm}$, corresponding to $I = 1.3 \times 10^{21}\,\mathrm{W/cm^2}$, with a relativistic electron beam of 20 GeV. In the reference frame of the relativistic electrons the laser field is enhanced by a factor $\gamma \approx 4 \times 10^4$, and then the threshold E_{qed} can in principle be overcome with existing laser technology [43]. Moreover, one should also notice that an electron initially at rest gains a velocity very close to the speed of light in a quarter of laser cycle. It means that for a very short time interval (of the order of a fs), it is subject to an acceleration approximately $10^{27}\,\mathrm{cm/s^2} \approx 10^{24}\,g$. These rough figures suggest that during the interaction of a petawatt class laser pulse, extreme conditions of matter are realized, which otherwise can be found in the universe only in very exotic objects like black holes or supernovae [44,45].

In order to illustrate in more detail few of the different interaction regimes made possible by the development of ultraintense lasers (what will be done in a forthcoming chapter), in the present Chapter the basic physical aspects of the interaction of relativistically intense EM radiation with a test electron and with a preformed plasma are reviewed. In particular, the orbit of a free electron in the combined electric and magnetic fields of a large amplitude laser pulse will be calculated, the theoretical models used to describe the interaction between intense laser pulses and plasmas will be presented, and the fundamental processes occurring in a plasma in the presence of strong EM radiation will be reviewed.

10.2 Interaction of Ultraintense Radiation and Plasmas

In this section several fundamental theoretical issues about the interaction between ultra-intense EM radiation and plasmas are discussed. In Sect. 10.2.1 the electron dynamics in the field of an intense EM plane wave is calculated to show the relevance of the relativistic effects when the radiation is intense enough. In Sect. 10.2.2, the main features of the non-linear relativistic interaction between ultra-intense and ultra-short laser pulses and plasmas are presented.

10.2.1 Relativistic Electron Dynamics in an EM Plane Wave

The electron dynamics under the action of electromagnetic fields depends essentially on the intensity of the radiation. Let us consider an electromagnetic plane wave in vacuum, described by the vector potential $\boldsymbol{A}(\boldsymbol{r},t) = \mathrm{Re}\{\boldsymbol{A_0} \exp(i\psi)\}$; if x is directed along the wave propagation direction, $\boldsymbol{A_0} = A_0\hat{e}_y$ for linear polarization, while $\boldsymbol{A_0} = A_0(\hat{e}_y \pm \mathrm{i}\hat{e}_z)$ in the case of circular polarization (+ and − are referred to right and left handed waves, respectively). In vacuum, the wave has only the transversal component and then $A_x = 0$; moreover, $\psi = \boldsymbol{k} \cdot \boldsymbol{r} - \omega t$ and $\omega = ck = 2\pi c/\lambda$. In the small velocity limit (electron velocity $|\boldsymbol{v}| \ll c$), the electron dynamics is described by the equation $m\mathrm{d}\boldsymbol{v}/\mathrm{d}t \approx -e\boldsymbol{E}$ whose integrals are

$$\boldsymbol{v} = -\frac{eA_0}{mc}(\cos\psi\hat{e}_y \pm \alpha\sin\psi\hat{e}_z) \qquad \boldsymbol{r} = \frac{eA_0}{mc\omega}(\sin\psi\hat{e}_y \mp \alpha\cos\psi\hat{e}_z) \quad (10.6)$$

where $\alpha = 0$ or $\alpha = 1$ for linear and circular polarization, respectively. By introducing the dimensionless amplitude $a_0 \equiv eA_0/mc^2$, it is evident from (10.6) that for radiation amplitude such that $a_0 = 1$ the electron velocity is of the order of c. This condition defines the radiation amplitude value above which a fully relativistic treatment of the electron dynamics is required. It is useful to express this condition in terms of the wave radiation intensity I, which is given by the modulus of the Poynting vector $\boldsymbol{S} = (c/4\pi)\boldsymbol{E} \times \boldsymbol{B}$; in this case $I = |\boldsymbol{S}| = (\omega k A_0^2/8\pi)(1-\alpha\cos 2\psi)$ and the mean intensity, averaged over a wave period, I_0, can be written as

$$I_0\lambda^2 = \frac{\pi}{2}cA_0^2 = 1.37 \times 10^{18}\, a_0^2\, \frac{\mathrm{W}}{\mathrm{cm}^2}\mu\mathrm{m}^2\,, \quad (10.7)$$

in the case of linear polarization, while for circular polarization it is twice this value.

The relativistic dynamics of a charged particle in the field of an electron plane wave of arbitrary amplitude can be solved exactly and the Lagrangian formulation provides a simple approach to this problem [46]. The same results can be obtained by means of the Hamilton–Jacobi equations [47], as well. The relativistic Lagrangian function of a particle with charge q in an electromagnetic field described by the potentials ϕ and $\boldsymbol{A}$ is:

$$L(\boldsymbol{r},\boldsymbol{v},t) = -mc^2\sqrt{1-\frac{v^2}{c^2}} + \frac{q}{c}\boldsymbol{v}\cdot\boldsymbol{A} - q\phi\,. \quad (10.8)$$

Equation (10.8) allows one to calculate the canonical momentum $\boldsymbol{P} = \partial L/\partial\boldsymbol{v} = \boldsymbol{p} + q\boldsymbol{A}/c$, where $\boldsymbol{p} = m\gamma\boldsymbol{v}$ is the particle momentum and $\gamma = 1/\sqrt{1-v^2/c^2}$ is its relativistic factor. The particle dynamics can be solved by determining the invariant quantities of the motion. By definition, a plane wave depends only on the coordinate which identifies the propagation direction, and consequently $\partial L/\partial\boldsymbol{r}_\perp = 0$. From the Euler–Lagrange equation, it gives $\partial L/\partial\boldsymbol{v}_\perp = \boldsymbol{p}_\perp + q\boldsymbol{A}_\perp/c = const$, that is the transverse component of the canonical momentum is constant during the particle motion. A second invariant follows from the observation that a plane wave depends on space and time through their combination $t - x/c$, *i.e.* $\boldsymbol{A} = \boldsymbol{A}(t - x/c)$, which implies $\partial L/\partial t = -c\partial L/\partial x$. It is well known that the particle energy E, given by its Hamiltonian $H(\boldsymbol{r}, \boldsymbol{p}, t)$, satisfies the relations $\mathrm{d}E/\mathrm{d}t = \mathrm{d}H/\mathrm{d}t = -\partial L/\partial t$. Therefore, using the fact that $A_x = 0$, it follows that $\mathrm{d}E/\mathrm{d}t = c\mathrm{d}p_x/\mathrm{d}t$ and consequently a second invariant of the motion is the quantity $E - cp_x = cost = mc^2$ (for an electron initially at rest in the laboratory frame). From these results it is possible to derive further important relationships among the physical parameters describing the particle dynamics. Starting from the expression of the kinetic energy $E_{\text{kin}} = E - mc^2 = mc^2(\gamma - 1)$, it follows immediately a relationship between the longitudinal momentum p_x and γ, $E_{\text{kin}} = cp_x \rightarrow \gamma = 1 + p_x/mc$. The general relativistic expression of the total particle energy, $E = mc^2\gamma = \sqrt{(mc^2)^2 + (p_x c)^2 + (p_\perp c)^2}$, combined with the above results, gives $E_{\text{kin}} = p_\perp^2/2m$ and consequently $p_\perp^2/2m = p_x c$. Then, the angle ϑ between the transverse and the longitudinal components of the momentum satisfies the relation $\tan^2\vartheta = (p_\perp/p_x)^2 \rightarrow \vartheta = \arctan[2/(\gamma - 1)]^{1/2}$. By introducing the dimensionless quantities $\boldsymbol{a} \equiv e\boldsymbol{A}/mc^2$ and $\tilde{\boldsymbol{p}} \equiv \boldsymbol{p}/mc$, the relativistic dynamics of an electron in a plane electromagnetic wave is therefore described by the following equations in the laboratory frame:

$$\tilde{\boldsymbol{p}}_\perp = \boldsymbol{a} = (0, a_y, a_z) \qquad \tilde{p}_x = \frac{\tilde{p}_\perp^2}{2} = \frac{a^2}{2} \qquad \gamma = 1 + \tilde{p}_x = 1 + \frac{a^2}{2}\,. \tag{10.9}$$

We can immediately observe that in the ultra-relativistic limit, i.e. for $a \gg 1$, $v_x/c = p_x/\gamma c \rightarrow 1$, $v_{y,z}/c = p_{y,z}/\gamma c \rightarrow 0$, $\tan\vartheta = p_\perp/p_{||} \rightarrow 0$. The electron motion, which for $a \ll 1$ is mainly in the transverse direction, becomes more and more directed along the wave propagation direction as the wave amplitude is increased towards relativistic values $a \gg 1$. Equations (10.9) can be easily integrated by introducing the variable $\tau = t - x(t)/c$, exploiting the fact that $\mathrm{d}\tau = \mathrm{d}t/\gamma$. Let us first consider a *linearly polarized* wave, for which $a_x = a_z = 0$ and $a_y = a_0\cos(\omega\tau)$. The trajectory of an electron initially in $(x, y, z) = 0$ is given by:

$$x = \frac{ca_0^2}{4}\left[\tau + \frac{\sin(2\omega\tau)}{2\omega}\right] = \frac{ca_0^2}{a_0^2 + 4}\left[t + \frac{\sin(2\omega\tau)}{2\omega}\right] \qquad y = \frac{ca_0}{\omega}\sin(\omega\tau)\,. \tag{10.10}$$

The motion is characterized by a component along the wave propagation direction, with constant drift velocity $v_d = ca_0^2/(a_0^2 + 4)$, with a superimposed oscillatory motion. In Fig. 10.1 the electron trajectory is shown for several values of the linearly polarized radiation amplitude a_0. In particular, the periods needed to travel a given longitudinal distance are visualized: it is clear that a larger number of cycles is necessary if the radiation amplitude is low, in accordance with the above discussion. It is instructive to study the properties of the oscillatory component of the motion in the reference frame moving with the velocity v_d of the electron, by applying to (10.10) the corresponding Lorentz transformation for the coordinates x and t; the phase $\omega\tau$ is a scalar invariant and consequently the frequency ω_d in the new frame is modified by relativistic Doppler effect and is related to ω by the relation

$\omega = \omega_d[1-\cos\alpha(v_d/c)]/\sqrt{1-v_d^2/c^2} = \omega_d\sqrt{1+a_0^2/2}$ (α is the angle measured in the laboratory frame between the radiation propagation direction and the direction of the moving frame: then, in this case $\alpha = 0$). In the moving system the electron trajectory is given by the equations

$$x_d = \frac{ca_0^2}{8\omega_d(1+a_0^2/2)}\sin(2\omega_d\tau_d) \qquad y_d = \frac{ca_0}{\omega_d\sqrt{1+a_0^2/2}}\sin(\omega_d\tau_d)\,, \quad (10.11)$$

where, $\omega_d\tau_d = \omega\tau$. In this reference frame the electron performs a typical figure-8 trajectory, parametrized by the quantity $\omega_d\tau_d$, as shown in Fig. 10.1. The relativistic factor of the electron in the moving system is given by $\gamma_d = \sqrt{1+a_0^2/2}$. In Fig. 10.1 this trajectory is shown for different values of a_0. The motion is mainly transverse for small a_0 values while it tends to a limit trajectory for ultra-relativistic radiation.

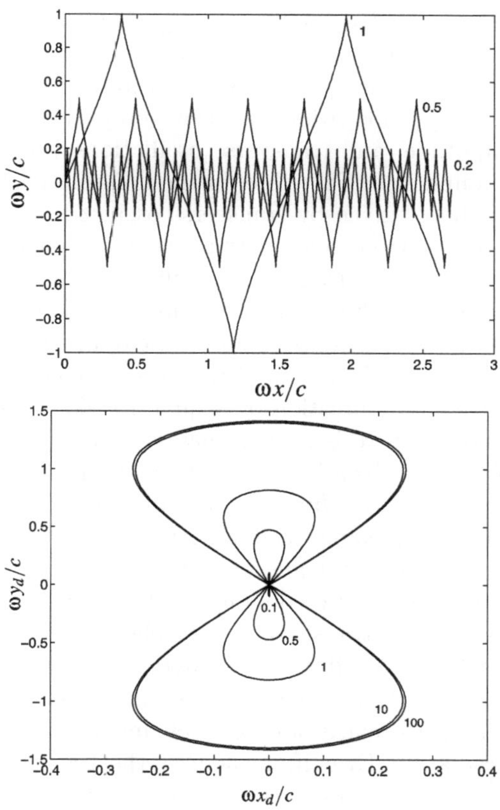

Fig. 10.1. (*Left*) Electron trajectory under the action of a plane linearly polarized EM wave, measured in the laboratory frame, for different values of the normalized wave amplitude, $a_0 = 0,2;\ 0,5;\ 1$. (*Right*) "Figure-8" trajectory in the reference frame moving with velocity v_d, for $a_0 = 0,1;\ 0,5;\ 1;\ 10;\ 100$.

The case of *circularly polarized* radiation can be solved analogously. In this situation $a_y = a_0 \cos(\omega\tau)$, $a_z = a_0 \sin(\omega\tau)$ e $a^2 = a_0^2$. The electron motion initially in $(x, y, z) = 0$ at $\tau = 0$ is now given by the following relations:

$$x = \frac{ca_0^2}{a_0^2 + 2}t, \qquad y = \frac{ca_0}{\omega}\sin(\omega\tau), \qquad z = -\frac{ca_0}{\omega}\cos(\omega\tau) \,. \tag{10.12}$$

The electron performs an helicoidal trajectory: there is a longitudinal motion with velocity $v_d = ca_0^2/(a_0^2 + 2)$, to which an orthogonal circular motion is superimposed. In the frame moving with velocity v_d along the propagation direction, the equations become $x_d(t_d) = 0$ and

$$y_d(\tau) = \frac{ca_0}{\omega_d\sqrt{1 + a_0^2}}\sin(\omega_d t_d), \qquad z_d(\tau) = -\frac{ca_0}{\omega_d\sqrt{1 + a_0^2}}\cos(\omega_d t_d) \,, \tag{10.13}$$

while the relativistic factor is now given by $\gamma_d = \sqrt{1 + a_0^2}$. This study clearly shows the relevance of the relativistic effects on the electron dynamics in an intense EM field. The overall motion is the result of a longitudinal drift along the plane wave propagation direction, proportional to the square of the radiation amplitude, and transverse oscillations which depend linearly on the radiation amplitude. Therefore, in the non-relativistic limit, the electron dynamics is almost given by transverse oscillations, while in the ultra-relativistic limit the electron moves with a velocity of the order of c along the wave propagation direction. To be more realistic, let's represent an ultraintense laser pulse of finite duration by means of a plane wave which exists only for a finite time τ_p. During the time τ_p the dynamics follows (10.10) (or (10.12)): in vacuum, after the action of the pulse, the electron will remain at rest displaced along the x direction by a quantity $\Delta x = c\tau_p a_0^2/(a_0^2 + 2)$ (for circular polarization). There is no net energy transfer between the particle and the pulse. If we describe a laser pulse by assuming a Gaussian envelope, $a = a_0 \exp[-(t/\tau_L)^2] \cos(\omega t - kx)$, it is not difficult to show that the main features of the electron dynamics is in agreement with the previous analysis: initially, the motion is mainly transverse since the amplitude is small, then it becomes more and more directed along the longitudinal direction. Even in this situation, no energy is transferred to the electron and after the pulse has overcome the electron, it is left at rest shifted by a certain amount Δx.

In conclusion, this study demonstrates the possibility of accelerating a charged particle up to relativistic energies under the action of an intense electromagnetic radiation. Even if, in vacuum, there is no net energy transfer from the radiation to the particle, the situation can be different in the presence of electron collisions with other plasma particles, or, generally speaking, when random processes affect the particle motion. Nevertheless, it is clear that the relativistic effects can strongly modify the interaction between ultra-intense electromagnetic radiation and plasmas, compared with the case of low-intensity radiation. From (10.7), it is seen that the transition to the relativistic regime becomes effective when the radiation intensity exceeds 10^{18} W/cm^2.

10.2.2 Fundamental Issues About the Interaction Between Ultra-Intense Laser Pulses and Plasmas

The theoretical investigation of the interaction between EM radiation with plasmas (more generally with matter) requires the calculation of the self-consistent EM fields generated by the interaction itself, determined by the Maxwell equations coupled with the model equations for the plasma. In this chapter, we will use the Maxwell equations written in the *Coulomb gauge*, obtained imposing a divergence-free vector potential, that is $\nabla \cdot \boldsymbol{A} = 0$, which reveals itself very useful in the study of a large class of phenomena. For example, the Coulomb gauge is naturally appropriate for studying spatially localized solutions of the Maxwell equations. In this gauge, Maxwell equations have the following form:

$$\nabla^2 \boldsymbol{A} - \frac{1}{c^2}\frac{\partial^2 \boldsymbol{A}}{\partial t^2} - \frac{1}{c}\nabla\frac{\partial \phi}{\partial t} = -\frac{4\pi}{c}\boldsymbol{j} \qquad \nabla^2 \phi = -4\pi\rho\,. \tag{10.14}$$

Notice that in the Coulomb gauge, the relationship between the scalar potential and its source, the electric charge, is instantaneous over all the space domain. Nevertheless, the corresponding EM fields do propagate at a finite velocity [48, 49]. In order to calculate the EM fields in a plasma, it is necessary to express the sources, namely the electric charge density ρ and current density $\boldsymbol{j}$ in terms of the EM field itself, using a plasma model which is suitable for our purposes. In plasmas it is possible to study a large class of phenomena (in particular those involving energetic particles) neglecting the Coulomb collisions between the particles. In this case, the kinetic properties of the plasma are derived from the *relativistic Vlasov equation*,

$$\frac{\partial f_j}{\partial t} + \frac{\boldsymbol{p}_j}{m_j\gamma_j}\cdot\nabla f_j + q_j\left[\boldsymbol{E}(\boldsymbol{r},t) + \frac{\boldsymbol{p}_j}{m_j c\gamma_j}\times\boldsymbol{B}(\boldsymbol{r},t)\right]\cdot\frac{\partial f_j}{\partial \boldsymbol{p}_j} = 0\,. \tag{10.15}$$

Here, f_j is the particle distribution function, m_j, q_j and $\boldsymbol{p}_j = m_j\gamma_j\boldsymbol{v}$ are the rest mass, the electric charge and momentum of the j-th species, $\gamma_j = (1 + p_j^2/m_j^2c^2)^{\frac{1}{2}}$ is the corresponding relativistic factor, $\boldsymbol{E}(\boldsymbol{r},t)$ and $\boldsymbol{B}(\boldsymbol{r},t)$ represent the self-consistent EM fields developed during the evolution of the system. Being a kinetic equation, (10.15) describes the microscopic behavior of the plasma. Once f_j has been determined, for each plasma species, in terms of the EM fields from (10.15), it is possible to construct the plasma electric charge and current densities,

$$\rho(\boldsymbol{r},t) = \sum_{j=e,i} q_j \int f_j \, \mathrm{d}^3\boldsymbol{p}_j \qquad \boldsymbol{j}(\boldsymbol{r},t) = \sum_{j=e,i} q_j \int \boldsymbol{v}\, f_j \, \mathrm{d}^3\boldsymbol{p}_j \tag{10.16}$$

that, after inserting in (10.14), allow the calculation of the self-consistent fields.

Instead of using (10.15), depending on the problem at hand, it is possible to model the plasma response to EM radiation in the frame of a relativistic hydrodynamic description. A consistent system of hydrodynamic equations for a warm relativistic plasma can be derived from the relativistic Vlasov equation by calculating the equations for the moments of the particle distribution functions [50]. An equivalent procedure consists in assuming the conservation of the particle number and of the energy-momentum for a fluid element of the j-th-species; then, one can construct the hydrodynamic equations governing the dynamics of a relativistic plasma with arbitrary temperature [51–53]. The specific forms of the fluid equations depend on the closure of the hydrodynamic system and on the particular choice of the energy-momentum tensor.

Following [51], the case of a perfect fluid, with the *adiabatic closure* $P_j/N_j^{\Gamma_j} = const.$, with $P_j = N_j T_j$, has been considered, where the equation of state relating the pressure P_j, the internal energy density ϵ_j, and N_j has been taken in the form

$$\epsilon - N_j m_j c^2 \approx \frac{P_j}{\Gamma_j - 1} .$$

With these choices, the momentum density and number density equations can be put in the form [54]

$$\frac{\partial \boldsymbol{P}_j}{\partial t} = \boldsymbol{u}_j \times \nabla \times \boldsymbol{P}_j - \nabla \left(q_j \phi + m_j^{eff} c^2 \right) , \tag{10.17}$$

$$\frac{\partial n_j}{\partial t} + \nabla \cdot (n_j \boldsymbol{u}_j) = 0 , \tag{10.18}$$

where $\boldsymbol{P}_j = \boldsymbol{p}_j R_j + q_j \boldsymbol{A}/c$ is the canonical momentum, $\boldsymbol{p}_j = m_j \boldsymbol{u}_j \gamma_j$ the particle momentum, $n_j = N_j \gamma_j$ the number density in the laboratory reference frame (being N_j the density in the local frame), m_j the rest mass, q_j the (invariant) electric charge, $m_j^{eff} = m_j \gamma_j R_j$ the *effective mass*. Moreover,

$$R_j = 1 + \alpha_j \frac{P_j}{m_j c^2 N_j}$$

is a factor which combines the relativistic thermal motion, represented by the pressure P_j, and the ordered particle motion, contained in $N_j = n_j/\gamma_j$; $\alpha_j = \Gamma_j/(\Gamma_j - 1)$, Γ_j is the adiabatic index (see also [55]). Notice that, in the adiabatic regime, the temperature is a function of space, $T(\boldsymbol{r})$. The momentum equation (10.17) can be cast in the convenient form

$$\frac{\partial \boldsymbol{\Omega}_j}{\partial t} = \nabla \times (\boldsymbol{u}_j \times \boldsymbol{\Omega}_j) , \tag{10.19}$$

where $\boldsymbol{\Omega}_j = \nabla \times \boldsymbol{P}_j$ is the generalized vorticity.

When the typical macroscopic velocities and the wave phase velocity are much larger than the thermal random motion of the particles, that is $u_j, \omega/k \gg \sqrt{T_j/m_j}$, thermal effects can be neglected and the cold hydrodynamic model can be used. In addition in the investigation of ultra-fast processes, on the scales of tens or hundreds fs, ions can be assumed to remain at rest, and only electrons respond to the EM fields. In this case (10.17), (10.18) can be reduced to the set of equations for the electron dynamics, only

$$\frac{\partial \boldsymbol{P}}{\partial t} = \left(\boldsymbol{P} + \frac{e}{c}\boldsymbol{A}\right) \times \frac{\nabla \times \boldsymbol{P}}{m\gamma} + \nabla\left(e\phi - mc^2\gamma\right) , \tag{10.20}$$

$$\frac{\partial n}{\partial t} + \nabla \cdot \left[\frac{n}{m\gamma}\left(\boldsymbol{P} + \frac{e}{c}\boldsymbol{A}\right)\right] = 0 , \tag{10.21}$$

where, $\boldsymbol{P} = \boldsymbol{p} - e\boldsymbol{A}/c$ is the canonical momentum, $\boldsymbol{p} = m\gamma\boldsymbol{u}$ is the electron momentum, m is the electron rest mass, $\gamma = \left[1 + (\boldsymbol{P} + e\boldsymbol{A}/c)^2/(m^2c^2)\right]^{1/2}$ the relativistic factor. Here the subscript e is eliminated. The electric charge density and the current density take the form $\rho = -e(n - n_0)$ and $\boldsymbol{J} = -(en/m\gamma)(\boldsymbol{P} + e\boldsymbol{A}/c)$, respectively, and $n_i = n_0/Z$ is the background ion density, which is considered uniform, and independent of time. The momentum equation (10.20) can be cast in the form given by (10.19), which shows that the generalized electron vorticity, $\boldsymbol{\Omega} = \boldsymbol{\nabla} \times \boldsymbol{P}$, is frozen into the electron fluid. If at $t = 0$, the electrons are at rest and the laser field is absent, that is $\boldsymbol{P}(t = 0) = 0$, then (10.19) states that $\boldsymbol{\nabla} \times (\boldsymbol{p} - e\boldsymbol{A}/c) = 0$ should hold at any time, that is $\boldsymbol{B} = (c/e)\boldsymbol{\nabla} \times \boldsymbol{p}$. Applying the *curl* operator to (10.20), we derive the equation governing the magnetic field in the laser-plasma interaction:

$$\Delta \boldsymbol{B} - \frac{1}{c^2}\frac{\partial^2 \boldsymbol{B}}{\partial t^2} - \frac{\omega_p^2}{c^2}\frac{n}{n_0\gamma}\boldsymbol{B} = -\frac{4\pi e}{mc}\left(\boldsymbol{P} + \frac{e}{c}\boldsymbol{A}\right) \times \nabla\frac{n}{\gamma} = \boldsymbol{S} . \tag{10.22}$$

where $\boldsymbol{S}$ is the source of $\boldsymbol{B}$. Equation(10.22) is a general equation for the magnetic field generated during the laser-plasma interaction. It describes both the laser magnetic field and the magnetic field induced in the plasma [20]. The magnetic field is nonlinearly coupled to the electron dynamics, through the last term in the l.h.s. and the source term, in (10.22). It can be worth separating the contributions to the source term along the propagation direction of the laser pulse

$$\boldsymbol{S}_{||} = \frac{4\pi e}{mc}\boldsymbol{p}_{\perp} \cdot \left(\hat{e}_{||} \times \nabla_{\perp}\frac{n}{\gamma}\right) , \tag{10.23}$$

and transverse to it

$$\boldsymbol{S}_{\perp} = \frac{4\pi e}{mc}\left\{\hat{e}_{||} \times \boldsymbol{p}_{\perp}\nabla_{||}\frac{n}{\gamma} - \boldsymbol{p}_{||}\hat{e}_{||} \times \nabla_{\perp}\frac{n}{\gamma}\right\} . \tag{10.24}$$

According to (10.22–10.24), a nonvanishing source term implies the space inhomogeneity either of the electron density or of the electron mass, through the relativistic factor γ. However, this is just a necessary condition for a nonzero source. Generally speaking, the transverse and the parallel components of $\boldsymbol{p}$ are related to the laser pulse and to the plasma wakefield produced behind the pulse, respectively. By inspection of (10.23) we can argue that the source of the longitudinal magnetic field is localized in the laser pulse region, and it depends on the transverse non uniformities. No longitudinal field can be generated in a one-dimensional plasma. On the contrary, the source of the transverse component of the magnetic field is present either in the laser pulse region (first term in the r.h.s. of (10.24)), or outside it (second term). Again, in the latter case transverse gradients are needed. Magnetic field generation during the strong laser-plasma interaction has been investigated analytically and numerically in [18–20, 56–61]. Depending on the polarization of the laser and on the particular geometry of the problem, rapidly oscillating (on the laser period time scales) magnetic fields, as well as "quasistatic" magnetic fields can be generated.

The theoretical models of strong laser-plasma interaction, which have been discussed above, predict that in such extreme physical conditions a large variety of physical processes are likely to take place, due to the complex non linear nature of the coupling between EM radiation and matter.

Let us review the main features of the propagation of EM radiation in plasmas in the non-relativistic regime. In this case, the dispersion relation for a wave of frequency ω and wave vector k in a plasma of density n is $kc = (\omega_{pe}^2 - \omega^2)^{1/2}$, where $\omega_{pe} = (4\pi e^2 n/m)^{1/2}$. Thus, there exists a *critical density* $n_c = m\omega^2/4\pi e^2$, for which $\omega_{pe} = \omega$, which defines two well distinct regimes of interaction. A plasma with density below the critical value is called *underdense*, while the plasma is *overdense* with respect to the considered radiation if $n > n_c$. EM waves can propagate only in underdense plasmas: then, for a given density the corresponding plasma frequency ω_{pe} defines the minimum frequency of an EM wave propagating in the plasma. In non-uniform plasmas an EM wave can propagate only until it reaches the region at the critical density, where it is partly reflected and partly absorbed by the plasma, decaying exponentially inside the overdense region on distances of the order of the so-called *classical skin depth* c/ω_{pe}. The propagation of EM radiation in non-uniform underdense plasmas can be characterized by the onset of varius forms of instabilities, depending on the kind of non-uniformity of the system. The relativistic effects alter this picture in many ways. First of all, the relativistically correct dispersion relation becomes $kc = (\omega_{pe}^2/\gamma - \omega^2)^{1/2}$, where γ is the relativistic factor of an electron in the field of the wave: therefore, the critical density in the relativistic regime becomes

$$n_c = \gamma \frac{m\omega^2}{4\pi e^2} = 1.1 \times 10^{21} \frac{\gamma}{\lambda^2(\mu\mathrm{m})}\,\mathrm{cm}^{-3}\,. \tag{10.25}$$

An EM wave can then propagate also in plasmas that are overdense in the non-relativistic regime, an effect called *relativistic induced transparency.* Even the propagation and the instabilities in underdense plasmas, as well as absorption phenomena in overdense plasmas, are influenced by relativistic and non-linear effects, and it is often impossible to treat separately the different issues. From an analytical point of view, it is necessary to introduce approximations in the description, for example restrict to 1D geometry and/or perform asymptotic expansions, while the use of as realistic as possible numerical simulations becomes essential for the study of the more general cases. Let us now consider some of the features associated with the interaction between intense EM radiation and plasmas, both underdense and overdense.

While propagating in an underdense plasma, an ultraintense laser pulse interacts with the electron population and can induce a large charge separation which ultimately generates large intensity waves in the plasma; in this way, in the interaction zone behind the pulse, a large amplitude plasma wave is formed, usually called *wake field.* If compared with what happens with low-intense waves, the intense wave propagation is significantly more complex, because of the non-linear and relativistic effects. Studying the propagation of relativistically intense longitudinal and transverse EM waves in cold plasmas, neglecting the ion motion and Coulomb collisions, it has been demonstrated [62,63] that the frequency of the radiation depends on the wave amplitude (contrary to the linear small amplitude case). There exists a maximum value E_m for the amplitude of the electric field which can be sustained by the plasma, given by the *Akhiezer and Polovin limit,* $eE_m = m\omega_{pe}c[2(\gamma_\varphi - 1)]^{1/2}$, where γ_φ is the relativistic factor associated to the wave velocity v_φ. When the field reaches this value, the fluid description, which predicts a singular behavior of the electron density, is no longer applicable and a kinetic description becomes necessary. This phenomenon is called *wave breaking* [64–66]: when the wave intensity is sufficiently high to induce the *wave breaking*, an effective energy transfer to the plasma takes place, and a generation of highly energized relativistic electron population is possible [65]. Several studies have been performed to include the thermal effects and understand their influence on the *wake field* and *wave breaking* [67–70].

As already said, the propagation of intense EM waves in non-uniform plasmas can be accompanied by the onset of various kind of instabilities [8]. The non-homogeneity could be due to the presence of a density gradient in the plasma: in this case in the vicinity of the critical density the so called *resonant absorption* takes place, leading to an efficient absorption of the wave. On the other side, a spatial variation of the density can be also due to the presence of electron and/or ion plasma waves. In this case, the coupling between these density fluctuations and the EM wave can give rise to mechanisms which eventually lead to the growth of the plasma waves themselves. Generally speaking, processes of this kind are named *parametric instabilities.* An important class of these instabilities involves the coupling between a large amplitude EM wave and fluctuations like an electron plasma wave, giving rise to a process called *Stimulated Raman Scattering*, SRS, or an ion-acoustic wave, the so called *Stimulated Brillouin Scattering*, SBS. In these processes an intense EM wave (*pump wave*), with frequency ω_0 and wave vector $\boldsymbol{k_0}$, couples with an electron plasma wave (or ion acoustic wave) with frequency ω and wave vector $\boldsymbol{k}$, generates a scattered EM wave with frequency ω_s and wave vector $\boldsymbol{k_s}$ which, interacting with the original EM wave, produces an amplification of the electron plasma (ion acoustic) wave, giving rise to the instability. Depending on the direction of the scattered EM wave we can speak about *backward-*, *forward-*, or *side-scattering*. The analysis of these phenomena [8] shows that the conditions to be satisfied by the wave parameters are $\omega_0 = \omega_s + \omega$ and $\boldsymbol{k_0} = \boldsymbol{k_s} + \boldsymbol{k}$. Therefore, these processes can be interpreted as the inelastic scattering of an incident photon, which results in a scattered photon and a plasmon (SRS) or ion-acoustic phonon (SBS): in this sense, the above conditions represent the energy and momentum conservation laws. Let us very briefly consider these important phenomena in further detail. For the SRS, since the minimum frequency for an EM wave in a plasma is ω_{pe}, from the energy conservation it follows that this instability can develop only if $\omega_0 \geq 2\omega_{pe}$, that is, if the plasma density n is such that $n \leq n_c/4$. The *Stimulated Backward Raman Scattering*, (for which $\boldsymbol{k_s} \approx -\boldsymbol{k_0}$ and $\boldsymbol{k} \approx 2\boldsymbol{k_0}$), is the fastest process in the interaction between ultraintense ultrashort laser pulses and underdense plasmas [15]: this instability is very important since it is one of the principal mechanisms which erodes the pulse energy and transfers it to the plasma in the relativistic regime. The SBS is instead characterized by the fact that the frequency of an ion-acoustic wave is always much less than the incident EM frequency: consequently, the instability can develop in every underdense plasma, without restriction imposed on the parameters. Moreover, for the same reason, almost all the incident energy can be transferred to the scattered wave. These features can be of great importance, especially in the framework of inertial-fusion studies, since Brillouin scattering can severely limit or negatively alter the absorption of the incident energy.

Consider now the interactions involving an overdense plasma. When an ultraintense laser pulse hits the surface of a solid target, the surface is immediately ionized and a non-uniform plasma with a maximum density largely exceeding the critical one is created. Therefore, in general, the properties of the interaction between EM radiation and overdense plasmas are of great importance because of its relevance to the study of the laser-solid interaction. The great interest in the laser-overdense plasma interaction comes from the fact that an intense EM wave, when reaching the critical density, can be very effectively absorbed, transferring much of its energy to the plasma electron, thus generating a population of relativistic electrons. The phenomenology of the various absorption mechanisms of an intense laser pulse from a solid target is very complicated: many interrelated processes are present, and the relative importance depends crucially on the intensity of the EM laser field [71]. In the case of non-relativistic intensity, *i.e.* from 10^{12} W μm^2/cm^2 up to $I\lambda^2 < 10^{17}$ W μm^2/cm^2, the principal absorption processes are the already mentioned *resonance absorption* and the so called *inverse Bremsstrahlung absorption* [8, 71]; they are present also in the overall absorption at higher intensies, but they are no longer the most dominant phenomena. Actually, when the intensity is greater than 10^{18} W/cm^2, the principal processes which regulate the absorption of ultraintense laser pulses are the so-called *vacuum heating* (or *Brunel effect*) and the *relativistic* $J \times B$ *heating.*

The *vacuum heating* [72–76] takes place at the vacuum-solid interface, when a very sharp density gradient is present. In this region, the electric field E of an ultraintense ultrashort laser pulse, of frequency ω, extract electrons from the plasma, accelerating them to velocity of the order of $eE/m\omega$, and putting them back into the overdense plasma; here the electric field is not present since the wave cannot penetrate and consequently the energy acquired by the electrons from the field is irreversibly transferred and eventually converted into thermal energy. It is interesting to note that in this process the presence of an underdense plasma is not required. When the density gradient is very sharp and the field intensity very high, the vacuum heating becomes much more effective than the resonant absorption and the absorption efficiency can be significant. In order to estimate it, let us consider a simple one dimensional non-relativistic model, made up of a perfect conductor (as a strongly overdense plasma) which occupies the region $x \geq 0$ and can emit electrons, while a uniform electric field $E_{\text{ext}} = E_0 \sin \omega t$ is present in the region $x < 0$. Starting from $t = 0$, electrons begin to be extracted from the solid into the vacuum from the electric field in such a way to let a zero electric field on the conductor surface. The electric field in $x < 0$ acting on an electron which is at $x_j(t)$, after being extracted at the time t_j ($x_j(t_j) = 0$), is $E(x_j) = E_{\text{ext}} + \Delta E_j$ where ΔE_j, the electric field generated by the extracted electrons, is given by the Poisson equation, $\Delta E_j = -4\pi e \int_{x_1(t)}^{x_j(t)} n dx$; here n is the electron density and x_1 is the position of the first extracted electron. In this simple model the extracted electrons never overtake and retain their initial order: this means that their contribution to the electric field is constant and can be evaluated by calculating E at the time t_j, that is at $x_j = 0$; however, $E(x_j = 0) = 0$, and from Poisson equation it follows $\Delta E_j = -E_{\text{ext}}(t_j)$. It is now possible to calculate the trajectory of the j-th electron, that is its velocity v_j and position x_j as a function of t and emission time t_j; then, by obtaining t_j as a function of t, imposing $x_j = 0$, and substituting these quantities in the expression of v_j, we can derive the velocity v_{j0} of the electron re-entering in the conductor, as a function of time. The corresponding kinetic energy $(1/2)mv_{j0}^2$ is absorbed and lost by the laser field, since inside the conductor the electric field is no longer active. The energy lost by the laser over one cycle, per unit area, is given by

$$W_{\rm abs} = \int \left(\frac{1}{2} m v_{j0}^2\right) n v_{j0} \mathrm{d}t \,. \tag{10.26}$$

The electron density n can be determined by means of the Poisson equation and the trajectory equation, giving $n = 2n_c/(\omega t - \omega t_j)$, where n_c is the critical density; dividing by the laser period $2\pi/\omega$, the absorbed intensity $I_{\rm abs}$ is finally calculated. By numerically integrating (10.26), the result is $I_{\rm abs} = (\eta/2\pi)(v_{osc}E_0^2/8\pi)$, where $v_{osc} = eE_0/m\omega$ and η is a numerical factor, equal to about 1.57 in this simple model. By collecting the velocities of the electrons coming back in the conductor, one can also determine the dependence of the thermal energy generated on the pulse intensity. The efficiency of this process is simply obtained by dividing $I_{\rm abs}$ by the intensity of the incident wave, $c(E_L^2/8\pi)\cos\vartheta$, where E_L is the electric field amplitude of the laser, forming an angle ϑ with the normal to the vacuum-target surface. In this way, we arrive at the following estimate for the efficiency f_{vh} of the process,

$$f_{vh} = \frac{\eta}{2\pi} \frac{v_{osc}^3}{v_L^2 c \cos\vartheta} \tag{10.27}$$

where $v_L = eE_L/m\omega$.

The relativistic $J \times B$ heating [77] is due to the action of the ponderomotive force associated with the laser pulse. At the interface between vacuum and an overdense plasma, the EM field penetrates over distances of the order of the skin depth. The gradient of the EM energy acts on the electrons and the potential associated with this force is equivalent to an effective pressure which gives rise to an effective heating of the electronic population. In order to show in the simplest way the essence of this effect, let us consider a relativistic cold plasma, where ions are assumed at rest. The eletron momentum (10.20) contains the relativistic ponderomotive effect in the term $-\nabla(mc^2\gamma)$, which can be formally viewed as an effective pressure-gradient term. Therefore the ponderomotive potential $\Phi_{pm} = mc^2(\gamma - 1)$ can be seen as an effective temperature T_{eff} associated with the electron population absorbing the laser radiation. From (10.7) this effective temperature T_h, which then characterizes the electron population generated by the laser ponderomotive potential, is related to the laser intensity,

$$T_h = \left(\sqrt{1 + \frac{I\lambda^2(\mu\mathrm{m})}{1.4 \times 10^{18}}} - 1\right) mc^2 \,. \tag{10.28}$$

The ponderomotive force depends on the gradient of the EM radiation which, for relativistically intense laser pulses, can be very large. In this simple derivation the ponderomotive potential still depends on the longitudinal momentum: the explicit calculation requires further developments and hypothesis and depends also on the effective structure of the laser EM field.

In conclusion, a laser pulse interacting with a solid target can generate a very energetic electron population, which propagates in the solid matter. The energetic spectrum of these electrons is characterized by an effective temperature of the order of MeV for radiation intensities exceeding $10^{18}\,\mathrm{W\,\mu m^2/cm^2}$.

10.3 Concluding Remarks

In this chapter the fundamental physical processes underlying the interaction of relativistically intense EM radiation with a preformed collisionless plasma have been discussed, introducing the most appropriate relativistic models which are currently used in the theoretical research: the kinetic approach, the multi-fluid model, and the electron-fluid model in a fixed-ion background. The role of relativistic nonlinearities for focused radiation intensities above $I\lambda^2 \approx 10^{18}\,\mathrm{W\mu m^2/cm^2}$ has been emphasized, reviewing several physical processes like relativistic laser pulse focusing, compression and splitting, harmonic generation, electron and ion acceleration, quasi-static magnetic field generation, X-ray production, laser-induced nuclear reactions, electron-positron production. All these issues demonstrate that the successful development of more and more intense lasers in the last decade has opened up several new fields of fundamental and applied science, which may results in applications unpredictable ten or twenty years ago.

The fundamental aspects of the relativistic dynamic of an electron in the relativistically intense laser pulse have been discussed, starting from the so-called "figure-of-8" trajectory. It is to be emphasized that in such interaction regimes, an appreciable elongation of the electron excursion in the direction of the pulse propagation develops, which may become of the same order as that in the transverse direction. In addition, in an increasingly strong field, the electron parallel momentum scales as a^2, while the perpendicular component scales as a. Therefore, in the ultra-relativistic regime the longitudinal electron dynamics can become more important than the transverse one. An important consequence is that a short and intense laser pulse, acting mainly on electrons, is able to create strong charge separations, which are at the basis of the effective charged particle acceleration, both of electrons and ions, and of the formation of relativistic electromagnetic solitons (RES). RES and laser-based ion-acceleration will be discussed in a forthcoming chapter.

The ultra-strong laser-pulse interaction in such extreme regimes can be considered as one of the most attractive fields of modern theoretical and experimental research and is destined to produce very interesting applications in the near future.[3]

[3] During the editorial process of this volume, two review articles on relevant subjects have been published, which are worth to be referenced [78, 79].

References

1. J.H. Eberly, P. Maine, D. Strickland et al.: Laser Focus **23**, 84 (1987)
2. P. Maine, D. Strickland, P. Bado et al.: IEEE Journal of Quantum Electronics **QE-24**, 398 (1988)
3. M.D. Perry and G. Mourou: Science **64**, 917 (1994)
4. G.A. Mourou, C.P.J. Barty, M.D. Perry: Physics Today **51**, 22 (1998)
5. J.D. Lindl, B.A. Hammel, B. Grant Logan et al.: Plasma Phys. Contr. Fus. **46**, B135 (2004)
6. C. Cavailler, N. Fleurot, T. Lonjaret, J.-M. Di-Nicola: Plasma Phys. Contr. Fus. **45**, A217 (2003)
7. A.V. Gaponov and M.A. Miller: Sov. Phys. JETP **7**, 168 (1958)
8. W.L. Kruer: *The Physics of Laser-Plasma Interaction* (Addison-Wesley, New York 1988)
9. P. Corkum: Phys. Rev. Lett. **71**, 1994 (1993)
10. M. Lontano, G. Lampis, A. Kim, A.M. Sergeev: Physica Scripta **T63**, 262 (1996)
11. A.M. Sergeev, M. Lontano, A.V. Kim et al.: Laser Part. Beams **17**, 129 (1998)
12. C.E. Max, J. Arons, A.B. Langdon: Phys. Rev. Lett. **33**, 2091 (1974)
13. W.B. Mori, C. Joshi, J.M. Dawson et al.: Phys. Rev. Lett. **60**, 1298 (1988)
14. A. Borisov, A.V. Borovskiy, O.B. Shiryaev et al.: Phys. Rev. A **45**, 5830 (1992)
15. S.V. Bulanov, F. Califano, G.I. Dudnikova et al.: Relativistic Interaction of laser pulses with Plasmas. In *Reviews of Plasma Physics*, edited by V.D. Shafranov (Kluwer Academic/Consultants Bureau New York 2001), vol. 22, pp 227–335
16. G. Zeng, B. Shen, W. Yu, Z. Xu: Phys. Plasmas **3**, 4220 (1996)
17. S.V. Bulanov, M. Lontano, T.Zh. Esirkepov et al.: Phys. Rev. Lett. **76**, 3562 (1996)
18. G.A. Askar'yan, S.V. Bulanov, G.I. Dudnikova et al.: Plasma Phys. Contr. Fus. **39**, A137 (1997)
19. G.A. Askar'yan, S.V. Bulanov, F. Califano et al.: Plasma Phys. Contr. Fus. **39**, B261 (1997)
20. D. Farina, M. Lontano, I.G. Murusidze, S.V. Mikeladze: Phys. Rev. E **63**, 056409 (2001)
21. T. Tajima, J.M. Dawson: Phys. Rev. Lett. **43**, 267 (1979)
22. Th. Katsouleas: Nature **431**, 515 (2004)
23. S.P.D. Mangles, C.D. Murphy, Z. Najimudin et al.: Nature **431**, 535 (2004)
24. C.G.R. Geddes, Cs. Toth, J. van Tilborg et al.: Nature **431**, 538 (2004)
25. J. Faure, Y. Glinec, A. Pukhov et al.: Nature **431**, 541 (2004)
26. S.V. Bulanov, I.N. Inovenkov, V.I. Kirsanov et al.: Phys. Fluids B **4**, 1935 (1992)
27. S.V. Bulanov, I.N. Inovenkov, V.I. Kirsanov et al.: Physica Scripta **47**, 209 (1993)
28. S.V. Bulanov, T.Zh. Esirkepov, N.N. Naumova et al.: Phys. Rev. Lett. **82**, 3440 (1999)
29. Y. Sentoku, T.Zh. Esirkepov, K. Mima et al.: Phys. Rev. Lett. **83**, 3434 (1999)
30. N.M. Naumova, S.V. Bulanov, T.Zh. Esirkepov et al.: Phys. Rev. Lett. **87**, 185004 (2001)

31. T. Esirkepov, K. Nishihara, S.V. Bulanov, F. Pegoraro: Phys. Rev. Lett. **89**, 275002 (2002)
32. M. Lontano, S.V. Bulanov, F. Califano et al.: Relativistic Electromagnetic Solitons produced by Ultrastrong Laser Pulses in Plasmas. In *Science of Superstrong Field Interactions.*, edited by K. Nakajima and M. Deguchi, AIP Conference Proceedings (American Institute of Physics New York 2002), **634**, pp 87–98
33. E.L. Clark, K. Krushelnick, J.R. Davies et al.: Phys. Rev. Lett. **84**, 670 (2000)
34. R.A. Snavely, M.H. Key, S.P. Hatchett et al.: Phys. Rev. Lett. **85**, 2945 (2000)
35. S. Fritzler, V. Malka, G. Grillon et al.: Appl. Phys. Lett. **83**, 3039 (2003)
36. T.E. Cowan, J. Fuchs, H. Ruhl et al.: Phys. Rev. Lett. **92**, 204801 (2004)
37. T.E. Cowan, A.W. Hunt, T.W. Phillips et al.: Phys. Rev. Lett. **84**, 903 (2000)
38. M.A. Stoyer, T.C. Sangster, E.A. Henry et al.: Rev. Sci. Instr. **72**, 767 (2001)
39. S. Fritzler, Z. Najimudin, V. Malka et al.: Rev. Sci. Instr. **89**, 165004 (2002)
40. M. Lontano and I.G. Murusidze: Optics Express **11**, 248 (2003)
41. C.N. Danson, P.A. Brummitt, R.J. Clarke et al.: Nucl. Fus. **44**, S239 (2004)
42. T. Tajima and G. Mourou: Superstrong Field Science. In *Superstrong Field in Plasmas*, edited by M. Lontano, G. Mourou, O. Svelto, T. Tajima, AIP Conference Proceedings (American Institute of Physics New York 2002), **611**, pp 423–436
43. E.P. Liang, S.C. Wilks, M. Tabak: Phys. Rev. Lett. **81**, 4887 (1998)
44. P. Chen, T. Tajima: Phys. Rev. Lett. **83**, 256 (1999)
45. S.V. Bulanov, T. Esirkepov, T. Tajima: Phys. Rev. Lett. **91**, 085001 (2003)
46. J. Meyer-ter-Vehn, A. Pukhov, Zh.-M. Sheng: Relativistic Laser Plasma Interaction. In *Atoms, Solids, and Plasmas in Super-Intense Laser Fields*, edited by D. Batani, Ch. Joachain, S. Martellucci, A.N. Chester (Kluwer Academic/Pleanum Publishers, New York 2001), pp 167–192
47. L.D. Landau and E.M. Lifshitz: *The Classical Theory of Fields*, 4th edn (Pergamon Press, Oxford 1980)
48. J.D. Jackson: *Classical Electrodynamics*, 3rd edn (John Wiley and Sons, New York 1998)
49. O.L. Brill, B. Goodman: Am. J. Phys. **35**, 832 (1967)
50. D.I. Dzhavakhishvili, N.L. Tsintsadze: Sov. Phys. JETP **37**, 666 (1973)
51. S. Weinberg: *Gravitation and Cosmology* (Wiley, New York 1972)
52. F.T. Gratton, G. Gnavi, R.M.O. Galvao, L. Gomberoff: Phys. Rev. E **55**, 3381 (1997)
53. C. Cercignani, G.M. Kremer: *The Relativistic Boltzmann Equation: Theory and Applications* (Birkhauser, Basel 2002)
54. M. Lontano, S.V. Bulanov, J. Koga: Phys. Plasmas **8**, 5113 (2001)
55. V.I. Berezhiani, S.M. Mahajan, Z. Yoshida, M. Ohhashi: Phys. Rev. E **65**, 047402 (2002)
56. L. Gorbunov, P. Mora, T.M. Antonsen Jr.: Phys. Rev. Lett. **76**, 2495 (1996)
57. S.V. Bulanov, T.Zh. Esirkepov, M. Lontano et al.: Physica Scripta **T63**, 280 (1996)
58. L. Gorbunov, P. Mora, T.M. Antonsen Jr.: Phys. Plasmas **4**, 4358 (1997)
59. A.M. Pukhov, J. Meyer-ter-Vehn: Phys. Plasmas **5**, 1880 (1998)
60. Z.M. Sheng, J. Meyer-ter-Vehn, A.M. Pukhov: Phys. Plasmas **5**, 3764 (1998)
61. L. Gorbunov, R.R. Ramazashvili: J. Exp. Theor. Phys. **87**, 461 (1999)
62. A.I. Akhiezer, R.V. Polovin: Sov. Phys. JETP **30**, 915 (1956)

63. A.I. Akhiezer, I.A. Akhiezer, R.V. Polovin et al.: *Plasma Electrodynamics* (Pergamon, Oxford 1975)
64. J.M. Dawson: Phys. Rev. **113**, 383 (1959)
65. S.V. Bulanov, V.I. Kirsanov, A.S. Sakharov: JETP Lett. **53**, 565 (1990)
66. S. Dalla, M. Lontano: Phys. Lett. A **173**, 456 (1993)
67. T.P. Coffey: Phys. Fluids **14**, 1402 (1971)
68. T. Katsouleas, W.B. Mori: Phys. Rev. Lett. **61**, 90 (1988)
69. J.B. Rosenzweig: Phys. Rev. A **38**, 3634 (1988)
70. Z.M. Sheng, J. Meyer-ter-Vehn: Phys. Plasmas **54**, 493 (1997)
71. S.C. Wilks, W.L. Kruer: IEEE J. Quantum Electronics **33**, 1954 (1997)
72. F. Brunel: Phys. Rev. Lett. **59**, 52 (1987)
73. F. Brunel: Phys. Fluids **31**, 2714 (1988)
74. G. Bonnaud, P. Gibbon, J. Kindel, E. Williams: Laser Part. Beams **9**, 339 (1991)
75. P. Gibbon, A. Bell: Phys. Rev. Lett. **68**, 1535 (1992)
76. S. Kato, B. Bhattacharyya, A. Nixhiguchi, K. Mima: Phys. Fluids B **5**, 564 (1993)
77. W.L. Kruer: Phys. Fluids **28**, 430 (1985)
78. G.A. Mourou, T. Tajima, S.V. Bulanov: Rev. Mod. Phys. **78**, 309 (2006)
79. M. Marklund, P.K. Shukla: Rev. Mod. Phys. **78**, 591 (2006)

11 Unusual Optical Properties of the Dense Nonequilibrium Plasma

G. Ferrante, M. Zarcone, and S.A. Uryupin

[1] Dipartimento di Fisica e Tecnologie Relative dell'Universitá di Palermo, Viale delle Scienze, edificio 18, 90128 Palermo, Italy
zarcone@unipa.it
[2] P.N. Lebedev Physics Institute, Leninsky pr. 53, 119991, Moscow, Russia
uryupin@sci.lebedev.ru

Summary. A concise overview of new optical properties of dense nonequilibrium plasma formed on the solid state target boundary is given. In this chapter, we describe phenomena such as the third harmonic generation in the skin layer, collisionless electron heating in the high frequency skin-effect regime, absorption, and reflection and transmission of radiation by a plasma with anisotropic electron distribution.

11.1 Introduction

In the context of powerful femtosecond laser pulse interaction with solid state targets, it is relatively simple to create conditions under which the interaction may be considered as occurring with a dense hot plasma with a sharp boundary. Consideration of the plasma surface as a sharp boundary is justified if, during the laser pulse action, it is legitimate to neglect the ion hydrodynamic motion. Radiation with frequency smaller than the electron plasma frequency penetrates into the plasma to distances of the order of the skin layer depth. As a consequence, the peculiar features of the radiation interaction with a dense plasma are determined, to a considerable extent, by the electron kinetics in the skin layer and depend essentially on the value of the electron mean thermal energy (see, for instance, [1–10]).

At relatively small thermal energy the electron–ion collision frequency ν is greater than or of the same order as the laser radiation fundamental frequency. Under such conditions the transition from normal to high-frequency skin-effect takes place and electron collisions determine plasma optical properties to a large degree. In particular, the electron–ion collisions in the skin layer are responsible for harmonic generation of a pump wave. Harmonic generation occurring as a result of electron–ion collisions in the presence of a strong high-frequency electromagnetic field has been attracted attention for many years [11–15]. As a rule, excluding the numerical calculations reported in paper [14], theoretical papers addressing this subject restrict their analysis to the conditions where the frequency of the generating field is much larger than the electron–ion collision frequency. The consideration of such conditions is of interest and appropriate when the high-frequency field interacts with a sufficiently hot plasma of relatively low density. In turn, the increase of plasma density causes the increase of electron–ion collision frequency ν, which may become comparable to the frequency ω of the generating field. So, to extend the investigation of the harmonics generation to the interesting domain of sufficiently dense plasmas, one needs to work out a theory for the case where the frequency ν is not small compared with ω. Such a theory is developed in [16] and is given here in Sect. 11.2.

The role of electron–ion collisions decreases with increasing temperature. In the conditions when the anomalous and the high-frequency skin-effect take place, the perturbations to the electron velocity distribution brought about by the high-frequency field significantly depend on the character of the electron interaction with the plasma surface. In the theoretical investigations available in literature [5–10], the electron response to the laser pulse action is described by using the simplest boundary condition at the plasma surface. Namely, it is assumed that the electrons are reflected by the plasma surface specularly. This boundary condition permits to give a relatively simple description of phenomena like reflection and absorption by a plasma undergoing heating. We shall show that the assumption of specular reflection is not the only one making the above processes easily treatable. Following [17] in Sect.11.3 we are using another assumption at the plasma boundary allowing more profitably to investigate the process of radiation interaction with a plasma surface. As matter of fact, in a hot plasma the charge density fluctuations are expected to reach a relatively high level, and these will yield a perturbation of the plasma surface. Electron reflection by a chaotically perturbed surface along any given direction is expected to have casual character. Chaotic electron reflection by such a plasma surface may be modeled introducing the parameter p giving the fraction of electrons diffusely reflected. Similarly to the definition familiar in the theory of metals (see, for instance, [18]), we call "diffuse reflection" the process when electrons are reflected in all directions with the same probability, and the reflected electron distribution coincides with the surface electron equilibrium distribution with temperature T_s. Specifically, in Sect. 11.3 we investigate plasma heating in the regime of high-frequency skin-effect in the conditions when a variable fraction of electrons, equal to p, is diffusely reflected by the plasma surface. It is assumed that the parameter p is a power function of surface electron temperature. Similarly to [5, 9], to describe the plasma heating we use the relatively simple model, in which the thermal flux in the plasma surface is determined by the corresponding radiation absorption coefficient, while the transfer of heat from the skin layer into the plasma interior is described by the classical heat transfer equation. Such a relatively simple approach to the plasma heating problem permits to clearly evidence the influence of the diffuse reflection on both the radiation absorption and the electron heating characteristics inside the plasma far from the skin layer.

Very intense, ultrashort laser pulses interacting with solid surfaces permit one to investigate the properties of hot dense plasmas in strongly nonequilibrium states. Thanks to the shortness of the laser pulse, it is possible to create the conditions, when the newly formed plasma has relatively sharp boundary and the influence of electron–electron and electron–ion collisions is negligible to a large extent due to the high electron kinetic energy. Another important issue concerning plasmas created by a powerful ultrashort ionizing laser pulse is the following. As a result of ionization, a photoelectron distribution function with large anisotropy in the velocity space is created. Such plasmas are formed both in the regime of tunnel ionization and in the regime of barrier suppression [19–23] and may exhibit unusual optical properties. Some of them are predicted in [24–26] and considered in Sect. 11.4.

11.2 Normal Skin-Effect in a Dense Plasma

11.2.1 Reflection and Absorption

Let us assume that on the plasma filling the half space $z > 0$, normally to its surface impinges an electromagnetic wave of the form

$$\frac{1}{2}\vec{E}_o \exp(-\mathrm{i}\omega t + \mathrm{i}kz) + c.c. , \tag{11.1}$$

linearly polarized along $0x$ axis. This wave penetrates inside the plasma to the skin-layer depth and is reflected by it. The reflected wave field is written as

$$\frac{1}{2}R\,\vec{E}_o \exp(-\mathrm{i}\omega t - \mathrm{i}kz) + c.c. , \tag{11.2}$$

where R is the complex reflection coefficient. We represent the field in the plasma as

$$\frac{1}{2}\vec{E}(z)\exp(-\mathrm{i}\omega t) + c.c. \equiv \vec{E}\cos\Psi , \tag{11.3}$$

where $\vec{E}(z) = (E(z), 0, 0)$, $\Psi = \omega t - \delta$, δ being the phase shift. In the conditions we are considering the magnetic field $\vec{B}(z,t) = (0, B(z,t), 0)$ both in vacuum and inside the plasma is connected to the electric one by the equation

$$-\frac{\partial}{\partial z}E(z,t) = \frac{1}{c}\frac{\partial}{\partial t}B(z,t) . \tag{11.4}$$

From (11.1)–(11.4) and the continuity requirements on the magnetic and electric fields on the plasma surface we have

$$E_o + RE_o = E(0) , \tag{11.5}$$

$$-E_o + RE_o = \mathrm{i}\frac{c}{\omega}E'(0) . \tag{11.6}$$

The field $\vec{E}\,(z)$ is determined by the Maxwell equations with the current density generated by the field at the frequency ω. In the conditions when the ratio of the electron–ion effective collision frequency to the field frequency is not assumed small nor large, as it is usually done in the theory of high-frequency ($\omega \gg \nu$) or normal ($\nu \gg \omega$) skin-effect, for the current density we have [16]

$$\delta \vec{j}_1 = J_1(\Omega) en \vec{v}_E \cos[\psi - \Delta_1(\Omega)] \,, \tag{11.7}$$

where $\vec{v}_E = e\,\vec{E}\,/m\omega$, e, m and n are, respectively, the electron charge, mass and density, $\Omega = \nu/\omega$, $\nu = 4\pi Z e^4 n \Lambda m^{-2} v_T^{-3}$, $-Ze$ the ion charge, Λ the Coulomb logarithm, v_T the electron thermal velocity. For small and large Ω values the functions $J_1(\Omega)$ and $\Delta_1(\Omega)$ yield known asymptotic expressions

$$J_1(\Omega) \simeq 1 \,, \qquad \Delta_1(\Omega) \simeq \frac{\pi}{2} - \frac{2\Omega}{3\sqrt{2\pi}} \,, \qquad \Omega \ll 1 \,, \tag{11.8}$$

$$J_1 \simeq \frac{32}{\sqrt{2\pi\Omega}} \,, \qquad \Delta_1(\Omega) \simeq \frac{315\sqrt{2\pi}}{32\Omega} \ll 1 \,, \qquad \Omega \gg 10 \,. \tag{11.9}$$

In Fig. 11.1 $J_1(\Omega)$ and $\Delta_1(\Omega)$ are plotted for intermediate Ω values. According to Fig. 11.1 both the current density phase shift and amplitude decrease monotonically increasing the ratio of the collision frequency ν to the field frequency ω.

For the function $E(z)$ from the Maxwell equations and (11.7) we have the equation

$$\frac{\mathrm{d}^2}{\mathrm{d}z^2} E(z) + \frac{\omega^2}{c^2} E(z) = -\mathrm{i} J_1(\Omega) \frac{\omega_L^2}{c^2} E(z) \exp[\mathrm{i}\Delta_1(\Omega)] \,. \tag{11.10}$$

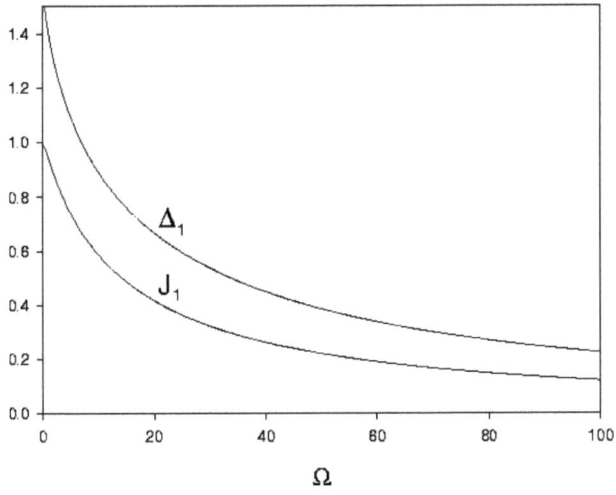

Fig. 11.1. Phase shift $\Delta_1(\Omega)$ and current third harmonic amplitude $J_1(\Omega)$ of the fundamental frequency versus $\Omega = \nu/\omega$

The expression (11.7) for the current density, determining the right-hand side of (11.10), has been obtained for arbitrary values of the ratio ν/ω. At the same time, in deriving (11.7) use has been made of a kinetic equation suited for weak electron–ion and electron–electron interactions, when $\nu \ll \omega_L$. It means that in the case $\nu \geq \omega$ is implied also the fulfilment of the inequality $\omega \ll \omega_L$. Thus, the field frequency is assumed to be smaller than the electron plasma frequency. A field with such a frequency penetrates inside the plasma only to the skin-layer depth. Let us assume that the inequality $\omega \ll \omega_L$ occurs also when $\nu \leq \omega$.

Further, we assume that

$$\omega \ll \omega_L \sqrt{J_1(\Omega)} \,. \tag{11.11}$$

When $\nu \ll \omega$ this last inequality is equivalent to $\omega \ll \omega_L$, while when $\nu \gg \omega$ from (11.11) we have $\nu\omega \ll \omega_L^2$. The inequality (11.11) allows one to neglect the second term proportional to the squared field frequency in the left-hand side of the equation for the field (11.10). Further, considering that the plasma fills the half space $z > 0$, from (11.10), we have approximately the following solution going to zero at $z \to \infty$:

$$E(z) = E(0) \exp(-\kappa z) \,, \tag{11.12}$$

$$\kappa = \frac{\omega_L}{c} \sqrt{J_1(\Omega)} \exp\left[\frac{\mathrm{i}}{2} \Delta_1(\Omega) - \mathrm{i}\frac{\pi}{4}\right] . \tag{11.13}$$

We now use the field distribution in the plasma given by (11.12), (11.13) to solve the problem of pump wave reflection and absorption.

From relations (11.5), (11.6), (11.12) and (11.13), we find the complex reflection coefficient

$$R = \frac{\omega - \mathrm{i}c\kappa}{\omega + \mathrm{i}c\kappa} \simeq -1 + \frac{2\omega}{\omega_L \sqrt{J_1(\Omega)}} \exp\left[-\frac{\mathrm{i}}{2} \Delta_1(\Omega) - \mathrm{i}\frac{\pi}{4}\right] \tag{11.14}$$

and the relation connecting the field inside the plasma to the electric field of the impinging wave

$$E(0) = \frac{2E_o}{1 + \mathrm{i}\kappa c/\omega} \simeq \frac{2\omega}{\omega_L \sqrt{J_1(\Omega)}} E_o \exp\left[-\frac{\mathrm{i}}{2} \Delta_1(\Omega) - \mathrm{i}\frac{\pi}{4}\right] . \tag{11.15}$$

From (11.14) follows a relatively simple expression for the absorption coefficient due to electron–ion collisions

$$A = 1 - |R|^2 \simeq \frac{4\omega}{\omega_L \sqrt{J_1(\Omega)}} \cos\left[\frac{1}{2} \Delta_1(\Omega) + \frac{\pi}{4}\right] . \tag{11.16}$$

Using the relations (11.8), (11.9) in two limiting cases from (11.16) we have

$$A \simeq \frac{2\nu_{ei}}{\omega} \,, \quad \nu_{ei} \ll \omega_L \,, \tag{11.17}$$

$$A \simeq \sqrt{\frac{3\pi}{4}} \frac{\sqrt{\nu_{ei}\omega}}{\omega_L} \,, \quad \nu_{ei} \gg \omega \,, \tag{11.18}$$

where use has been made of the notation $\nu_{ei} = \sqrt{2}\nu/3\sqrt{\pi}$ familiar in the theory of high frequency field absorption. For the general case, the dependence of the absorption coefficient A, give by formula (11.16), on the ratio ν/ω is reported in Fig. 11.2.

We conclude this section by reporting the relations that connect the strength and the phase shift of the field (11.3) to the electric field in vacuum. From (11.3), (11.13), and (11.15) we have approximately

$$E(z) \simeq \frac{2\omega e^{i\delta}}{\omega_L\sqrt{J_1(\Omega)}} E_o \exp\left\{-z\frac{\omega_L}{c}\sqrt{J_1(\Omega)}\cos\left[\frac{1}{2}\Delta_1(\Omega) - \frac{\pi}{4}\right]\right\}, \quad (11.19)$$

$$\delta \simeq -\frac{\pi}{4} - \frac{1}{2}\Delta_1(\Omega) - z\frac{\omega_L}{c}\sqrt{J_1(\Omega)}\sin\left[\frac{1}{2}\Delta_1(\Omega) - \frac{\pi}{4}\right]. \quad (11.20)$$

According to these relations the field strength and the phase shift depend on the space coordinate. Further, we neglect this dependence in dealing with the electron kinetics in the skin-layer. It is allowed if the distance covered by thermal electrons in the field period is much smaller than the effective depth of the skin-layer, which, according to (11.19), is $c/\omega_L\sqrt{J_1(\Omega)}$.

11.2.2 Third Harmonic Generation

The electric field (11.19) affects the electron–ion collisions in the skin-layer. As a result of such an effect, the generation of high order current density harmonics is possible. In particular, for the current density $\delta\vec{j_3}$ at the frequency 3ω we have [16]

$$\delta\vec{j_3} = 10^{-4} J_3(\Omega)\frac{\omega_L^2}{\omega}\frac{v_E^2}{v_T^2}\vec{E}\cos\left[3\psi - \Delta_3(\Omega)\right]. \quad (11.21)$$

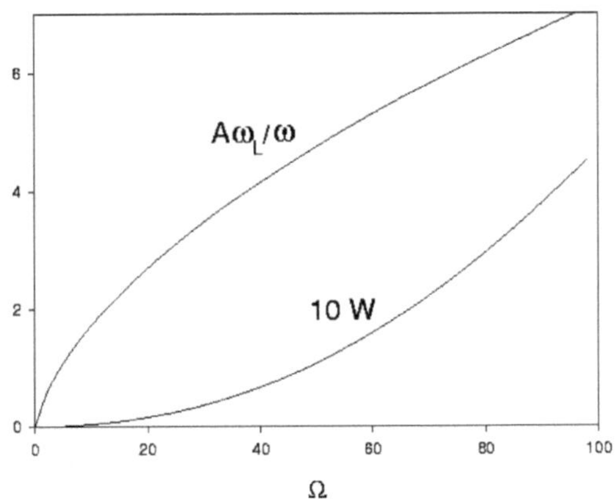

Fig. 11.2. Fundamental wave absorption coefficient A and third harmonic generation efficiency W, (11.37), in relative units versus $\Omega = \nu/\omega$

The functions $J_3(\Omega)$ and $\Delta_3(\Omega)$ take a simple form in the limits of small and large Ω. In the case of a high-frequency field, when $\omega \gg \nu$ up to terms linear in $\Omega = \nu/\omega \ll 1$ we find

$$J_3(\Omega) \simeq \frac{10^3}{24\pi\sqrt{2\pi}}\Omega\,, \qquad \Omega \ll 1\,, \tag{11.22}$$

$$\Delta_3(\Omega) \ll 1\,, \qquad \Omega \ll 1\,. \tag{11.23}$$

Obviously, the results (11.22), (11.23) follow also from the theory of harmonic generation based on the mechanism of electron–ion collisions (see [11], [15]), where from the beginning it is assumed that the electron–ion effective collision frequency is much smaller than the field frequency. In the opposite limiting case, when $\nu \gg \omega$ we approximately have

$$J_3(\Omega) \simeq \frac{105}{16\pi}\frac{10^4}{\Omega^2}\,, \qquad \Omega > 100\,, \tag{11.24}$$

$$\Delta_3(\Omega) \simeq -\frac{3\pi}{2}\,, \qquad \Omega > 100\,. \tag{11.25}$$

The results of numerical calculations of $J_3(\Omega)$ and $\Delta_3(\Omega)$ are shown in Fig. 11.3. Comparing the curves obtained numerically with the simple analytical dependencies (11.22)–(11.25), one can see that the domains of validity of the latter are restricted to very small and to very large values of the parameter Ω, respectively. In the important for applications range of Ω values, where the efficiency of current third harmonic generation is the largest,

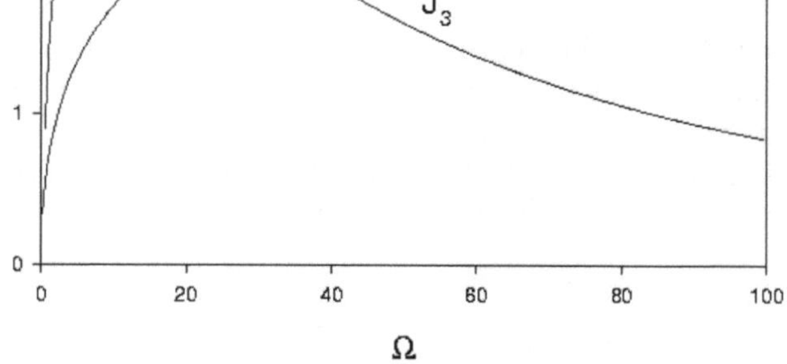

Fig. 11.3. Phase shift $-\Delta_3(\Omega)$ and current third harmonic amplitude $J_3(\Omega)$, in relative units, versus the ratio of the electron–ion collision frequency ν to the field frequency ω, $\Omega = \nu/\omega$

from Fig. 11.3, it is seen that the function $J_3(\Omega)$ exhibits a useful quantitative property. Namely, according to Fig. 11.3, the function $J_3(\Omega)$ changes by no more than a factor of three in the widest and most interesting region of the Ω values, when $1 \leq \Omega \leq 100$. We note also that the current third harmonic generation efficiency has its maximum at $\Omega \simeq 10$ with the value $\max[J_3(\Omega)] \simeq 3$. The phase shift in the maximum, as seen from Fig. 11.3, is close to $-\pi/2$. From Fig. 11.3 is also seen that by increasing Ω the phase shift is monotonically decreasing from zero to $-3\pi/2$. This last value of the phase shift $\Delta_3(\Omega)$ takes place at rather high Ω values, and corresponds to the situation when the current is generated in opposition of phase with the field. In concluding this section we observe that the non-monotonic dependence of the function $J_3(\Omega)$ on Ω established here is similar to the dependence on plasma density of the current third harmonic found in paper [14] as a result of numerical solution of a quantum kinetic equation.

Let us consider now the emission of the third harmonic by the plasma. The field inside the plasma at the frequency 3ω is written as

$$\frac{1}{2}\overrightarrow{E_3}(z)\exp(-3\mathrm{i}\omega t) + c.c. \equiv \overrightarrow{E_3}\cos(3\omega t - \delta_3)\ , \tag{11.26}$$

where $\overrightarrow{E_3}(z) = (E_3(z), 0, 0)$ and δ_3 is the corresponding phase shift. In accordance with the relation (11.7), the field (11.26) yields the current at 3ω with the density

$$\delta\overrightarrow{j_{3\omega}} = \frac{e^2 n}{3\omega m}\overrightarrow{E_3} J_1\left(\frac{\Omega}{3}\right)\cos\left[3\omega t - \delta_3 - \Delta_1\left(\frac{\Omega}{3}\right)\right]\ . \tag{11.27}$$

Taking into account the relation (11.7), similarly to (11.10), to determine the field $E_3(z)$ from the Maxwell equations, we find

$$\frac{\mathrm{d}^2}{\mathrm{d}z^2}E_3(z) + \frac{9\omega^2}{c^2}E_3(z) - \kappa_3^2 E_3(z) = F\exp\left(-3\kappa z\right)\ , \tag{11.28}$$

where F is given below by (11.30) and the notation

$$\kappa_3 = \frac{\omega_L}{c}\sqrt{J_1\left(\frac{\Omega}{3}\right)\exp\left[\frac{\mathrm{i}}{2}\Delta_1\left(\frac{\Omega}{3}\right) - \mathrm{i}\frac{\pi}{4}\right]} \tag{11.29}$$

is used. A substantial difference of this equation as compared to (11.10) consists in that its right-hand side contains a field source at frequency 3ω, due to nonlinear dependence of the electron–ion collision frequency on the field. According to relations (11.19)–(11.21), the current density value of the source F is proportional to the third power of the field strength and has the form

$$F = -\mathrm{i}12\pi 10^{-4} J_3(\Omega)\left(\frac{\mathrm{e}}{m\omega}\frac{\omega_L}{v_T}\frac{E_o}{c}\right)^2\left[\frac{2\omega}{\omega_L\sqrt{J_1(\Omega)}}\right]^3 \times$$
$$\times E_o\exp\left[-\mathrm{i}\frac{3\pi}{4} - \mathrm{i}\frac{3}{2}\Delta_1(\Omega) + \mathrm{i}\Delta_3(\Omega)\right]\ . \tag{11.30}$$

Further, together with the condition (11.11), assumed in analyzing the field at ω, we will consider that a similar condition takes place for the field at 3ω as well. Namely, we assume that

$$3\omega \ll \omega_L \sqrt{J_1\left(\frac{\Omega}{3}\right)} . \tag{11.31}$$

In this case, the solution to the nonhomogeneous (11.28), going to zero inside the plasma, has the form

$$E_3(z) = E(0)\exp\left(-\kappa_3 z\right) + \frac{F}{\kappa_3^2 - 9\kappa^2}\left[\exp\left(-\kappa_3 z\right) - \exp\left(-3\kappa z\right)\right] . \tag{11.32}$$

Next, in accordance with (11.4) the electric field (11.32) univocally determines the magnetic field in the plasma

$$B_3(z) = -\frac{\mathrm{i}}{3k}\frac{\mathrm{d}}{\mathrm{d}z}E_3(z) . \tag{11.33}$$

At the plasma boundary $z = 0$ the electromagnetic field (11.32), (11.33) transforms into the field of the wave irradiated by the plasma at 3ω. According to Maxwell equations, the field of the irradiated wave has the form

$$\frac{1}{2}\overrightarrow{E}_r \exp(-3\mathrm{i}\omega t - 3\mathrm{i}kz) + c.c. , \tag{11.34}$$

where $\overrightarrow{E_r} = (E_r, 0, 0)$. Besides, according to (11.4) $\overrightarrow{B_r} = (0, -E_r, 0)$. Further, using the field continuity conditions at $z = 0$, from (11.32)–(11.34) we find the strength of electric and magnetic fields of the irradiated wave:

$$-B_r = E_r = -\mathrm{i}\frac{F}{(3k + \mathrm{i}\kappa_3)(\kappa_3 + 3\kappa)} . \tag{11.35}$$

From here we find the energy flux density irradiated by the plasma at the frequency 3ω

$$\overrightarrow{S} = \frac{c}{4\pi T_r}\int_0^{T_r} \mathrm{d}t \left[\overrightarrow{E}(z,t)\overrightarrow{B}(z,t)\right] = -\overrightarrow{s} I_r , \tag{11.36}$$

where $\overrightarrow{s}$ is the unit vector along OZ axis, $I_r = c|E_r|^2/8\pi$ and $T_r = 2\pi/3\omega$ the corresponding period. As the energy flux density of the impinging on the plasma wave is $I_o = cE_o^2/8\pi$, using the inequality (11.31) from (11.30), (11.35) and (11.36) for the third harmonic generation efficiency in the skin-layer of a dense plasma we have the following result

$$\eta = \frac{I_r}{I_o} = \left(\frac{\mathrm{e}E_o}{m\omega v_T}\right)^4 \frac{\omega^6}{\omega_L^6} W\left(\frac{\nu}{\omega}\right) , \tag{11.37}$$

where for the function $W(\Omega)$ we have the simple asymptotic dependencies

$$W(\Omega) = \frac{5}{\pi} 10^{-3} \Omega^2 , \quad \Omega \ll 1 , \tag{11.38}$$

$$W(\Omega) \simeq \frac{\sqrt{2\pi}}{2+\sqrt{3}} \left(\frac{105\pi}{2048} \right)^2 \Omega , \quad \Omega \gg 10 . \tag{11.39}$$

In Fig. 11.2 we report the curve of the function $W(\Omega)$. From Fig. 11.2 it is seen that increasing $\Omega = \nu/\omega$ one has a monotonic increase of the third harmonic generation efficiency.

11.3 High-Frequency Skin-Effect

11.3.1 Collisionless Absorption

In this section the frequency ω of the incident wave is assumed to be much smaller than the electron plasma frequency ω_L, but much larger than the frequency $\omega_L v_T / c$, characterizing the time required to the thermal electrons to go through the skin layer. It is assumed, too, that ω is much larger than the effective electron–ion collision frequency ν_{ei}. Thus, below, the following inequalities must hold

$$\omega_L \gg \omega \gg \nu_{ei} , \frac{\omega_L v_T}{c} . \tag{11.40}$$

The inequalities (11.40) are compatible, provided an ideal plasma is considered, for which $\omega_L \gg \nu_{ei}$ and a nonrelativistic temperature is assumed, $v_T \ll c$.

In the linear theory of high-frequency radiation interaction with a hot plasma the inequalities (11.40) define the conditions for the high-frequency skin-effect regime. Then, the field in the skin layer is written as

$$\frac{1}{2} \vec{E}(0) \exp\left(-\frac{z}{d}\right) \exp(-\mathrm{i}\omega t) + c.c. , \tag{11.41}$$

where $\vec{E}(0)$ is the electric field at $z = +0$, while the skin layer dept is given by the London's length

$$d = \frac{c}{\omega_L} . \tag{11.42}$$

Let us consider the absorption of the field (11.41) by plasma electrons starting with collisionless absorption, when the influence of the electron–ion collisions on the electron motion in the skin layer may be neglected (see, below, inequality (11.56)).

To obtain the corresponding absorption coefficient, we must define the work done by the field (11.41) on the plasma electrons. To this aim, we use the kinetic equation allowing to find the small perturbations to the electron distribution function

$$\frac{1}{2}\delta f(z)\exp(-\mathrm{i}\omega t) + c.c.\,, \tag{11.43}$$

where $\delta f = \delta f(z)$ is found from

$$-\mathrm{i}\omega\delta f + v_z\frac{\mathrm{d}}{\mathrm{d}z}\delta f = -\frac{\mathrm{e}}{m}E(z)\frac{\partial f}{\partial v_x}\,. \tag{11.44}$$

In (11.44) $E(z) = E(0)\exp(-z/d)$ and it has been assumed that the OX axis is directed along the electric field direction. The electron distribution function f is assumed to be Maxwellian.

The solution to (11.44), for electrons moving towards the plasma surface $v_z < 0$, going to zero for $z \to \infty$

$$\delta f(z \to \infty,\ v_z < 0) = 0\,, \tag{11.45}$$

is given by

$$\begin{aligned}\delta f(z, v_z < 0) &= \frac{\mathrm{e}}{mv_z}\int_z^{\infty}\mathrm{d}z'\exp\left[\mathrm{i}\frac{\omega}{v_z}(z-z')\right]E(z')\frac{\partial f}{\partial v_x} = \\ &= \frac{\mathrm{e}}{m}\left(\mathrm{i}\omega + \frac{v_z}{d}\right)^{-1}E(z)\frac{\partial f}{\partial v_x} = \delta f^-(z, v_z)\,.\end{aligned} \tag{11.46}$$

The distribution of electrons reflected by the plasma surface depends on the choice of boundary condition at $z = 0$. At variance with the frequently used in recent times boundary condition of specular reflection (see, for instance, [5–10]), we assume that a fraction of electrons is reflected diffusely. This assumption is justified, in particular, if we consider that as a consequence of thermal fluctuations of the electron charge density, the plasma surface is not sharp nor smooth. More likely, the plasma surface is a fluctuating boundary at which the electron reflection along any given arbitrary direction is a casual event. The fraction of diffusely reflected electrons is characterized by the parameter $p < 1$, which is a function of temperature. Then, for the non-equilibrium part of the distribution function we have the following boundary condition

$$\delta f(z = 0,\ v_z > 0) = (1-p)\delta f^-(z = 0, -v_z)\,. \tag{11.47}$$

Taking into account (11.47), for electrons moving away from the plasma surface ($v_z > 0$), from (11.44) we have

$$\delta f(z, v_z > 0) = -\frac{\mathrm{e}}{m v_z} \int_0^z \mathrm{d}z' \exp\left[\mathrm{i}\frac{\omega}{v_z}(z - z')\right] E(z') \frac{\partial f}{\partial v_x} + \\ + (1-p) \exp\left(\mathrm{i}\frac{\omega}{v_z} z\right) \delta f^-(z=0, -v_z) = \\ = -\frac{\mathrm{e}}{m}\left(\mathrm{i}\omega + \frac{v_z}{d}\right)^{-1} \left[\exp\left(\mathrm{i}\frac{\omega}{v_z} z\right) - \exp\left(-\frac{z}{d}\right)\right] E(0) \frac{\partial f}{\partial v_x} + \\ + \frac{\mathrm{e}}{m}(1-p)\left(\mathrm{i}\omega - \frac{v_z}{d}\right)^{-1} \exp\left(\mathrm{i}\frac{\omega}{v_z} z\right) E(0) \frac{\partial f}{\partial v_x} \,. \quad (11.48)$$

Relations (11.46) and (11.48) allow one to calculate the high-frequency current density in the skin layer

$$\frac{1}{2}\delta j(z) \exp(-\mathrm{i}\omega t) + c.c. \,, \quad (11.49)$$

$$\delta j(z) = e \int_{v_z < 0} \mathrm{d}\vec{v}\, v_x \delta f(z, v_z < 0) + e \int_{v_z > 0} \mathrm{d}\vec{v}\, v_x \delta f(z, v_z > 0) \,, \quad (11.50)$$

and the average over a period of the high-frequency field Joule heat, transferred to plasma electrons

$$Q = \frac{1}{4} \int_0^\infty \mathrm{d}z \left[\delta j(z) E^*(z) + c.c.\right] . \quad (11.51)$$

Taking into account (11.41), (11.46), (11.48) and (11.50), from (11.51) we find

$$Q = \frac{v_T}{(2\pi)^{3/2}} \left(\frac{\omega_L^2}{4\omega^2} |E(0)|^2\right) \beta \times \\ \times \left\{-p e^\beta Ei(-\beta) - 2(1-p)\left[1 + (1+\beta) e^\beta Ei(-\beta)\right]\right\} \,, \quad (11.52)$$

where $\beta = 1/2\delta^2$, $\delta = \omega_L v_T / \omega c$ is the skin-effect anomaly parameter and $Ei(-\beta) = -\int_\beta^\infty \mathrm{d}x e^{-x}/x$ the integral exponential function. In the high-frequency skin-effect regime (11.40) $\delta \ll 1$ or $\beta \gg 1$. Taking into account the definition of the absorption coefficient A,

$$A = \frac{Q}{I_o} \,, \quad (11.53)$$

from (11.52) we have

$$A = \sqrt{\frac{8}{\pi}} \frac{v_T}{c} \left[p + 4(1-p)\left(\frac{\omega_L v_T}{\omega c}\right)^2\right] . \quad (11.54)$$

In (11.54), the term proportional to p gives the radiation absorption by electrons diffusely reflected, while that proportional to $(1-p)$ gives the absorption by specularly reflected electrons. In the case of absence of diffusely reflected electrons $p=0$ and the expression (11.54), as expected, goes over the known in the literature coefficient of the collisionless absorption (see, for instance, [8–10]).

In the case, instead, when the fraction of diffusely reflected electrons is sufficiently large, such as to give rise to the condition

$$p > 4(1-p)\left(\frac{\omega_L v_T}{\omega c}\right)^2 , \tag{11.55}$$

the absorption coefficient becomes considerably larger than that which results in the specular electron reflection regime. It is worth to note that, in the conditions under consideration here, when the anomaly parameter is very small (high-frequency skin-effect, $\omega c \gg \omega_L v_T$), a large absorption enhancement takes place even when the fraction of diffusely reflected electrons is relatively small. Expression (11.54) gives the collisionless Landau absorption in the high-frequency skin-effect regime. In such a regime the conditions are also possible when the absorption is due to electron–ion collisions. The corresponding absorption coefficient does not depend on the particular electron-plasma surface interaction mechanism and is given by (11.17). The same absorption coefficient (11.17) occurs also when the thermal electron mean free path is larger than the skin layer depth. In this last case, the absorption is basically determined by slow electrons which, in spite of being a relatively small fraction (of the order of $n(d/l)^{3/4} \ll n$), have an effective mean free path $v/\nu(v) \approx (v/v_T)^4 l$ smaller than the skin layer depth d, $l = v_T/\nu$. Comparing (11.17) and (11.54) it may be seen that absorption is collisionless provided the plasma temperature is high enough to fulfil the condition

$$p + 4(p-1)\left(\frac{\omega_L v_T}{\omega c}\right)^2 \gg \sqrt{\frac{\pi}{2}}\frac{c}{\omega_L}\frac{\nu_{ei}}{v_T} = \frac{d}{3l} . \tag{11.56}$$

It is also seen that the condition (11.56) is more simply fulfilled if electron diffuse reflection takes place. We note that the analysis of the peculiarities of radiation absorption reported above rests significantly on the use of the Maxwellian electron velocity distribution function. In other words, we have excluded from our analysis the cases, when the electron distribution differs appreciably from the Maxwellian. In particular, to the class of non-Maxwellian distributions belong distributions exhibiting a relatively large number of hot electrons. The presence of hot electrons witnesses that different plasma heating mechanisms are acting, and that the peculiarities of the radiation absorption too are different as compared to those analyzed above. At the same time, in the conditions of normal incidence of the laser radiation on the plasma surface with practically fixed ions, the possibility to generate hot electrons implies the presence of laser radiation of such a high intensity, that becomes necessary to take into account relativistic effects on the electron motion. Accordingly the peculiarities of radiation absorption reported above are meaningful provided the laser intensities are such to make negligible relativistic effects.

11.3.2 Heat Flux Propagation

Electrons undergoing heating in the skin layer transfer heat into plasma interior. Heat flux propagation outside the skin layer is described by the heat transfer equation

$$\frac{3}{2}\frac{\partial}{\partial t}(nT) + \frac{\partial}{\partial z}q = 0 , \tag{11.57}$$

where T is the electron temperature in energy units, and q is the heat flux density. To the aim of demonstrating the effects due to the electron diffuse reflection, we confine our analysis to a regime of heat transfer, when the Fick law is valid,

$$q = -\frac{128}{\sqrt{2\pi}} n v_T l \frac{\partial}{\partial z} T . \tag{11.58}$$

In doing so, in particular, we disregard any consideration concerned with the effects of nonlocality in the heat transfer. According to [9, 27] the use of (11.58) is legitimate, provided the characteristic scale of electron temperature variation L is sufficiently larger than the mean free path length l,

$$L = \left| \frac{\partial}{\partial z} \ln T \right|^{-1} > 50l . \tag{11.59}$$

Further, we assume that the mean free path length is larger than the skin layer depth

$$l > d = \frac{c}{\omega_L} . \tag{11.60}$$

In such a case, in looking for the solution to (11.57) and (11.58), we may disregard the finite value of the skin layer, and take that the thermal flux in the plasma surface is given and is described by the absorption coefficient

$$q(z = 0, t) = I_0 \, A(t) . \tag{11.61}$$

Equations (11.57) and (11.58) jointly with the boundary condition (11.61) allows one to investigate the characteristics of thermal flux propagation, depending on which physical mechanism controls the field absorption in the skin layer.

Let us analyze the essential features of heat propagation in the regime of collisionless absorption, when the condition (11.56) is fulfilled.

Case of Diffuse Reflection

First, we consider the case when the inequality (11.55) is fulfilled and the field absorption is controlled by diffusely reflected electrons. We assume that the fraction of diffusely reflected electrons depends on temperature as follows:

$$p = \left(\frac{T_s}{W}\right)^{\alpha} < 1\,, \tag{11.62}$$

where W is the energy characteristic of the given degree of nonspecularity of the plasma boundary and α a number controlling the diffuseness parameter dependency on the temperature. If the diffuse reflection and its degree are traced back to the level of charge density thermal fluctuations, then the value $\alpha = 1$ is, to a considerable extent, the most appropriate one, as it establishes a linear dependency between the level of fluctuations and the electron temperature (see, for instance, [28]). In what follows we analyze the plasma heating characteristics for $\alpha = 1$. One can find consideration for arbitrary α in [17]).

In the case of collisionless absorption the boundary condition for (11.57) has the form

$$q(z=0,t) = I_0\sqrt{\frac{8}{\pi}}p\frac{v_T}{c} = I_0\left(\frac{8W}{\pi mc^2}\right)^{1/2}\left(\frac{T_s}{W}\right)^{3/2}\,. \tag{11.63}$$

Using this condition we look for a solution to (11.57) and (11.58) in the self-similar form

$$T(z,t) = T_s(t)\frac{\Psi\left[z/z_f(t)\right]}{\Psi(0)}\,, \tag{11.64}$$

$\Psi(0) = \Psi(\xi = 0)$. The functions $T_s(t)$ and $z_f(t)$ have the forms

$$\begin{aligned} T_s(t) &= mv_E^2\Psi(0)\left(\frac{mv_E^2}{W}\right)^{4/3}\left(\frac{\nu t}{48\sqrt{2\pi}}\right)^{2/3} \equiv \\ &\equiv mv_E^2\Psi(0)\left(\frac{p(t)}{\Psi(0)}\right)^{4/7}\left(\frac{\nu t}{48\sqrt{2\pi}}\right)^{2/7}\,, \end{aligned} \tag{11.65}$$

$$\begin{aligned} z_f(t) &= 64\frac{v_E}{\nu}\left(\frac{mv_E^2}{W}\right)^{5/3}\left(\frac{\nu t}{48\sqrt{2\pi}}\right)^{4/3} \equiv \\ &\equiv 64\frac{v_E}{\nu}\left(\frac{p(t)}{\Psi(0)}\right)^{5/7}\left(\frac{\nu t}{48\sqrt{2\pi}}\right)^{6/7}\,. \end{aligned} \tag{11.66}$$

The equation for the function $\Psi(\xi)$ has a form

$$\frac{\mathrm{d}}{\mathrm{d}\xi}\left[\Psi^{5/2}\frac{\mathrm{d}}{\mathrm{d}\xi}\Psi\right] = -\frac{4}{3}\xi^{3/2}\frac{\mathrm{d}}{\mathrm{d}\xi}\left[\xi^{-1/2}\Psi\right]\,, \tag{11.67}$$

and the boundary conditions

$$\Psi(0)\frac{d\Psi}{d\xi}|_{\xi=0} = -1 \; , \qquad \Psi(\xi \to \infty) = 0 \; . \tag{11.68}$$

The numerical solution to (11.67), (11.68) gives

$$\Psi(0) = 1.43 \; , \qquad \xi_m = 0.84 \; , \tag{11.69}$$

where ξ_m defines the boundary position of the heat front $\Psi(\xi_m) = 0$. Figure 11.4 shows a plot of the function $\Psi(\xi)$. The portion of the $\Psi(\xi)$ curve around $\xi \approx \xi_m$ in Fig. 11.4 is shown dotted. Dots are meant to underline that in the neighborhood $\xi \approx \xi_m$ the derivative absolute value $\Psi'(\xi)$ is large and, accordingly, the temperature gradient (11.59), too, is large and

$$L = \frac{z_f}{|\Psi'(\xi)|} \ll z_f \; . \tag{11.70}$$

We note that, due to the smallness of the temperature variation space scale, around $z \approx z_f \xi_m$ effects of heat transfer nonlocality are likely to come into play, and it will act to smooth the $\Psi(\xi)$ curve around $\xi \approx \xi_m$. At the same time, from Fig. 11.4 it is seen that the region, where the effect of heat transfer nonlocality may affect the reported solution, is very small. It allows one to say that this effect weakly influences the plasma heating depth and the radiation absorption coefficient, which are the most important results of the present analysis.

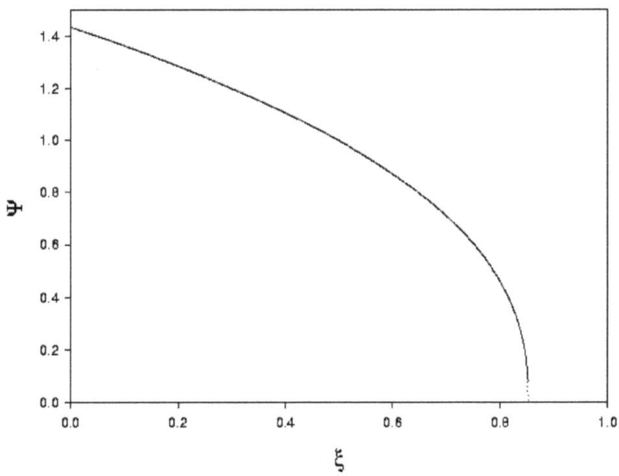

Fig. 11.4. Spatial electron temperature profile inside the plasma as a function of the distance from the boundary for the case of absorption due to electron diffuse reflection by the plasma boundary, $\xi = z/z_f(t)$

Case of Specular Reflection

In this case the absorption coefficient is given by (11.54), where we need to let $p = 0$. Further, the squared skin-effect anomaly parameter $\delta^2 = \omega_L^2 v_T^2/\omega^2 c^2$ depends linearly on the electron temperature on the plasma surface, i.e., similarly to the parameter p (11.62) for $\alpha = 1$. It means that in writing the solution corresponding to the electron specular reflection, it is sufficient to substitute W by $mc^2\omega^2/4\omega_L^2$ in the (11.65) and (11.66). As a result we obtain

$$T_s(t) = mv_E^2 \Psi(0) \left(\frac{\omega_L v_E}{\omega c}\right)^{8/3} \left(\frac{\nu t}{3\sqrt{2\pi}}\right)^{2/3} \equiv$$

$$\equiv mv_E^2 \Psi(0) \left(\frac{4\delta_s^2(t)}{\Psi(0)}\right)^{4/7} \left(\frac{\nu t}{48\sqrt{2\pi}}\right)^{2/7}, \qquad (11.71)$$

$$z_f(t) = 64 \frac{v_E}{\nu} \left(\frac{\omega_L v_E}{\omega c}\right)^{10/3} \left(\frac{\nu t}{12\sqrt{\pi}}\right)^{4/3} \equiv$$

$$\equiv 64 \frac{v_E}{\nu} \left(\frac{4\delta_s^2(t)}{\Psi(0)}\right)^{5/7} \left(\frac{\nu t}{48\sqrt{2\pi}}\right)^{6/7}, \qquad (11.72)$$

where $\delta_s^2 = \omega_L^2 T_s(t)/m\omega^2 c^2$. The temperature profile inside the plasma in the case of specular reflection is given by the same curve in Fig. 11.4. Comparing (11.65) and (11.66) to (11.71) and (11.72) we see that, when the fraction of diffusely reflected electrons is not very small ($p(t) > 4\delta_s^2(t)$, with $\delta_s(t) \ll 1$), the electron diffuse reflection yields a significant increase of electron temperature on the plasma surface (by a factor of $[p(t)/4\delta_s^2(t)]^{4/7} \gg 1$), and even a greater enlargement of the region where plasma undergoes heating (by a factor of $[p(t)/4\delta_s^2(t)]^{5/7} \gg 1$). It means that in the regime of diffuse reflection it is legitimate to say that the effect of enhancement of plasma heating is taking place, and that a more efficient radiation absorption too is occurring. The new features corresponding to the enhancement of the collisionless absorption are described by the relations (11.54), (11.62), (11.65), and (11.66).

11.4 Anomalous and High Frequency Skin-Effect in a Nonequilibrium Plasma

11.4.1 Anomalies in Absorption and Reflection

Let us investigate a plasma with the bi-Maxwellian electron distribution function (EDF)

$$F = \left(\frac{m}{2\pi}\right)^{3/2} \frac{n}{T_\perp \sqrt{T_x}} \exp\left[-\frac{mv_x^2}{2T_x} - \frac{m}{2T_\perp}\left(v_y^2 + v_z^2\right)\right], \qquad (11.73)$$

where the two effective temperatures T_x and $T_\perp$ are given in energy units. EDF like (11.73) is formed as a result of tunnel ionization of atoms [20]. We consider the interaction of such a plasma with a weak linearly polarized electromagnetic wave, impinging normally on the plasma surface

$$\frac{1}{2}\vec{E}\exp(-\mathrm{i}\omega t+\mathrm{i}kz)+c.c.\,, \quad z<0\,, \tag{11.74}$$

with $\vec{E}=(E_x,E_y,0)=E(\cos\varphi,\sin\varphi,0)$, φ the angle between the electric field polarization direction $\vec{E}$ and the OX axis. The wave frequency ω is assumed much smaller than the electron plasma frequency ω_L. According to (11.73), the OX axis is also the symmetry axis of the EDF. Thus we have two new parameters characterizing the process as compared to the case of an isotropic plasma. One is the angle φ, the other is the degree of temperature anisotropy $\Delta=1-T_x/T_\perp$. We will show below that these new parameters effectively control the process new features. The wave (11.74) is reflected by the plasma surface and partially penetrates into the plasma, where it is absorbed as a result of interaction with the electrons. In the geometry of interaction chosen here, only two components of the electric field are different from zero. As the incident wave has a low intensity, we can describe independently the interaction of each component with the plasma. First of all we consider the interaction of the E_x component. Then for the reflected component we have

$$\frac{1}{2}E_xR_x\exp(-\mathrm{i}\omega t-\mathrm{i}kz)+c.c.\,, \quad z<0\,, \tag{11.75}$$

where R_x is the complex reflection coefficient. In its turn, the field inside the plasma has the form

$$\frac{1}{2}E_x(z)\exp(-\mathrm{i}\omega t)+c.c.\,, \quad z>0\,. \tag{11.76}$$

To define the field $E_x(z)$, from the Maxwell equations we have

$$\frac{\mathrm{d}^2}{\mathrm{d}z^2}E_x(z)+\frac{\omega^2}{c^2}E_x(z)=-\frac{4\pi\mathrm{i}e\omega}{c^2}\int \mathrm{d}\vec{v}\,v_x\delta f\,. \tag{11.77}$$

The perturbation δf to the EDF (11.73) is found from the kinetic equation

$$\begin{aligned}-\mathrm{i}\omega\delta f+v_z\frac{\mathrm{d}}{\mathrm{d}z}\delta f=&-\frac{\mathrm{e}}{m}\left\{E_x(z)\frac{\partial F}{\partial v_x}+\right.\\ &\left.+\frac{1}{c}B_y(z)\left[v_x\frac{\partial F}{\partial v_z}-v_z\frac{\partial F}{\partial v_x}\right]\right\}\\ \equiv&-S_x(z,v_z)\,,\end{aligned} \tag{11.78}$$

where $B_y(z)$ is the magnetic field component in the plasma, created by the impinging wave

$$B_y(z)=-\mathrm{i}\frac{c}{\omega}\frac{\mathrm{d}}{\mathrm{d}z}E_x(z)\,. \tag{11.79}$$

In plasmas with isotropic EDFs, the term in (11.78) containing $B_y(z)$ goes to zero. In other words, in isotropic plasmas, the magnetic field does not affect the electron kinetics in the skin layer. At the contrary, in plasmas with anisotropic EDFs like (11.73), the term containing $B_y(z)$ is not zero. Physically, it implies that the anisotropic distribution over velocities creates the conditions for the magnetic field $B_y(z)$ to rotate the electrons from one degree of freedom to the other. Because of this the magnetic field contributes to determine δf, i.e., to influence significantly the electron kinetics. Besides, in the skin-effect conditions, the magnetic field in absolute value considerably exceeds the electric field, according to the inequality $c/\omega \gg |d \ln E_x(z)/\mathrm{d}z|^{-1}$. The joint manifestation of both these causes, as is shown below, is responsible for the appearance of new optical properties in plasmas with anisotropy in the EDF. Further, we consider the simplest boundary conditions on the plasma surface. Namely, we assume that electrons are specularly reflected by the plasma boundary. Provided these conditions are fulfilled, (11.78) gives

$$\begin{aligned}\delta f &= \frac{1}{v_z}\int_z^{\infty} \mathrm{d}z' S_x(z',v_z)\exp\left[\mathrm{i}\frac{\omega}{v_z}(z-z')\right], \quad v_z<0\,;\\ \delta f &= -\frac{1}{v_z}\int_0^{z} \mathrm{d}z' S_x(z',v_z)\exp\left[\mathrm{i}\frac{\omega}{v_z}(z-z')\right] -\\ &\quad -\frac{1}{v_z}\int_0^{\infty} \mathrm{d}z' S_x(z',-v_z)\exp\left[\mathrm{i}\frac{\omega}{v_z}(z+z')\right], \quad v_z>0\,. \qquad (11.80)\end{aligned}$$

Substituting the perturbation δf (11.80) into the right-hand side of (11.77) and performing the Fourier transform over z, as is usually done when the specular reflection conditions are assumed (see, for instance, [29]), from (11.77) we obtain

$$E_x(z) = -\mathrm{i}\frac{k}{\pi}\frac{E_x(+0)}{Z_x}\int_{-\infty}^{\infty}\frac{\mathrm{d}q}{q^2-k^2\varepsilon_x(q)}\exp(\mathrm{i}qz)\,, \quad z>0\,, \qquad (11.81)$$

where $E_x(+0)$ is the electric field on the plasma boundary. In writing the expression (11.81), we have used the notation Z_x to indicate the component of the surface impedance

$$Z_x = \frac{E_x(+0)}{B_y(+0)}\,, \qquad (11.82)$$

giving the ratio between the electric and magnetic fields on the plasma surface; besides, $\varepsilon_x(q)$ represents the x-component of the plasma dielectric susceptibility. If the electrons have a velocity distribution like (11.73), the function $\varepsilon_x(q)$ is given by

$$\epsilon_x(q) = 1-\frac{\omega_L^2}{\omega^2}J_+\left(\frac{\omega}{qv_T}\right)-\left(1-\frac{T_\perp}{T_x}\right)\frac{\omega_L^2}{\omega^2}\left[1-J_+\left(\frac{\omega}{qv_T}\right)\right], \qquad (11.83)$$

where $v_T = \sqrt{T_\perp/m}$, and the function J_+ has the form [28]

$$J_+(x) = J'_+(x) + \mathrm{i}J''_+(x) \,, \tag{11.84}$$

$$J'_+(x) = x \exp\left(-\frac{x^2}{2}\right) \int_o^x \mathrm{d}t \exp\left(\frac{t^2}{2}\right) \,, \tag{11.85}$$

$$J''_+(x) = -\sqrt{\frac{\pi}{2}} x \frac{q}{|q|} \exp\left(-\frac{x^2}{2}\right) . \tag{11.86}$$

Further, using the continuity condition of the electric and magnetic fields on the plasma surface, from (11.74), (11.75) and (11.79), we find R_x and Z_x. The interaction of the component E_y with the plasma is considered in a completely similar way. The plasma response to this component is described by the functions R_y and Z_y, which formally are similar to R_x and Z_x, but Z_y is determined through $\varepsilon_y(q)$, whose expression follows from (11.83) letting $T_x = T_\perp$. As a final result, for the surface impedance components Z_α and the complex reflection coefficients R_α we find

$$Z_\alpha = -\frac{2\mathrm{i}}{\pi} k \int_0^\infty \frac{\mathrm{d}q}{q^2 - k^2 \epsilon_\alpha(q)} \,, \quad \alpha = x, y \,, \tag{11.87}$$

$$R_\alpha = \frac{(Z_\alpha - 1)}{(Z_\alpha + 1)} = |R_\alpha| \exp\left(\mathrm{i}\Psi_\alpha\right) \,, \tag{11.88}$$

where the functions $|R_\alpha|$ and Ψ_α are expressed through the real and imaginary parts Z'_α and Z''_α of the surface impedance components $Z_\alpha = Z'_\alpha + iZ''_\alpha$. As a consequence of the Landau damping, the intensity of the reflected wave is smaller than the intensity of the incident one. The intensity decrease of the reflected wave is measured by the decrease of the function $|R_\alpha|$ with respect to unity. Another important effect is related to the circumstance that different components of the incident wave are reflected by the anisotropic plasma with different phase-shifts, $\Psi_x \neq \Psi_y$. Due to the difference between Ψ_x and Ψ_y, the reflected wave polarization differs as compared to that of the incident wave. Namely, the linearly polarized incident wave (11.74) is reflected by the anisotropic plasma as an elliptically polarized wave (11.75). Under these conditions, the ellipticity degree of the reflected wave is characterized by the phase-shift difference

$$\Psi_r = \Psi_x - \Psi_y \,. \tag{11.89}$$

In the physical conditions under consideration here, the real and imaginary parts of the impedance components in absolute value show small departures from unity. It allows one to write the following approximate expressions for the absorption coefficient A,

$$A = 1 - R, R = |R_x|^2 \cos^2\varphi + |R_y|^2 \sin^2\varphi \,,$$
$$A \simeq 4Z'_x \cos^2\varphi + 4Z'_y \sin^2\varphi \,, \tag{11.90}$$

and the phase-shift Ψ_r

$$\Psi_r \simeq 2Z''_y - 2Z''_x \,. \tag{11.91}$$

According to (11.90) the absorption coefficient is basically determined by the real parts of the impedance components. In contrast, the phase-shift is determined by the difference of the imaginary parts. The functions A (11.90) and Ψ_r (11.91) are studied in [24, 25]. Here we show only basic results of numerical calculations.

In Figs. 11.5 and 11.6 we report the results of calculations of the collisionless absorption coefficient A (11.90), as a function of the angle φ, formed by the polarization vector of the absorbed wave with the EDF symmetry axis. Figure 11.5 shows the curves corresponding to $\delta = 9$ and three values of the temperature anisotropy $T_x = 5T_\perp$; $T_x = T_\perp$; and $T_\perp = 4T_x$. According to Fig. 11.5, for the chosen plasma and laser parameters, a relative decrease of absorption takes place as compared to the case of an isotropic plasma. Besides, the absorption coefficient in a plasma with $T_x \neq T_\perp$ is found to grow monotonously with the angle φ increase. The dependencies of A versus φ, for $\delta = 0.3$ and the same values of temperatures as in Fig. 11.5, are shown in Fig. 11.6. We note that the curve with $T_\perp = 4T_x$ is qualitatively similar to the corresponding curve of Fig. 11.5. A different behavior, instead, is shown by the curve with $T_x = 5T_\perp$. In fact, a significant increase of absorption in an anisotropic plasma is observed. The dependency of A on the angle φ changes as well. For $\delta = 0.3$ and $T_x = 5T_\perp$ the absorption maximum occurs at $\varphi = 0$, when the wave field is polarized along the EDF symmetry axis. Figure 11.7 reports plots of the phase-shift Ψ_r versus δ for three values of the temperature anisotropy: $T_x = 2T_\perp$; $T_x = 5T_\perp$; $T_\perp = 4T_x$. For small δ all the curves are close to zero. Far from $\delta = 0$, the behavior of the function Ψ_r depends on the

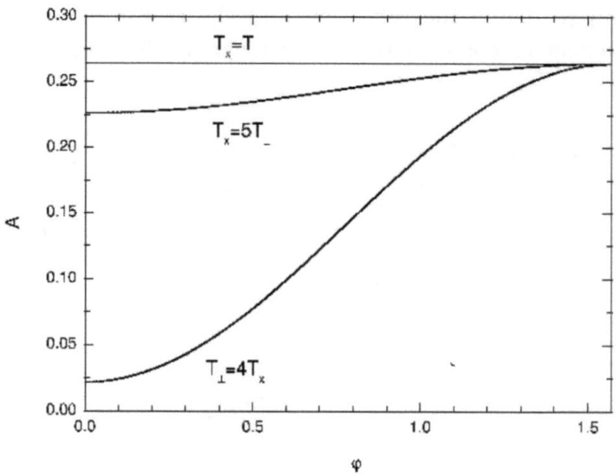

Fig. 11.5. Collisionless absorption coefficient A versus φ (in radians) the angle between the polarization vector of the electromagnetic wave and the EDF symmetry axis, in a plasma with $\delta = v_T\omega_L/\omega c = 9$ and for three values of the temperature anisotropy: $T_x = 5T_\perp$; $T_\perp = T_x$; $T_\perp = 4T_x$

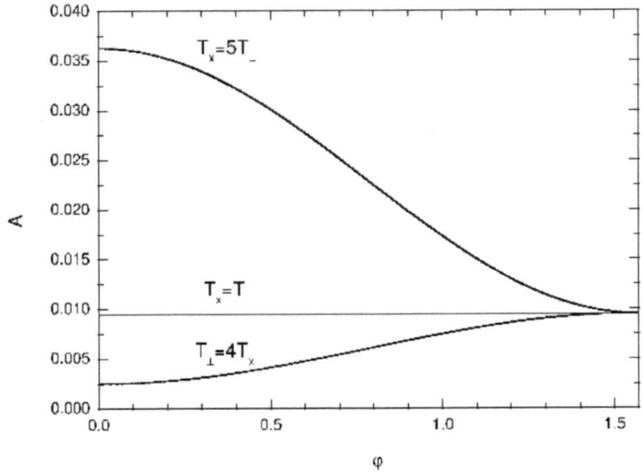

Fig. 11.6. The same function as in Fig. 11.5, but for $\delta = 0.3$

temperature anisotropy. When $T_\perp = 4T_x$, the function Ψ_r depends weakly on the parameter δ, if $\delta \leq 9$. If $T_x = 2T_\perp$ or $T_x = 5T_\perp$ the curves in Fig. 11.7 demonstrate essential decrease of the function Ψ_r with δ increase. As a whole, Fig. 11.7 shows that the larger δ and the ratio $T_x/T_\perp$, the larger the absolute value of the function Ψ_r too.

11.4.2 Transmission Through a Thin Foil

With the aim to define the reflection and the transmission coefficients by a plasma layer let us consider the electromagnetic field in the plasma layer.

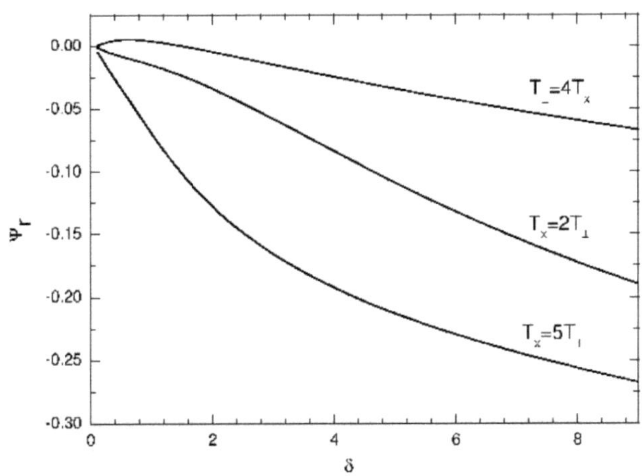

Fig. 11.7. The phase-shift of the reflected wave Ψ_r versus $\delta = v_T\omega_L/\omega c$ for three values of the electron temperature anisotropy: $T_\perp = 4T_x$; $T_x = 2T_\perp$; $T_x = 5T_\perp$

According to Maxwell's equations for the electric and magnetic components of the radiation field inside the plasma, we have (11.77)–(11.79). In solving (11.78), we take advantage of the circumstance that the average electron kinetic energy along the z axis $T_z = m\int \mathrm{d}v \cdot v_z^2 F/n$ is a small quantity. Namely, we assume that $D \gg \sqrt{T_z/m\omega^2}$ and

$$l_s\omega \gg \sqrt{\frac{T_z}{m}}\,, \tag{11.92}$$

where l_s is the effective skin-layer depth. The inequality (11.92) allows one to treat the spatial derivative in the left-hand side of (11.78) as a perturbation.

We consider the case, when F does not depend on the test wave electric field. This evidently takes place for a prepared plasma. Then, taking into account (11.92) and keeping only the corrections containing the field derivative up to the second order, from (11.78) we obtain

$$\delta f = -\frac{\mathrm{i}e}{m\omega}\left\{E_x(z)\frac{\partial F}{\partial v_x} - \frac{\mathrm{i}}{\omega}\frac{\mathrm{d}E_x}{\mathrm{d}z}v_x\frac{\partial F}{\partial v_z} - \frac{1}{\omega^2}\frac{\mathrm{d}^2E_x(z)}{\mathrm{d}z^2}v_xv_z\frac{\partial F}{\partial v_z}\right\}\,. \tag{11.93}$$

This expression permits one to find the current density in the right hand side of (11.77). Further, when $\omega \ll \omega_L$ and

$$T_x = m\int \mathrm{d}v \cdot v_x^2\frac{F}{n} \gg \frac{\omega^2}{\omega_L^2}mc^2\,, \tag{11.94}$$

(11.77) becomes

$$\frac{\mathrm{d}^2}{\mathrm{d}z^2}E_x(z) - \frac{E_x(z)}{l_s^2} = 0\,. \tag{11.95}$$

In (11.95), the length l_s characterizes the field penetration depth inside a highly anisotropic plasma and is found as

$$l_s = \left(\frac{T_x}{m\omega^2}\right)^{1/2}\,. \tag{11.96}$$

This result implies that the inequality (11.92) amounts to $T_x \gg T_z$. We note also that, thanks to inequality (11.94), the length l_s, given by (11.96), is much larger than the skin-layer dimension resulting when the EDF is isotropic. It is true, both in the case of high-frequency and anomalous skin effects. In fact, in the first case, one has $l_s \sim c/\omega_L$, while in the second case $l_s \sim \left(c^2\sqrt{T_x/m}/\omega_L^2\omega\right)^{1/3}$. Equation (11.95) has the solution

$$E_x(z) = \frac{1}{\sinh(D/l_s)}\left\{E(0)\sinh\left(\frac{D-z}{l_s}\right) + E(D)\sinh\left(\frac{z}{l_s}\right)\right\}\,, \tag{11.97}$$

where $E(0)$ and $E(D)$ are the fields in the plasma at the boundaries. The magnetic field is described by (11.79) and (11.97). In its turn, the electric fields of the incident, reflected and transmitted waves have, respectively, the forms

$$\begin{aligned} E_i &= \frac{1}{2} E_0 \exp(-\mathrm{i}\omega t + \mathrm{i}kz) + c.c. \,, \\ E_r &= \frac{1}{2} E_0 R \exp(-\mathrm{i}\omega t - \mathrm{i}kz) + c.c. \,, \\ E_t &= \frac{1}{2} E_0 T \exp(-\mathrm{i}\omega t + \mathrm{i}kz - \mathrm{i}kD) + c.c. \,, \end{aligned} \tag{11.98}$$

where R and T are the complex reflection and transmission coefficients. We assume that all electric fields are directed along the OX axis. Using (11.76), (11.97) and (11.98) and the continuity requirement of the magnetic and electric fields at the foil boundaries $z = 0$ and $z = D$, we find for R and T:

$$R = (1 + k^2 l_s^2) \sinh\left(\frac{D}{l_s}\right) \left[2\mathrm{i}kl_s \cosh\left(\frac{D}{l_s}\right) + (k^2 l_s^2 - 1) \sinh\left(\frac{D}{l_s}\right)\right]^{-1} , \tag{11.99}$$

$$T = 2\mathrm{i}kl_s \left[2\mathrm{i}kl_s \cosh\left(\frac{D}{l_s}\right) + (k^2 l_s^2 - 1) \sinh\left(\frac{D}{l_s}\right)\right]^{-1} . \tag{11.100}$$

From (11.99) and (11.100) follows the relation

$$|R|^2 + |T|^2 = 1 \,, \tag{11.101}$$

meaning that the incident wave energy is partly reflected and partly transmitted. In (11.101) the absorption coefficient is absent, as a result of the adopted approximate account of the spatial dispersion in (11.78). In other words, we have disregarded the small energy losses due to the field Cherenkov absorption. According to (11.94) and (11.100), the ratio of the transmitted energy density flux to the incident one is given by

$$|T|^2 = \left\{1 + \frac{1}{4}\left(\sqrt{\frac{T_x}{mc^2}} + \sqrt{\frac{mc^2}{T_x}}\right)^2 \sinh^2\left(\frac{2\pi D}{\lambda}\sqrt{\frac{mc^2}{T_x}}\right)\right\}^{-1} , \tag{11.102}$$

where $\lambda = 2\pi c/\omega$ is the wavelength of the incident radiation. According to (11.102) for $mc^2 \gg T_x$, and relatively thin foil, such that

$$D < \frac{\lambda}{\pi} \frac{T_x}{mc^2} \,, \tag{11.103}$$

the transmission coefficient is close to unity. The foil transparency increase is due to the relatively large depth of field penetration inside the plasma, which possesses a highly anisotropic EDF. In its turn, the large penetration depth is a consequence of the wave magnetic field action on the electron motion in the skin-layer. It is just thanks to the magnetic field that a fast electron motion appears also in the direction perpendicular to the electric field yielding the penetration depth increase.

Acknowledgments

This work is part of the research activity of the Italian–Russian Forum of Laser Physics and Related Technologies and was supported by INTAS (project 05-51-5037) and the Russian–Italian Agreement for Scientific Collaboration 2003–2004 (projects B26 and B28 of ENCL. 3). The authors wish to acknowledge also the support by the Palermo University through the International Relations Fund. Also acknowledged is the support from the Italian Ministry of education and Science through the PRIN project N. 2004 023130-005.

References

1. D.F. Price, R.M. More, R.S. Walling, G. Guethlein, R.L. Shepherd, R.E. Stewart, W.E. White, Phys. Rev. Lett. **75**, 252 (1995).
2. M. Fraenkel, A. Zigler, Z. Henis, S. Eliezer, N.E. Andreev, Phys. Rev. E **61**, 1899 (2000).
3. M. Schnürer, R. Nolte, A. Rousse, G. Grillon, G. Cheriaux, M.P. Kalachnikov, P.V. Nickles, W. Sandner, Phys. Rev. E **61**, 4394 (2000).
4. L.M. Chen, J. Zhang, H. Teng, Q.L. Dong, Z.L. Chen, T.J. Liang, L.Z. Zhao, Z.Y. Wei, Phys. Rev. E **63**, 036403 (2001).
5. W. Rozmus, V.T. Tikhonchuk, Phys. Rev. A **42**, 7401 (1990).
6. A.A. Andreev, E.G. Gamaly, V.N. Novikov, A.N. Semakhin and V.T. Tikhonchuk, JETP **74**, 963 (1992).
7. W. Rozmus, V.T. Tikhonchuk, Phys. Rev. A **46**, 7810 (1992).
8. T.-Y. Brian Yang, W.L. Kruer, R.M. More, A.B. Langdon, Phys. Plasmas **2**, 3146 (1995).
9. W. Rozmus, V.T. Tikhonchuk, R. Cauble, Phys. Plasmas **3**, 360 (1996).
10. T.-Y. Brian Yang, W.L. Kruer, A.B. Langdon, T.W. Johnston, Phys. Plasmas **3**, 2702 (1996).
11. V.P. Silin, Sov. Phys. JETP **20**, 1510 (1965).
12. V.P. Silin, Sov. Phys. JETP **87**, 486 (1998).
13. G. Ferrante, M. Zarcone, S.A. Uryupin, J. Opt. Soc. Am. B **17**, 1383 (2000).
14. H. Haberland, M. Bonitz, and D. Kremp, Phys. Rev. E **64**, 026405 (2001).
15. G. Ferrante, M. Zarcone, S.A. Uryupin, Laser Part. Beams **20**, 177 (2002).
16. G. Ferrante, M. Zarcone, S.A. Uryupin, Phys. Rev. E **70**, 016403 (2004).
17. G. Ferrante, M. Zarcone, and S.A. Uryupin, Phys. Plasmas **9**, 4560 (2002).
18. H. Reuter, E. Zondheimer, Proc. Roy. Soc. **155**, 336 (1949).
19. P.B. Corkum, N.H. Burnett, F. Brunel, Phys. Rev. Lett. **62**, 1259 (1989).

20. N.B. Delone, V.P. Krainov, J.Opt. Soc. Am. B **8**, 1207 (1991).
21. W.P. Leemans, C.E. Clayton, W.B. Mori, K.A. Marsh, A. Dyson, C. Joshi, Phys. Rev. Lett. **68**, 321 (1992).
22. J.C. Kieffer, J.P. Matte, H. Pepin, M. Chaker, Y. Beaudoin, T.W. Johnston, C.Y. Chien, S. Coe, G. Mourou, J. Dubau, Phys. Rev. Lett. **68**, 480 (1992).
23. V.P. Silin, JETP **90**, 805 (2000).
24. G. Ferrante, M. Zarcone, S.A. Uryupin, EPJ D **19**, 349 (2002).
25. G. Ferrante, M. Zarcone, S.A. Uryupin, EPJ D **22**, 109 (2003).
26. G. Ferrante, M. Zarcone, S.A. Uryupin, Phys. Rev. Lett. **91**, 085005 (2003).
27. E.M. Epperlein, R.W. Short, Phys. Fluids B **3**, 3092 (1991).
28. A. F Alexandrov, L.S. Bogdankevich and A.A. Rukhadze, "Principles of Plasma Electrodynamics", (Springer-Verlag, Berlin, 1984).
29. E.M. Lifshitz, L.P. Pitaevskii, Physical Kinetics (Pergamon, Oxford, 1981), Chap. 9.

12 Radiative Recombination in a Strong Laser Field

Saverio Bivona, Riccardo Burlon, Gaetano Ferrante, and Claudio Leone

Dipartimento di Fisica e Tecnologie Relative, Universita di Palermo, Viale delle Scienze, Edificio 18, 90128 Palermo, Italy
leone@unipa.it

Summary. Recent advances of radiative recombination in the presence of strong laser fields are reported. The intense laser radiation is found to introduce new relevant features, among which enhancement and control of the emitted X-ray spectra are the most important. The influence of the plasma medium in which the process generally takes place is considered as well. The results of the recent investigations shed new light on the laser assisted radiative recombination physics and give relevant indications concerning the possibilities to have effective slow electrons and to balance the plasma heating, as needed in important applications.

12.1 Introduction

In recent years particular attention has been paid to the generation of high frequency coherent radiation, as it is a topic of great importance in different fields of science and technology. Different processes may be considered for the production of high-energy photons such as: high order harmonic generation, laser induced bremsstrahlung, laser assisted X-ray-atom scattering, and laser assisted radiative recombination (LARR). This chapter is concerned with just the last process. We note that in the context of strong field radiation-matter interaction this process is of special interest, as the LARR may be viewed as a step in the three-step scheme usually invoked in the high-order harmonic generation (HHG). The third step in HHG (that of recombination) provides a mechanism for the generation of attosecond pulses (Drescher et al. 2001 [9]; Paul et al. 2001 [21]), so the study of the LARR may yield a better understanding of the elementary process. The recombination of a free electron with a positive ion followed by the spontaneous emission of a photon has a long story and has been studied extensively in the past few years, because the results of such studies are relevant to the analysis of astrophysical and laboratory plasmas (Hahn 1997 [12]). As the photon emission time is very long as compared to the collision time, the probability of the process is very small and its direct experimental observation difficult. In the experimental investigation great progress has been achieved by using merged electron and ion beams (Quinteros et al. 1995 [23]).

The recent story of radiative recombination (RR) is mainly concerned with the inclusion into the process of the laser radiation to take advantage of the new possibilities offered by the properties of these coherent sources of e.m. radiation. Schematically, advances in RR in the presence of laser radiation have taken place along the following directions: a) RR in the presence of a weak laser field tuned into resonance with an optical transition of the recombining system. Only one photon participates into the elementary process. b) RR in the presence of an moderate intense nonresonant laser field. Several photons participates to the process, but one-photon and few-photon channels are largely dominating. c) RR in the presence of very intense nonresonant laser fields, when a large number of photons may take part into the process, and many different multiphoton channels have comparable probabilities of occurrence. It is intuitive that in this latter case the physics of field-free RR will be strongly modified.

This chapter is basically meant to present a concise overview of just the new physics of RR made possible by the presence of very intense laser radiation fields. Before confining ourselves to the issue related to the strong field RR (direction c)), it is worth mentioning some results concerning the directions a) and b).

Stimulated radiative recombination experiments have been carried out in ion storage ring by merging electrons and ions with a weak laser field tuned into resonance with an optical transition from the continuum state of the free electron to a bound state of the recombined system. In such experiments several groups have observed significant enhancements of the recombination rate for the case when one photon is emitted at the laser frequency (Neumann et al. 1983 [20]; Schramm et al. 1991 [26]; Yousif et al. 1991 [28]; Borneis et al.1994 [8]; Schramm et al. 1996 [27]; Pastuszka et al. 1996 [22]; Asp et al. 1996 [1]; Rogelstad et al. 1997 [24]).

In particular Rogelstad et al. 1997 [24] have observed stimulated recombination in $e^- + H^+$ and $e^- + He^+$ and have found enhancement of the stimulated recombination over the spontaneous one by a gain factor of about three orders of magnitude in the case of final states with the principal quantum numbers $n = 11, 12$ and 13. The gain factor represents the induced recombination rate normalized to the spontaneous one. Similar enhancement has been observed also during the recombination of free electrons with free ions within an ion trap using a pulsed electric field (Wesdorp et al. 2000 [29]). In the stimulated resonant radiative recombination the laser field is weak and only one-photon process occurs. The availability of intense laser fields allows to observe multiphoton processes in radiative recombination and it has prompted investigations i) in the regime of moderate intense fields, when more than one photon is exchanged, but the number of exchanged photons remains still small (Kuchiev and Ostrovsky 2000 [16]; Milosevic and Ehlotzky 2002 [19]); and ii) in the strong field regime, when a large number of photons participates to the process (Leone et al. 2002 [18]; Bivona et al. 2003 [3]; Bivona et al. 2004a [4]; Bivona et al. 2004b [5]; Bivona et al. 2004c [6]; Hu and Collins 2004 [13]).

In the strong field context, the recombination of an electron in a laser-dressed continuum state into an atomic bound state generally results in the emission of photons with energy given, in atomic units, by

$$\omega_X = \varepsilon_q + |\varepsilon_0| + \Delta + n\omega_L \ , \tag{12.1}$$

where ε_0 is the quasienergy of the final state, $\Delta = E_L^2/4\omega_L^2$ the ponderomotive shift, ε_q the electron average translation energy, n the number of exchanged laser photons, and E_L and ω_L, respectively, the electric field amplitude and the photon energy of the laser field.

Kuchiev and Ostrovsky 2000 [16] addressed the multiphoton LARR by employing a Keldysh-type approximation, i.e., neglecting the interaction of the incoming electron with the atomic core in the initial continuum state, while Jaron et al. 2000 [15] working within the same theoretical scheme took approximately into account the interaction between the incoming electron and the atomic ion. Definite results concerning the emitted photon spectrum as a function of the laser intensity and of the incoming electron average translation energy (Kuchiev and Ostrovsky 2000 [16]; Jaron et al. 2000 [15]) are restricted to the regime in which the amplitude of the quiver velocity V_E of the incoming electron is much smaller than the average translation velocity V_q related to the average momentum q. In particular for this few-photon regime, Kuchiev and Ostrovsky [16] have found that the total cross section of electron recombination summed over all the channels of exchanged laser photons exhibits negligible dependence on the laser intensity, and its value is close to that of the spontaneous process. Accordingly, the presence of the laser field does not modify the total yield of emitted photons, but only redistributes the photons in different channels, with energies given by (12.1).

It is worth noting that this "sum rule" is restricted to the case when the energy exchanged between the electrons and the laser field is a very small fraction of the initial translation energy of the incoming electron and clearly does not apply in the strong field regime (corresponding to the situation when the amplitude of the electron quiver velocity V_E is equal to or larger than the translation one V_q). It is worth stressing that in entering a truly multiphoton regime the RR physics undergoes significant changes. In particular, for instance, the electron medium and laser radiation parameters may be chosen in such a way to approximately match the two characteristic velocities V_E and V_q to have effective slow electrons and thus large recombination cross sections (Leone et al.2002 [18]; Bivona et al. 2003 [3]). The treatment concerning the strong field LARR, on which all the results and calculations reported below are based, is outlined in the next section. Below we report also on a number of LARR issues first addressed by the present authors in the strong field context; namely, we report on i) the possibility of the process coherent control; ii) an accurate low-frequency (LF) treatment of the process; and iii) the effects due to the plasma medium.

An interesting possibility to generate effective slow electrons is offered by using two radiation fields and varying their relative phase. Here we report on the main results concerning just the coherent control of multiphoton radiation recombination in the presence of two laser fields, having commensurate frequencies ω_L and $2\omega_L$.

The results reported in Bivona et al. 2005 [7] have shown that the emitted power spectra may be calculated accurately in the LF approximation. The basic physical content of this approximation is that the elementary process takes place instantaneously in the presence of a quasistatic field, and that the influence of the latter on the process is accounted for by means of the appropriate modification of the electron momentum and energy. Since the instant of recombination is not observed, the instantaneous result needs to be averaged over the ensemble of the laser field phases in order to obtain observable quantities. The conceptual simplicity and accuracy of the LF approximation allow to gain a new useful insight on the LARR physics in the region of the laser parameters, in which strong enhancement is found in the emitted radiation power.

In the RR context, it is important to account for the influence of the medium (a plasma), in which the process takes place. This influence here is accounted for by performing the appropriate ensemble averages over the relevant process physical quantities. The electron velocity distribution functions (EDFs) required for the ensemble averages are chosen in such a way to be consistent, within the given model, as much as possible with the plasma and the laser field parameters characterizing the process. In other words, we use laser-modified EDFs, which specific investigations have shown to exist in plasmas subject to the laser action (Ferrante et al. 2001 [11]).

12.2 The Elementary Process in the Presence of a Monochromatic Laser Field

12.2.1 Basic Formulas

To introduce notions and notations needed below, we recall briefly the basics of our model and treatment. Let us consider the recombination of a free electron with an hydrogenic ion with charge Z^+ in the presence of a spatially and temporally homogeneous laser field, taken in dipole approximation and linearly polarized along the z-axis

$$\boldsymbol{E}_L(t) = E_{0L}\hat{\boldsymbol{z}}\cos\omega_L t\,, \tag{12.2}$$

where $\hat{\boldsymbol{z}}$ is a unit vector. The transition amplitude of emitting a high-frequency photon by an electron recombining from a continuum state $\Psi_q^+(\boldsymbol{r},t)$ of average momentum $\boldsymbol{q}$ into a bound state $\Psi_0(\boldsymbol{r},t)$ is equal to

$$T_{fi} = -\mathrm{i}\int_{-\infty}^{+\infty} \mathrm{d}t \left\langle \Psi_0(\boldsymbol{r},t) \left| \boldsymbol{E}_H \cdot \boldsymbol{r} \right| \Psi_q^+(\boldsymbol{r},t) \right\rangle \tag{12.3}$$

with

$$\boldsymbol{E}_H = -\sqrt{\frac{2\pi\omega_X}{V}}\hat{\boldsymbol{\varepsilon}}_X \exp\left[-\mathrm{i}\left(\boldsymbol{k}_X \cdot \boldsymbol{r} - \omega_X t\right)\right] . \tag{12.4}$$

In (12.4) ω_X is the energy of the emitted X-ray photon, $\hat{\boldsymbol{\varepsilon}}_X$ the polarization vector, $\boldsymbol{k}_X = (\omega_X/c)\hat{\boldsymbol{n}}_X$ the momentum, $\hat{\boldsymbol{n}}_X$ the propagation direction, and V the quantization volume of the high-frequency radiation.

In nonrelativistic conditions the wavefunctions $\Psi_q^+(\boldsymbol{r},t)$ and $\Psi_0(\boldsymbol{r},t)$ are exact solutions of the time-dependent Schroedinger equation

$$\mathrm{i}\frac{\partial}{\partial t}\Psi(\boldsymbol{r},t) = \left[-\frac{\nabla^2}{2} - \frac{Z}{r} + \boldsymbol{r}\cdot\boldsymbol{E}_L\right]\Psi(\boldsymbol{r},t) . \tag{12.5}$$

Assuming $E_{0L} < ZE_{at}$ (E_{at} the infratomic electric field of the hydrogen ground state), $\Psi_0(\boldsymbol{r},t)$ is approximated by the unperturbed ground state of the hydrogenic ion with $(Z-1)$ charges

$$\Phi_i(\boldsymbol{r},t) = u_0(r)\exp\left(-\mathrm{i}I_0 t\right) \tag{12.6}$$

$$u_0(r) = \frac{Z^3}{\sqrt{\pi}}\exp\left(-Zr\right) \tag{12.7}$$

and $I_0 = 0.5Z^2$. The continuum state $\Psi_q^+(\boldsymbol{r},t)$ is approximated by the Coulomb–Volkov ansatz:

$$\Psi_q^+(\boldsymbol{r},t) = \chi_q(\boldsymbol{r},t)u_q^+(\boldsymbol{r},t) . \tag{12.8}$$

$\chi_q(\boldsymbol{r},t)$ is the familiar Volkov wavefunction

$$\chi_q(\boldsymbol{r},t) = \exp\left\{\mathrm{i}\left[\boldsymbol{q} + \boldsymbol{k}_L(\omega_L t)\right]\cdot\boldsymbol{r} - \frac{1}{2}\int^t \mathrm{d}t'\left[\boldsymbol{q} + \boldsymbol{k}_L(\omega_L t)\right]^2\right\} \tag{12.9}$$

$\boldsymbol{k}_L(\omega_L t)$ the quiver momentum imparted to the electron by the radiation field

$$\boldsymbol{k}_L(\omega_L t) = \frac{E_{0L}}{\omega_L}\hat{\boldsymbol{z}}\sin\omega_L t \tag{12.10}$$

and

$$u_q^+(\boldsymbol{r}) = \frac{\exp\left(1/2\pi\nu\right)}{(2\pi)^{3/2}}\Gamma(1-\nu)_1F_1\left[-\mathrm{i}\nu, 1, \mathrm{i}(\boldsymbol{q}\cdot\boldsymbol{r} - qr)\right] , \tag{12.11}$$

with $\nu = Z/q$, $\Gamma(x)$ and ${}_1F_1$, respectively, the gamma and the confluent hypergeometric functions. The ansatz (12.8) and slightly modified versions of it (Jain and Tzoar 1978 [14], Basile et al. 1988 [2]) are in use since many years in dealing with elementary atomic and electrodynamics processes in the presence of strong laser fields. Their use has allowed to explain the measurements of angular distribution of photoelectron ionised by a strong low frequency elliptically polarized laser field (Basile et al. 1988 [2]), and to calculate energy spectra of ejected electrons during the ionisation of hydrogen atoms irradiated by extreme-ultraviolet laser pulses in the subfemtosecond regime, which are in good agreement with the ones obtained by using a numerical code based on the integration of the time-dependent Schrödinger equation (Duchateau et al. 2002 [10]).

By using the above approximate wavefunctions for $\Psi_q^+(\boldsymbol{r},t)$ and $\Psi_0(\boldsymbol{r},t)$ and applying the Fermi's golden rule, one can derive the probability of the X-ray photon

$$W = \sum_n W_n \,, \tag{12.12}$$

where

$$W_n = (2\pi)^2 \omega_X \left| \frac{1}{T} \int_0^T \mathrm{d}t M(\omega_L t) \exp\left\{ \mathrm{i}(\omega_X - |I_0|)t - \frac{1}{2}\int^t \mathrm{d}t' \left[\boldsymbol{q} + \boldsymbol{k}_L(\omega_L t')\right]^2 \right\} \right|^2 \times \delta\left[\omega_X - \tilde{\boldsymbol{\omega}}_X(n)\right] \tag{12.13}$$

is the probability for the single event in which n laser photons are exchanged,

$$\tilde{\boldsymbol{\omega}}_X(n) = \varepsilon_q + \frac{k_L^2}{4} + Z^2 |I_0| + n\omega_L \,, \tag{12.14}$$

and

$$M(\omega_L t) = \left\langle u_0(r) \left| \hat{\boldsymbol{\varepsilon}}_X \cdot \boldsymbol{r} \exp\left\{ -\mathrm{i}\frac{\omega_X}{c} \hat{\boldsymbol{n}}_X \cdot \boldsymbol{r} \right\} \right| \exp\left\{ \mathrm{i}\left[\boldsymbol{q} + \boldsymbol{k}_L(\omega_L t)\right]\right\} \right\rangle \,. \tag{12.15}$$

According to the normalization of the coulomb function, (12.11), the density of the free electrons is

$$\mathrm{d}^3 p = p \mathrm{d}\Omega_q \mathrm{d}\varepsilon_q \tag{12.16}$$

while the density of states of the X-ray radiation is given by

$$\frac{\mathrm{d}^3 k_X}{(2\pi)^3} V = \frac{V \omega_X^2}{(2\pi c)^3} \mathrm{d}\omega_X \mathrm{d}\Omega_X \,. \tag{12.17}$$

Proceeding in the usual way, one finds that the differential rate (DR) of the X-ray photon emission in the solid angle $\mathrm{d}\Omega_X$, during the recombination of an electron impinging within the solid angle $\mathrm{d}\Omega_q$

$$R(\varepsilon_q, \Omega_q) = \frac{\partial^3 R}{\partial\varepsilon_q \partial\Omega_X \partial\Omega_q} = \sum_n R_n(\varepsilon_q, \Omega_q) \,. \tag{12.18}$$

In (12.18) $R_n(\varepsilon_q, \Omega_q)$ is the single-channel DR in which the incoming electron exchanges n laser photon; it is equal to

$$R_n(\varepsilon_q, \Omega_q) = \int_{-\infty}^{+\infty} \mathrm{d}\omega_X \frac{\omega_X^3 q}{(2\pi)^4 c^3} \left|T_n(\varepsilon_q, \Omega_q)\right|^2 \delta\left[\omega_X - \tilde{\omega}_X(n)\right] \tag{12.19}$$

with $T_n(\varepsilon_q, \Omega_q)$ the corresponding transition amplitude

$$T_n(\varepsilon_q, \Omega_q) = \frac{1}{2\pi} \int_{-\pi}^{\pi} \mathrm{d}\phi \exp\left\{\mathrm{i}\left[n\phi + \frac{k_L}{\omega_L} \boldsymbol{q} \cdot \hat{\boldsymbol{\varepsilon}}_L \cos\phi + \frac{k_L^2}{8\omega_L} \sin(2\phi)\right]\right\} M(\phi) . \tag{12.20}$$

The quantity $M(\phi)$ is defined in (12.15), $\phi = \omega_L t$ and $\boldsymbol{k}_L(\phi) = k_L \hat{\boldsymbol{z}} \sin\phi$ is the electron quiver momentum at the field phase ϕ and $k_L = E_L/\omega_L$. The rates integrated over the impinging directions of the free electron, for a given n-th channel and summed over the channels, are given respectively by:

$$R_n(\varepsilon_q) = \int R_n(\varepsilon_q, \Omega_q) \mathrm{d}\Omega_q \tag{12.21}$$

and

$$R(\varepsilon_q) = \int R(\varepsilon_q, \Omega_q) \mathrm{d}\Omega_q \tag{12.22}$$

12.2.2 Low Frequency Approximation

In multiphoton regime the results reported in Bivona et al. 2003 [3] have shown that the transition probability of LARR may be calculated very profitably in the adiabatic approximation if the laser frequency is sufficiently small (see below). Within this approximation, the LARR exhibits a large probability in the domain of emitted photon frequencies for which emission is allowed classically, while it is strongly suppressed in the classical forbidden domain.

In the LF domain, where it is valid the adiabatic approximation, a very large number of the processes contributes to the rate, so that a relevant computer time would be necessary to obtain the required convergence of the sum over the rates of all the channels. By resorting to a LF approximation, largely used in the last two decades in dealing with laser assisted atomic processes, it is possible to perform analytically the sum, appearing in the rate (12.18). In other words we assume that the laser photon energy is much lower than the energies involved in the process ($\varepsilon_q/\omega_L \gg 1$ and $\omega_X/\omega_L \gg 1$).

Here it is well suited the procedure adopted in Rosenberg 1979 [25]; therefore we introduce the integral representation of the δ-function in (12.19), so that the DR, (12.18), becomes

$$R(\varepsilon_q, \Omega_q) = \sum_{n=-\infty}^{+\infty} \int_0^{\infty} \mathrm{d}\omega_X \int_{-\pi}^{+\pi} \mathrm{d}\phi \int_{-\pi}^{+\pi} \mathrm{d}\phi' \int_{-\infty}^{+\infty} \mathrm{d}s \left(\frac{\omega_X^3}{(2\pi)^7 c^3} \right) \times$$
$$\times \exp\left\{ \mathrm{i}n(\phi - \phi' - \omega_L s) + \mathrm{i}\frac{k_L}{\omega_L} \boldsymbol{q} \cdot \hat{\boldsymbol{z}}(\cos\phi - \cos\phi') + \right.$$
$$\left. + \mathrm{i}\frac{k_L^2}{8\omega_L^3}(\sin 2\phi - \sin 2\phi') \right\} \times$$
$$\times \exp\left\{ \mathrm{i}\left(\omega_X - \frac{k_L^2}{4} - \varepsilon_q - Z^2 |I_0| \right) s \right\} M(\phi) M^*(\phi) . \quad (12.23)$$

By using the relation

$$\sum_{n=-\infty}^{+\infty} \exp\left[\mathrm{i}n(\phi - \phi' - \omega_L s)\right] = 2\pi\delta(\phi - \phi' - \omega_L s) \quad (12.24)$$

and treating $\omega_L s$ as a first-order quantity, with the magnitude of the parameter s effectively limited by the convergence of the integral in which it appears. The integration over ϕ and s may be carried out, obtaining the following expression for the DR in the LF approximation

$$R_{\mathrm{LF}}(\varepsilon_q, \Omega_q) = \frac{1}{2\pi} \int_{-\pi}^{+\pi} \mathrm{d}\phi \int_0^{\infty} \mathrm{d}\omega_X \frac{\omega_X^3}{(2\pi)^4 c^3} |M(\phi)|^2 \delta\left[\omega_X - \bar{\omega}_X(\phi, \Omega_q)\right] \quad (12.25)$$

with

$$\bar{\omega}_X(\phi, \Omega_q) = \frac{1}{2}(\boldsymbol{q} + k_L \hat{\boldsymbol{z}} \sin\phi)^2 + Z^2 |I_0| . \quad (12.26)$$

After the integration over ω_X in (12.25), the final expression of the DR in LF approximation is

$$R_{\mathrm{LF}}(\varepsilon_q, \Omega_q) = \frac{1}{2\pi} \int_{-\pi}^{+\pi} R_{\mathrm{LF}}(\varepsilon_q, \Omega_q, \phi) \mathrm{d}\phi \quad (12.27)$$

where

$$R_{\mathrm{LF}}(\varepsilon_q, \Omega_q, \phi) = \frac{\bar{\omega}_X^3(\phi, \Omega_q)}{(2\pi)^4 c^3} |M(\phi)|^2 . \quad (12.28)$$

The rate integrated over the impinging directions of the free electron is given by

$$R_{\mathrm{LF}}(\varepsilon_q) = \int R_{\mathrm{LF}}(\varepsilon_q, \Omega_q) \mathrm{d}\Omega_q \quad (12.29)$$

Formulae (12.27) and (12.29) are the LF counterparts of formulae (12.18) and (12.22). If in (12.23) the integration over ϕ' is carried out, we may introduce the DR density, defined as the DR for unit energy of the X-ray photon.

$$\begin{aligned} DR_{\mathrm{LF}}(\varepsilon_q, \Omega_q, \omega_X) &= \frac{\partial}{\partial \omega_X}\left[R\left(\varepsilon_q, \Omega_q\right)\right] = \\ &= \frac{\omega_X^3 q}{(2\pi)^4 c^3} \sum_l \frac{\left|M(\phi_l)\right|^2}{\left|k_l \cos\phi_l \sqrt{2\left(\omega_X - Z^2|I_0|\right) - q^2 \sin^2\vartheta}\right|} \end{aligned} \tag{12.30}$$

where ϑ is the angle between the average translation momentum $\boldsymbol{q}$ of the incoming electron and the z-axis. The index l numbers the real solutions of equation $\bar{\omega}_X(\phi, \Omega_q) = 0$ in the interval $-\pi < \phi < \pi$. The DR density, (12.30), exhibits a divergence when $\cos\phi_l = 0$, so that $\omega_L s$ cannot be treated as a first order quantity, and terms of higher order must be considered to obtain an appropriate expression. For the purposes of the present chapter we do not need to analyse in detail this situation. However, the above consideration allows us to predict that the DR spectrum will exhibit large values when the energy ω_X of the emitted photon is about $(q\cos\vartheta \pm k_L)^2/2 + (q\sin\vartheta)^2/2 + Z^2 I_0$, and it occurs when the electric laser field instantaneous value is close to zero and the instantaneous absolute value of the quiver velocity is close to its maximum.

The formulae (12.25)–(12.30) are the basic results of the LF approximation and suggest the following two step picture of the LARR: in the first step, the free electron propagates towards the ion and its motion is described classically with motion changes mainly due to the laser field; in the second step the free electron recombines with an ion instantaneously at a given value of the laser field phase ϕ, having instantaneous momentum $\boldsymbol{p}(\phi, \Omega_q)$ and energy $\varepsilon(\phi, \Omega_q)$, respectively:

$$\boldsymbol{p}(\phi, \Omega_q) = \boldsymbol{q} + \boldsymbol{k}_L(\varphi) \tag{12.31}$$

and

$$\varepsilon(\phi, \Omega_q) = \frac{1}{2}\left[\boldsymbol{q} + \boldsymbol{k}_L(\varphi)\right]^2 . \tag{12.32}$$

According to the energy conservation, the energy of the emitted photon is given by (12.26), and (12.28) may be interpreted as the instantaneous DR. Since the recombination instant is not observed, $R_{\mathrm{LF}}(\varepsilon_q, \Omega_q, \phi)$ must be averaged over the laser field phase in order to obtain an observable quantity. This average corresponds to the low-frequency DR, (12.27), which is the counterpart of the DR reported in (12.18). According to the above discussion, in the LF regime the discrete picture is given up and the LARR is described classically in terms of a field picture. As a consequence, in the LF regime the emitted radiation spectrum is a continuum and the DR density depends on the laser parameters only through the amplitude k_L of the quiver momentum imparted to the incoming electron by the laser.

Wishing to compare in detail the predictions of the LF treatment vs those of the treatment in which the discrete photon picture is retained, we have to discretize the DR density. The most natural procedure consists in integrating the formula (12.30) from $\bar{\omega}_X(n) - \omega_L/2$ to $\bar{\omega}_X(n) + \omega_L/2$. The quantities

$$R_{\mathrm{LF},n}(\varepsilon_q, \Omega_q) = \int_{\bar{\omega}_X(n)-\omega_L/2}^{\bar{\omega}_X(n)+\omega_L/2} DR_{\mathrm{LF}}(\varepsilon_q, \Omega_q, \omega_X)\mathrm{d}\omega_X \tag{12.33}$$

$$R_{\mathrm{LF},n}(\varepsilon_q) = \int \mathrm{d}\Omega_q R_{\mathrm{LF},n}(\varepsilon_q, \Omega_q) \tag{12.34}$$

may then be considered respectively as the counterparts in the LF approximation of the (12.19) and (12.21).

By concluding this section we observe that, according to (12.26), large values of the DR, (12.19), are expected in the range of emitted photon energy where the (12.26) gives real values of $\bar{\omega}_X(\vartheta, \phi)$, that is when the emission is classically allowed. Below, in considering various LARR issues, we will take advantage of the LF approximation simplicity.

12.2.3 Calculations

In Fig. 12.1 we report the LARR rate integrated over all the electron incidence directions and summed over all the channels $R(\varepsilon_q)$, (12.22), normalized to the field free rate, as a function of the average translation momentum q. The laser frequency is $\omega_L = 0.043$ a.u., the $\hat{\boldsymbol{\varepsilon}}_X$ and $\hat{\boldsymbol{\varepsilon}}_L$ polarization vectors are parallel to the z-axis, the laser intensity I_L and the corresponding amplitude of the quiver velocity V_E are: $I_L = 10^{14}\,\mathrm{W/cm^2}$ and $V_E = 1.24$ a.u. (continuous line), $I_L = 3 \times 10^{14}\,\mathrm{W/cm^2}$ and $V_E = 2.1$ a.u (dotted line), $I_L = 6 \times 10^{14}\,\mathrm{W/cm^2}$ and $V_E = 3$ a.u (dashed line). The reported curves show that the rate exhibits an enhancement when the amplitude of the quiver velocity V_E matches the average translation velocity V_q.

In Fig. 12.2 we report the LARR rate integrated over all the electron incidence directions and summed over all the channels $R(\varepsilon_q)$, (12.22), and the low frequency counterpart, (12.29), as a function of the average translation momentum q. For comparison we show also the field free rate. The parameters are: laser frequency $\omega_L = 0.043$ a.u., laser intensity $I_L = 6 \times 10^{14}\,\mathrm{W/cm^2}$, and the $\hat{\boldsymbol{\varepsilon}}_X$ and $\hat{\boldsymbol{\varepsilon}}_L$ polarization vectors parallel to the z-axis. The rates calculated by (12.29) are in excellent agreement with those obtained with (12.22). We note that calculations with formula (12.29) are performed without the need of large computer time, which is required instead when (12.22) is used.

In Fig. 12.3 we report the DR as a function of the angle ϑ between the average translation momentum $\boldsymbol{q}$ of the incoming electron and the z-axis, and

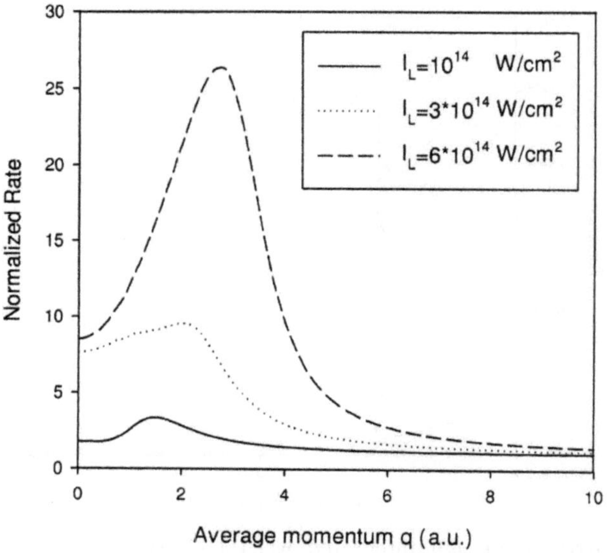

Fig. 12.1. The laser-assisted radiative recombination (LARR) rate, in a.u., integrated over all the electron incidence directions and summed over all the channels $R(\varepsilon_q)$, (12.22), normalized to the field free rate, as a function of the average translation momentum q for different values of the laser intensity I_L. Laser frequency $\omega_L = 0.043$ a.u.; ion charge $Z = 1$; $\hat{\varepsilon}_X$ and $\hat{\varepsilon}_L$ polarization vectors parallel to the z-axis

of the angle ϑ_L between $\hat{\varepsilon}_L$ and the z-axis for different values of the laser intensity, by assuming a laser linearly polarized and both the polarization vector $\hat{\varepsilon}_L$ and the average translation momentum $\boldsymbol{q}$ lying in the plane (z, x). As the simultaneous inversion of the vectors $\boldsymbol{q}$, $\hat{\varepsilon}_X$ and $\hat{\varepsilon}_L$ is a symmetry operation, at a given value of ϑ_L the DR as a function of ϑ shows a twofold symmetry, i.e., when $\vartheta \longrightarrow \vartheta + \pi$ the same value of the DR is obtained. When ϑ_L is equal to zero or $\pi/2$, the operation of the reflection of the above specified vectors about the x or the z-axis is an additional symmetry operation which manifests in fourfold symmetry of the DR as function of ϑ, at a given value of ϑ_L. We note that the asymmetries of the DR are reduced when the laser intensity decreases, as the influence of the laser field is reduced and the DR approaches the field-free value. Besides, the DR values decrease by increasing ϑ_L from 0 to $\pi/2$ for any value of the laser intensity, as the multiphoton radiative recombination is more favoured when the polarization vectors of the laser field and the emitted high frequency radiation are parallel.

The results reported in Fig. 12.1–12.3 suggest that the LARR is favoured when $\hat{\varepsilon}_X$ and $\hat{\varepsilon}_L$ are parallel and the average translation momentum is of the same order of magnitude as or lower than the quiver momentum amplitude k_L.

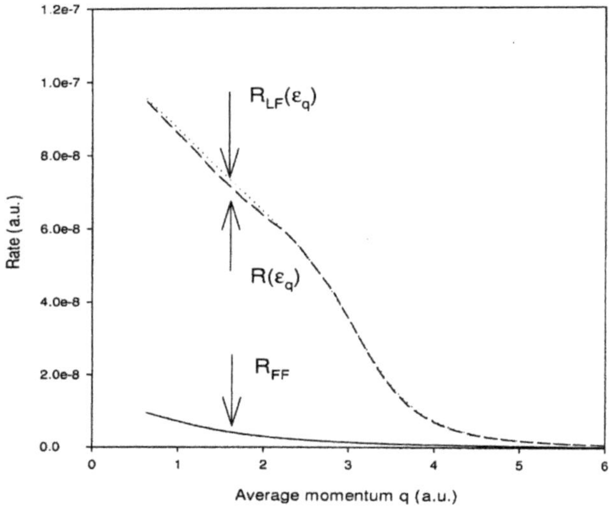

Fig. 12.2. LARR rate in a.u. integrated over the electron incoming directions and summed over all the channels: $R(\varepsilon_q)$ (*dashed line*), (12.22), and the low frequency counterpart $R_{\rm LF}(\epsilon_q)$ (*dotted line*), (12.29), as a function of the average translation momentum $\boldsymbol{q}$. Ion charge $Z = 1$, laser frequency $\omega_L = 0.043$ a.u., laser intensity $I_L = 6\times10^{14}\,\mathrm{W/cm^2}$. The *solid line* corresponds to the field free rate

In Fig. 12.4 we report $R_n(\varepsilon_q, \Omega_q)$, (12.19), and the counterpart in the LF approximation $R_{\mathrm{LF},n}(\varepsilon_q, \Omega_q)$, (12.33), as a function of the emitted photon energy ω_X, assuming the vectors $\hat{\boldsymbol{\varepsilon}}_X$, $\hat{\boldsymbol{\varepsilon}}_L$ and $\boldsymbol{q}$ all parallel to the z-axis ($\vartheta = 0$). The values of the laser parameters are the same as in Fig. (12.2), and the electron energy ε_q is chosen equal to 2 a.u. in order to have $q < k_L$ ($q = 2$ a.u. and $k_L = 3.04$ a.u.). According to the LF predictions, both spectra are practically confined in a range of ω_X where the emission is classically allowed. By inspecting the figure, we remark that this range may be divided in two regions: the first of them defined by $-I_0 < \omega_X < \omega_1$ and the second by $\omega_1 < \omega_X < \omega_2$ with $\omega_1 = \omega_{ff} + k_L(k_L - 2q)/2$; $\omega_2 = \omega_{ff} + k_L(k_L + 2q)/2$ and ω_{ff} the emitted photon energy when the recombination occurs in the absence of laser. With the choice of the above parameters, $\omega_1 = 1.04$ a.u.; $\omega_2 = 13.2$ a.u. and $\omega_{ff} = 2.5$ a.u. if $I_0 = -0.5$ a.u.

Within the first region, both spectra exhibit larger values (as compared to the second region); this occurs as the electron instantaneous momentum may be either parallel or antiparallel to the average translation momentum and consequently the electron recombines with the ion having an instantaneous momentum close to zero for a large interval of the laser phases. The broadness of this region in which the rate enhancement reaches its maxima values contracts to zero when $q > k_L$. The calculations show that the DR exhibits the maxima at the spectrum edges, as ω_1 and ω_2 are the values of $\bar{\omega}_X(\vartheta, \phi)$ at $\phi = 0$, where the LARR is enhanced as remarked in the previous section. Finally,

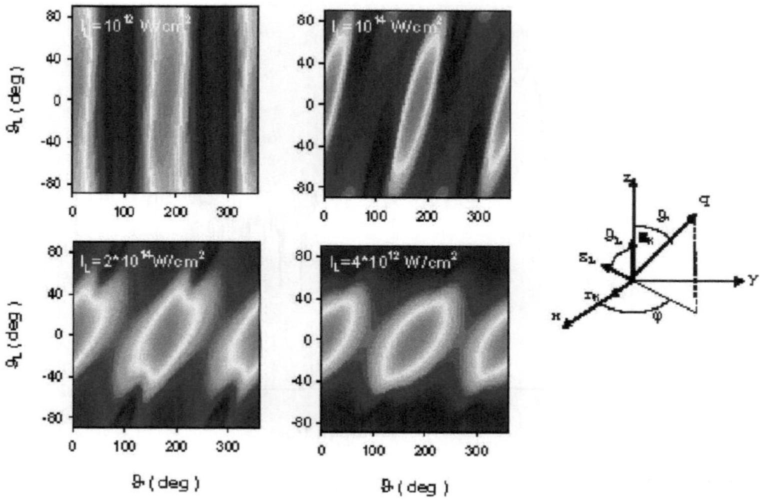

Fig. 12.3. Contour plot of the differential rate (DR) as a function of the angle ϑ between the average translation momentum $\boldsymbol{q}$ of the incoming electron and the z-axis, and of the angle ϑ_L between $\hat{\varepsilon}_L$ and the z-axis for different values of the laser intensity, by assuming a laser linearly polarized and both the polarization vector $\hat{\varepsilon}_L$ and the average translation momentum $\boldsymbol{q}$ lying in the plane $(z,\, x)$. Ion charge $Z = 1$ and laser frequency $\omega_L = 0.043$ a.u.

we observe that the spectrum calculated by $R_{\mathrm{LF},n}(\varepsilon_q, \Omega_q)$ does not reproduce the oscillations appearing in the spectrum evaluated by $R_n(\varepsilon_q, \Omega_q)$. It occurs as in the LF approximation the recombination takes place instantaneously and the transition amplitudes at different values of the laser phase, giving emitted photons with the same energy, are summed incoherently (see (12.30)). The presence of the oscillations in the spectrum calculated by $R_n(\varepsilon_q, \Omega_q)$ is discussed in Bivona et al. 2003 [3]. Here we remark only that the oscillations are due to quantum mechanical interferences, as the main contributions to the transition amplitude $T_n(\varepsilon_q, \Omega_q)$ appearing in (12.19) are summed coherently.

In Fig. 12.5 we report the DR spectrum, assuming the incoming electron direction perpendicular to the polarization vectors $\hat{\varepsilon}_X$, $\hat{\varepsilon}_L(\vartheta = \pi/2)$. The values of ε_q, ω_L and I_L are equal to the ones chosen in Fig. 12.4. On the basis of the LF picture, the spectrum is confined within the interval $\omega_{ff} < \omega_X < \omega_{ff} + \frac{k_L^2}{2} = 7.12$ a.u., where the emission is classically allowed. Here the spectrum exhibits a maximum at the upper edge, as the X-ray photon is emitted at the values of the laser phase giving $\cos\phi = 0$, which optimises the LARR emission. However, more interesting is that, within the interval $\omega_{ff} < \omega_X < \omega_3 = \omega_{ff} + \frac{k_L^2}{2}$, the values of the DR rate are larger than the ones reported in Fig. 12.4. In the inset we report the ratio between the values of the differential rates, calculated within the LF approximation, for $\vartheta = \pi/2$ and $\vartheta = 0$.

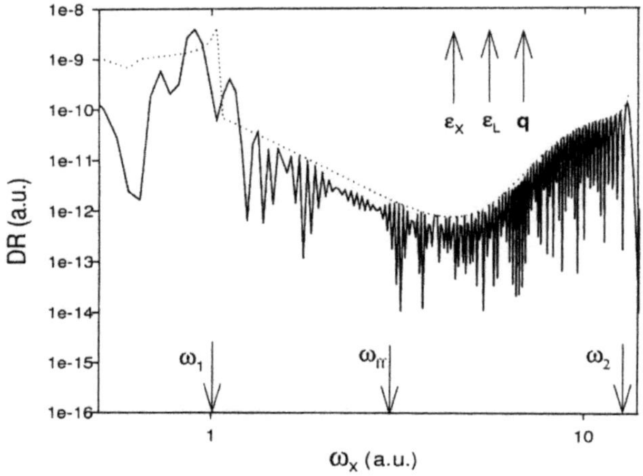

Fig. 12.4. Differential rate (DR) $R_n(\varepsilon_q, \Omega_q)$ (*continuous line*), (12.19), and the counterpart in the LF approximation $R_{\mathrm{LF},n}(\varepsilon_q, \Omega_q)$ (*dotted line*), (12.33), as a function of the emitted photon energy ω_X. The vectors $\hat{\varepsilon}_X$, $\hat{\varepsilon}_L$ and $\boldsymbol{q}$ are all parallel and along the z-axis ($\vartheta = 0$). The values of the ion charge, the laser parameters and the electron energy are the same as in Fig. 12.2. The arrows indicate the emitted photon energies which mark off the two regions in which the spectrum may be divided according to the LF approximation: $\omega_1 = \omega_{ff} + k_L(k_L - 2q)/2 = 1.04$ a.u. and $\omega_2 = \omega_{ff} + k_L(k_L + 2q)/2 = 13.2$ a.u; $\omega_{ff} = 2.5$ a.u. is the emitted photon energy when the laser field is off

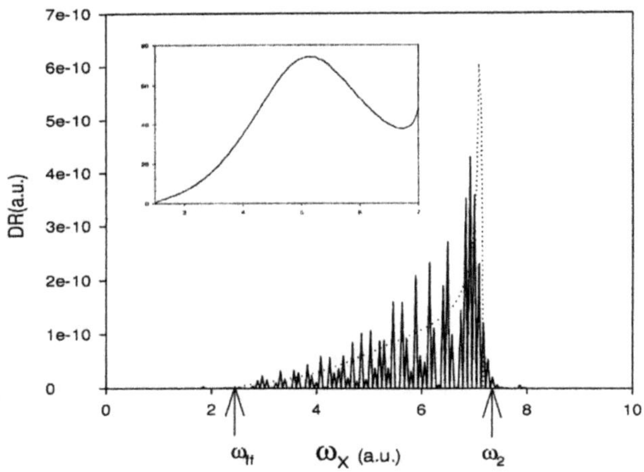

Fig. 12.5. As in Fig. (12.4) by assuming the incoming electron direction perpendicular to the polarization vectors $\hat{\varepsilon}_X$, $\hat{\varepsilon}_L$ ($\vartheta = \pi/2$). The *arrows* indicate the emitted photon energies which mark off the spectrum according to the LF approximation. In the *inset* we report the ratio between the values of the differential rates calculated within the LF approximation, for $\vartheta = \pi/2$ and $\vartheta = 0$

12.3 The Elementary Process in the Presence of a Bichromatic Laser Field

The new features of the radiative recombination in the presence of a strong monochromatic laser field may be further enriched by varying the relative phase δ of two radiation fields, having commensurate frequencies ω_L and $2\omega_L$ (Bivona et al. 2004b [5] and 2004c [6]):

$$E_L = [E_{1L} \cos(\omega_L t) + E_{2L} \cos(2\omega_L t + \delta)]\, \hat{z} \,. \tag{12.35}$$

Here we remember that the bichromatic radiation (as the monochromatic one) field induces virtual states in the continuum, so that the radiative recombination may be described, loosely speaking, as a process in which the incoming electron with energy ε_q recombines in a bound state from a laser embedded continuum state $|\varepsilon_q, n\rangle$, in which $n\omega_L$ is the energy that the incoming electron exchange with the laser radiation. As shown in Fig. 12.6, owing to the commensurability of the frequencies of the two radiation fields, the incoming electron with energy ε_q can follow different pathways to reach the same dressed state of the continuum $|\varepsilon_q, n\rangle$. In each of these routes the electron may exchange different numbers n_1 and n_2 of energy quanta ω_L and $2\omega_L$. The resulting LARR transition amplitudes, in which the incoming electron recombines in a bound state emitting a single photon of energy ω_X and exchanging the energy $n\omega_L$ with the two laser fields, turns out to be a coherent sum of single-channel transition amplitudes T_{n_1,n_2} pertaining to all the possible pathways in which the numbers n_1 and n_2 combine to give $n_1 + 2n_2 = n$. By varying the relative phase δ of the two fields, the phase

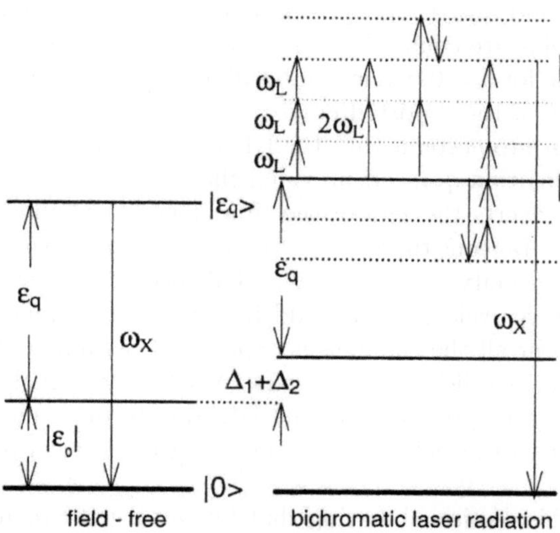

Fig. 12.6. Scheme of radiative recombination in the presence of a bichromatic laser radiation

factors relative to each single-channel transition amplitude T_{n_1,n_2} change and the DR for a given channel, in which the electron exchange the amount of energy $n\omega_L$ with the two laser fields, is modified. Consequently, the DR summed over the channels is expected to be strongly affected.

When the LARR takes place in the presence of a monochromatic field, the probability amplitude, (12.20), describing the process is an even function of the average translation momentum $\boldsymbol{q}$ of the incoming electron if an odd number of laser photons is exchanged, provided that the bound state has an even parity. Instead, the probability amplitude is an odd function of $\boldsymbol{q}$ when the number of photons is even. Hence, the DR turns out to be invariant under the translation $\boldsymbol{q} \rightarrow -\boldsymbol{q}$. When a second field of double frequency $2\omega_L$ is added, the LARR amplitude will be a coherent sum of partial transition amplitudes and each of them is an odd or an even function of $\boldsymbol{q}$. Therefore, the DR may turn out to be not invariant under the transformation $\boldsymbol{q} \rightarrow -\boldsymbol{q}$.

Here we report on selected calculations that show the effect of the relative phase δ on the LARR in the highly nonlinear regime, in which the incoming electron may exchange a large amount of energy $n\omega_L$ with the bichromatic radiation. In all the calculations the polarization vector of the high frequency photon $\hat{\boldsymbol{\varepsilon}}_X$ is directed along the z-axis, its momentum $\boldsymbol{k}_X$ along the x-axis, $Z = 1$, and both the amplitudes of the bichromatic radiation are chosen at the same value ($E_{L1} = E_{L2} = E_{0L}$).

Some effects of the relative phase δ on the LARR are shown in Fig. 12.7, where we report the DR summed over the channels as a function of the average momentum q, divided by field-free DR, assuming the incoming electron momentum $\boldsymbol{q}$ parallel or antiparallel to, for two different values of the relative phase ($\delta = 0$ and $\delta = \pi/2$). $\hat{\boldsymbol{\varepsilon}}_L$ is parallel to $\hat{\boldsymbol{\varepsilon}}_X$ and the $\hat{\boldsymbol{\varepsilon}}_L$ intensities of the two laser fields are equal. The curves corresponding to the two above geometries coincide for $\delta = 0$, while exhibit different shapes for $\delta = \pi/2$. The different behaviour of the reported curves by changing the relative phase δ is explained taking into account that the DR for a given channel is invariant under the transformation $\boldsymbol{q} \rightarrow -\boldsymbol{q}$ provided that it is possible to find a time translation which inverts the sign of the time dependent quiver velocity. In fact, under the simultaneous time translation and the transformation $\boldsymbol{q} \rightarrow -\boldsymbol{q}$ the instantaneous velocity inverts its sign. This occurs only when $\delta = 0$, as in this case the quiver velocity is an odd function of the time t. Therefore, the DR summed over all the channels is expected to assume different values for $\boldsymbol{q}$ parallel or antiparallel to except when $\delta = 0$. Accordingly we may assert that the symmetry properties of the DR, due to quantum interference effects, are related to the behaviour of the electron classical instantaneous velocity.

We conclude this section observing that by varying the relative phase of the two radiation fields it is possible to control enhancement, broadening and symmetry properties of the DR summed over the channels.

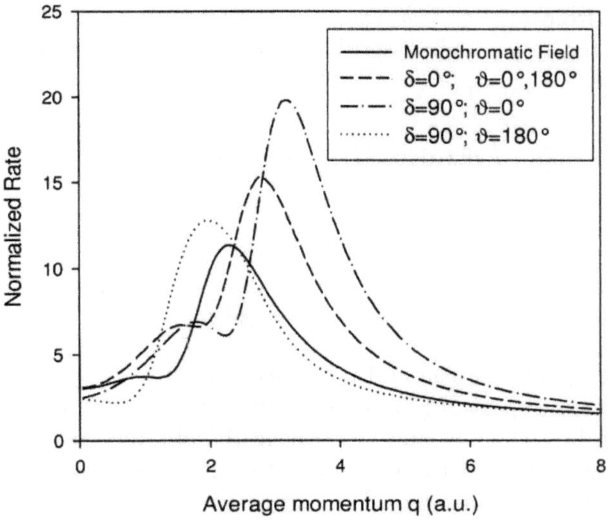

Fig. 12.7. Differential rate (DR) summed over the channels as a function of the average momentum q, normalized to the field-free one, assuming the incoming electron momentum $\boldsymbol{q}$ parallel ($\vartheta = 0$) or antiparallel to $\hat{\varepsilon}_L$, for two different values of the relative phase ($\delta = 0$ and $\delta = \pi/2$). $\hat{\varepsilon}_L$ is parallel to $\hat{\varepsilon}_X$ and the intensities of the two laser fields are equal, $I_{L1} = I_{L2} = 2 \times 10^{14}\,\mathrm{W/cm^2}$. The frequencies of the bichromatic laser radiation are $\omega_L = 0.018$ a.u. and $2\omega_L$, and the electron average translation energy is $\varepsilon_q = 6$ a.u. The DR for a monochromatic field is calculated at the frequency ω_L and the same intensity of the two fields

Unreported calculations show that by decreasing the laser intensity, the polar asymmetry reduces for any value of δ until it practically vanishes in the weak field regime. This results by the fact that, when the laser intensity decreases, the quiver velocity decreases and becomes a small fraction of the average translation velocity of the incoming electron. As the instantaneous velocities of the electrons incoming in the forward and backwards directions are almost equal, so are the recombination rates. Moreover, in weak field regime, the DR becomes almost equal to the field-free one.

12.4 Influence of the Plasma

Account of the plasma influence is taken by averaging the LARR physical quantities of interest over the translation velocity or equivalently over the free electron energy distributions. To this end, an appropriate EDF $f(\boldsymbol{V})$ is required. Additionally an average over the field phase $\phi = \omega_L t$ is required as well, because $\boldsymbol{V} = \boldsymbol{q} + \hat{\boldsymbol{z}} E_{0L}/\omega_L \sin\phi$.

Starting from (12.18), the rate per steradian and unit frequency of the emitted photon in a plasma is given by Bivona et al. 2004 [4–6]:

$$R(\omega_L, \Omega_X) = N_i N_e J(\omega_X) = \\ = \sum_n \int \mathrm{d}\Omega_q \mathrm{d}\varepsilon_q \mathrm{d}\phi f(\varepsilon_q, \phi) \frac{\omega_X^3 q}{(2\pi)^5 c^3} |T_n(\varepsilon_q, \Omega_q)|^2 \delta(\varepsilon_q - \bar{\varepsilon}_q(n)) \tag{12.36}$$

with

$$\bar{\varepsilon}_q(n) = \omega_X + \frac{k_L^2}{4} + Z^2|I_0| + n\omega_L \tag{12.37}$$

and N_e and N_i the electron and ion concentrations.

The LF counterpart of (12.34) is obtained by multiplying the DRLF density, (12.30), by the EDF, and integrating the obtained result over the solid angle and the energy of the incoming electron, and averaging the obtained result over the field phase ϕ:

$$R_{\mathrm{LF}}(\omega_X, \Omega_X) = N_i N_e J_L(\omega_X) \\ = \frac{N_i N_e}{2\pi} \int \mathrm{d}\Omega_q \mathrm{d}\varepsilon_q \mathrm{d}\phi f(\varepsilon_q) DR_{\mathrm{LF}}(\varepsilon_q, \Omega_q, \omega_X) \tag{12.38}$$

Before proceeding to calculations, some remarks are in order. In a plasma interacting with a strong laser field, thanks to collisions and various plasma processes, the electrons not only get heated but their velocity distribution may as well undergo shape modifications as compared to the initial distribution. Besides, depending on the physical mechanism yielding the ionised medium formation, the initial EDF may well be very different as compared to the familiar Maxwellian. In other words, in wishing to take into account the plasma influence on the LARR process, we are faced with the additional task of knowing the laser modified EDF of the recombining electrons and its time evolution for the given set of laser-plasma interaction parameters. Thus, rigorously speaking, we need to solve the pertinent kinetic equation for the EDF containing all the processes able to shape it, including the radiative recombination. Here we will follow instead a much simpler procedure that is expected to hold well provided that judicious advantage is taken of the existing literature on the subject. As a matter of fact, in the last two decades many investigations have been carried out to understand which EDFs are established in plasmas interacting with strong laser fields, and how they evolve with time (Ferrante et al. 2001 [11]). For sake of simplicity, here we confine ourselves to consider only the situation of a very strong field when the electron–electron collisions dominate over the electron–ion ones and, as a result the EDF is approximately Maxwellian:

$$f(\varepsilon_q) = \frac{1}{(2\pi T)^{3/2}} \exp\left(-\frac{\varepsilon_q}{T}\right) \tag{12.39}$$

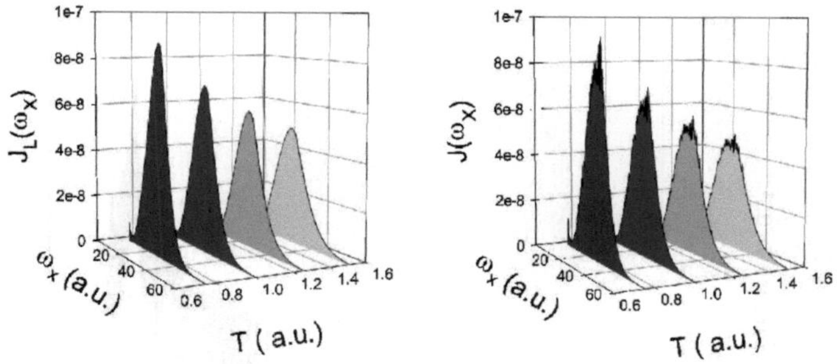

Fig. 12.8. Normalized rates per steradian and frequency units of the emitted photon in a plasma $J(\omega_X)$, and $J_L(\omega_X)$, in a.u., as a function of ω_X and for different values of the plasma temperature T. Ion charge $Z = 3$, laser frequency and intensity, respectively, $\omega_L = 0.057$ a.u. and $I_L = 6 \times 10^{15}$ W/cm^2

In Fig. 12.8 we report the normalized rates $J(\omega_X)$ and $J_L(\omega_X)$ as a function of ω_X and of the plasma temperature by assuming $Z = 3$, the laser frequency and intensity, respectively, $\omega_L = 0.057$ a.u. and $I_L = 6 \times 10^{15}$ W/cm^2. The LF formula ((12.38)) reproduces the patterns of the $J(\omega_X)$ spectra for any value of the plasma temperature, but as before it does not give the oscillations appearing in $J(\omega_X)$.

The reported spectra, calculated in the LF approximation, are characterized by the presence of a large maximum at the same value of $\omega_X = \omega_{ff} + k_L^2/2 = 24,66$ a.u., independently of the plasma temperature. These maxima are also present in the oscillating spectra predicted by (12.37). As expected, by increasing the temperature, the maximum value of the spectrum decreases.

The features of the reported spectra may be explained by considering that the EDF exhibits its maximum at $\varepsilon_q = 0$, and after the maximum a decreasing behaviour with decreasing slope by increasing the plasma temperature. So the main features of the spectra reported in Fig. 12.9 are determined by the recombination of very slow electrons. In Fig. 12.9 we report the integrated rate as a function of the emitted photon energy calculated by (12.19), choosing $Z = 3$, $\varepsilon_q = 0.1$ a.u and the same laser parameters as in Fig. 12.8. The reported spectrum exhibits a maximum at $\omega_X = \omega_{ff} + k_L^2/2$ and this explains the presence of a large maximum in the curves of Fig. 12.8 at same ω_X value independently of the plasma temperature. However, the plasma temperature controls the maximum height and the slope of the spectrum after the maximum. In fact, by increasing the temperature, the height of the maximum decreases as the number of very slow electrons decreases, while the slope after the maximum decreases as the number of more energetic electrons increases. For a more complete discussion on the issue of this section see Bivona et al. 2004a [4].

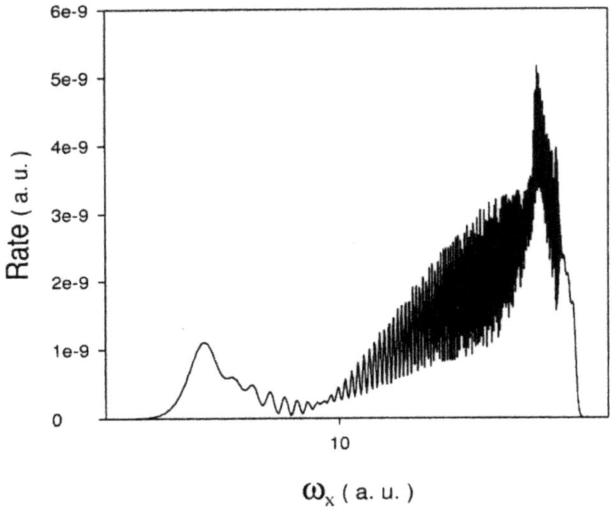

Fig. 12.9. Recombination rates, in a.u., integrated over all the electron incoming directions as a function of the emitted photon energy ω_X with $Z = 3$, by choosing the same laser parameters as in Fig. 12.8 and $\varepsilon_q = 0.1$ a.u.

12.5 Concluding Remarks

We have reported on the influence of a strong laser field on the outcome of the radiative electron–ion recombination. Such presence is found to introduce new relevant features, which may be explained in a simple way resorting to the LF approximation. Within this approximation the radiative recombination in the presence of an intense laser field is thought of as a two step process: in the first step, the free electron propagates towards the ion and its motion may be described classically and motion changes are mainly due to the laser field; in the second step, the free electron recombines with an ion instantaneously at a given value of the laser field phase ϕ. Since the instant of recombination is not observed, the instantaneous result must be averaged over the laser field phase in order to obtain observable quantities.

The main results in the presence of a monochromatic laser field are summarized as follows: The LARR is favoured when $\hat{\varepsilon}_X$ and $\hat{\varepsilon}_L$ are parallel and the average translation momentum is of the same order magnitude as or lower than the quiver momentum amplitude k_L (see Figs. 12.1–12.3).

The DR spectrum exhibits large values in the range of X-ray photon energy where the emission is classically allowed and the DR maxima occur when the electric laser field instantaneous value is close to zero and the instantaneous absolute value of the quiver velocity is close to its maximum, as expected according to (12.30) (see Figs. 12.4–12.5).

The presence of a bichromatic laser field introduces new relevant features in the LARR process. The most interesting is the possibility to control the enhancement and the broadening of the emitted X-ray spectra (Fig. 12.7).

As radiative recombination in plasmas is a prospective candidate as a source of coherent radiation in the high-frequency domain, we have reported also on results when the influence of the medium (a plasma) is taken into account. The spectra are characterized by the presence of a large maximum at the same value of $\omega_X = \omega_{ff} + k_L^2/2 = 24,66$ a.u., independently of the plasma temperature. Besides, these maxima decrease with the increase of the temperature. These features of the spectra, reported in Fig. 12.8, are explained in a simple way, by resorting to the conceptual simplicity of the LF approximation.

Acknowledgments

This work was supported in part by the Italian Ministry of University and Scientific Researches and by the European Community under Contract no. HPRN-CT-2000-00156

References

1. S. Asp, R. Schuch, D.R. DeWitt, C. Biedermann, H. Gao, W. Zong, G. Andler and E. Justiniano: Nucl. Instrum. Methods: Phys. Res. B **117**, 31–37 (1996),"Laser-induced recombination of D^+"
2. S. Basile, F. Trombetta and G. Ferrante: Phys. Rev. Lett. **61**, 2435–2437(1988),"Twofold Symmetric Angular Distributions in Multiphoton Ionization with Elliptically Polarized Light"
3. S. Bivona, R. Burlon, G. Ferrante and C. Leone: Laser Physics **13**, 1077–1082 (2003),"Strong field effects of multiphoton radiative recombination"
4. S. Bivona, R. Burlon, G. Ferrante and C. Leone: Laser Phys. Letters **1**, 86–92 (2004a),"Influence of a plasma medium on laser assisted radiative recombination"
5. S. Bivona, R. Burlon, G. Ferrante and C. Leone: Laser Phys. Letters **1**, 118–123 (2004b),"Control of multiphoton radiative recombination through the action of two-frequency fields"
6. S. Bivona, R. Burlon, G. Ferrante and C. Leone: Appl. Phys. B **78**, 809–812 (2004c),"Control of radiative recombination by a strong laser field"
7. S. Bivona, R. Burlon, G. Ferrante and C. Leone: to be published in IOSA , (2005),"Radiative Recombination in a Strong Laser Field. Low Frequency Approximation"
8. S. Borneis, F. Bosch, T. Engel, M. Jung, I. Klaft, O. Klepper, T. Kuhl, D. Marx, R. Moshammer, R. Neumann, S. Schroder, P. Seelig and Volker L.: Phys. Rev. Lett. **72**, 207–209 (1994),"Laser-stimulated two-step recombination of highly charged ions and electrons in a storage ring"
9. M. Drescher, M. Hentschel, R. Kienberger, G. Tempea, C. Spielmann, G.A. Reider, P.B. Corkum and F. Krausz: Science **291**, 1923–1927 (2001),"X-ray pulses approaching the attosecond frontier"
10. G. Duchateau, E. Cornier and R. Gayet: Phys. Rev. A **66**, 023412 (2002),"Coulomb–Volkov approach of ionization by extreme-ultraviolet laser pulses in the subfemtosecond regime"

11. G. Ferrante, M. Zarcone and S.A. Uryupin: Plasma Sources Sci Technol. **10**, 318–328 (2001),"Electron distribution functions in laser fields"
12. Y. Hahn: Rep. Prog. Phys. **60**, 691–759 (1997),"Electron–ion recombination process–an overview"
13. S.X. Hu and L.A. Collins: Phys. Rev. A **70**, 013407 (2004),"Phase control of the inverse above-threshold-ionization processes with few-cycle pulses"
14. J. Manoj and T. Narkis : Phys. Rev A **18**, 538–545 (1978),"Compton scattering in the presence of coherent electromagnetic radiation"
15. A. Jaron, J.Z. Kaminski, and Ehlotzky F. : Phys. Rev. A **61**, 023404 (2000),"Stimulated radiative recombination and X-ray generation"
16. Y.M. Kuchiev and V.N. Ostrovsky: Phys. Rev. A **61**, 033414 (2000),"Multiphoton radiative recombination of electron assisted by a laser field"
17. C. Leone, S. Bivona, R. Burlon and G. Ferrante: Phys. Rev A **38**, 5642–5651 (1988),"Two-frequency multiphoton ionization of hydrogen atoms"
18. C. Leone, S. Bivona, R. Burlon and G. Ferrante: Phys. Rev. A **66**, 051403(R) (2002),"Strong-field and plasma aspects of multiphoton radiative recombination"
19. D.B. Milosevic and F. Ehlotzky: Phys: Rev. A **65**, 042504 (2002),"Rescattering effects in soft-X-ray generation by laser-assisted electron–ion recombination"
20. R. Neumann, H. Poth, A. Winnacker and A. Wolf: Z. Physik. A **313**, 253–262 (1983),"Laser-enhanced electron–ion capture and antihydrogen formation"
21. P.M. Paul, E.S. Toma, P. Breger, G. Mullot, F. Auge, P. Balcou, H.G. Muller and P. Agostini: Science **292**, 1689–1692 (2001),"Observation of a train of attosecond pulses from high harmonic generation"
22. S. Pastuszka, U. Schramm, M. Grieser, C. Broude, R. Grimm, D. Habs, J. Kenntner, H.J. Miesner, T. Schussler, D. Schwalm and A. Wolf: Nucl. Instrum. Methods Phys. Res. A **369**, 11–22 (1996),"Electron cooling and recombination experiments with an adiabatically expanded electron beam"
23. T. Quinteros, H. Gao, D.R. DeWitt, R. Schuch, M. Pajek, S. Asp, and Dz. Belkic: Phys. Rev. A **51**, 1340–1346 (1995),"Recombination of D^+ and He^+ ions with low-energy free electrons"
24. M.L. Rogelstad, F.B. Yousif, T.J. Morgan and J.B.A. Mitchell: J. Phy. B **30**, 3913–3931 (1997),"Stimulated radiative recombination of H^+ and He^+"
25. L. Rosenberg: Phys. Rev. A **20**, 1352–1358 (1979),"Sum rule and classical limit for scattering in a low-frequency laser field"
26. U. Schramm, J. Berger, M. Grieser, D. Habs, E. Jaeschke, G. Kilgus, D. Schwalm, A. Wolf, R. Neumann and R. Schuch: Phys. Rev. Lett. **67**, 22–25 (1991),"Observation of laser-induced recombination in merged electron and proton-beams"
27. U. Schramm, T. Schussler, D. Habs, D. Schwalm and A. Wolf: Hyperfine Interact. **99**, 309–316 (1996),"Laser-induced recombination studies with the adiabatically expanded electron beam of the Heidelberg TSR"
28. F.B. Yousif, P. Vanderdonk, Z. Kucherovsky, J. Reis, E. Brannen, J.B.A. Mitchell and T.J. Morgan: Phys. Rev. Lett. **67**, 26–29 (1991),"Experimentalobservation of laser-stimulated radiative recombination"
29. C. Wesdorp, F. Robicheaux and L.D. Noordam: Phys. Rev. Lett. **84**, 3799–3802 (2000),"Field-induced electron–ion recombination: A novel route towards neutral (anti-) matter"

13 Femtosecond Filamentation in Air

A. Couairon[1] and A. Mysyrowicz[2]

[1] Centre de Physique Théorique, École Polytechnique, CNRS UMR 7644, F-91128, Palaiseau Cedex, France
couairon@cpht.polytechnique.fr

[2] Laboratoire d'Optique Appliquée, École Nationale Supérieure des Techniques Avancées – École Polytechnique, CNRS UMR 7639, F-91761 Palaiseau Cedex, France
mysyr@ensta.fr

13.1 Introduction

A peculiar propagation of intense femtosecond laser pulses is observed in air without apparent diffraction or dispersion, during which the pulse is severely reshaped in the spatial, temporal and spectral domains [1]. The beam first shrinks upon itself to form a narrow intense core of about 100 μm diameter, surrounded by extended region of laser radiation which constitute an energy reservoir able to feed the intense core and sustain its propagation over long distances reaching several hundreds of meters. This phenomena is called filamentation in spite of the fact that the longitudinal and the transverse dimensions of the core have the same order of magnitude. The filament core has an intensity of a few 10^{13} cm^{-3} sufficient to ionize air molecules. Therefore a long plasma channel is left in the wake of the light filament. The term of filamentation which has been adopted in the literature refers not only to this trailing plasma column but also to the cumulated tracks of the intense core along the propagation. Among the important physical effects responsible for this spectacular propagation, there are diffraction, the optical Kerr effect which leads to self-focusing and collapse if the beam peak-power exceeds a critical value, multiphoton absorption, ionization of the gas and associated plasma defocusing. The effect of these phenomena is not only visible in the transverse diffraction plane but also in the temporal and spectral domains. For example the Kerr effect induces self-phase modulation which broadens the spectrum of the pulse by generating red frequencies in the leading and blue frequencies in the trailing parts of the pulse. For specific pulse parameters or experimental conditions, additional physical effects such as group velocity dispersion, or self-steepening occasionally become important in the filamentation process itself or in the interpretation of associated phenomena, as e.g. the conical emission of white light in the form of colored rings [2,3].

One of the remarkable properties of light filaments is their ability to carry high intensities over long distance with little attenuation. To understand this, it is useful to consider the loss mechanism induced by long and short pulses. The losses during the propagation of an intense laser pulse of long duration in air are mainly due to the interaction between the free electrons generated by the forward part of the pulse (or preexistent free electrons) and the laser field. This interaction involves an acceleration of the electrons (inverse Bremsstrahlung) which leads to their multiplication by avalanche, when the energy of the electrons becomes sufficient to produce secondary electrons. The major part of the energy of a long pulse is consumed at the beginning of avalanche, by local absorption in the plasma (dielectric breakdown of air). Avalanche processes, however, do not have time to develop with fs pulses. The electrons are generated mainly by multiphoton ionization, an effect which occurs over time scales of a few fs and involves only a fraction of the energy of a mJ pulse. Once triggered, multiphoton absorption, a highly nonlinear process, immediately depletes the intensity peaks which has two consequences: it stops the energy absorption until new intense peak are generated and it can even drive an energy flux refilling the intense peaks from the low intensity reservoir [4–9]. For ultrashort pulses, multiphoton absorption therefore occurs in steps of small amount of energy. This explains the propagation over long distances of an intense fs laser pulse.

To summarize, the most spectacular properties of light filaments are:

a) Their propagation over long distances from the meter scale in the laboratory to the kilometric scale outdoor [10–13].
b) The generation of thin plasma columns in their wake.
c) The generation of a supercontinuum [14, 15] which constitutes the main property used in atmospheric LIDAR applications such as the detection of pollutants [16].
d) The generation of an electromagnetic pulse (EMP) at the interface between the plasma channel and the gas. This EMP contains frequency components up to the GHz and beyond [17–20].
e) Filamentation also occurs in other media such as transparent liquids or solids [21–24]. These systems provide a compact model which exhibits a similar physics although at lower energies and on reduced scales. Filaments in fused silica play an important role in the formation of buried guiding structures, and find applications in the writing of optical elements in general.
f) Filamentation is accompanied by a self-shortening of the pulse duration, a process which can lead to isolated single cycle optical pulses in an adequate gas pressure gradient [25–27].
g) At large powers, multiple filaments are usually obtained, a process potentially interesting for applications since the multiple filamentation patterns can be controlled under certain conditions.
h) The regeneration of filaments when they encounter obstacles such as water droplets [5–8]

This chapter will review and discuss femtosecond filamentation and some of their properties.

13.2 Modeling Light Filamentation

13.2.1 Propagation

Brabec and Krausz [28] have derived an envelope equation modeling the propagation of laser pulses with durations of a few to many optical cycles in a nonlinear dispersive medium. It was called the nonlinear envelope equation and is obtained from a wave equation by assuming that the pulse envelope is slowly varying in the propagation direction z, but not necessarily in time. Their model associated with specific physical effects occurring in filamentation became rapidly a standard in the field. In the frequency domain corresponding to the retarded time $t \equiv t_{lab} - z/v_g$, this model reads [21, 28–33]:

$$U\frac{\partial \hat{\mathcal{E}}}{\partial z} = \mathrm{i}[\frac{1}{2k}\Delta_\perp + \frac{k}{2}L_+L_-]\hat{\mathcal{E}} + \mathrm{FT}\{N(\mathcal{E})\mathcal{E}\} \tag{13.1}$$

Here $\hat{\mathcal{E}}(x,y,z,\omega) = \mathrm{FT}\{\mathcal{E}(x,y,z,t)\}$, where FT denotes Fourier transform, $U \equiv 1 + (\omega - \omega_0)/kv_g$, $L_\pm \equiv n(\omega)\omega/kc \pm U$. The first term on the right hand side (rhs) of Eq. (13.1) accounts for diffraction within the transverse plane with $\Delta_\perp \equiv \partial^2/\partial x^2 + \partial^2/\partial y^2$. It can be readily seen from a small $\omega - \omega_0$ expansion that $kL_+L_-/2 \simeq k''(\omega-\omega_0)^2/2 + k'''/6(\omega-\omega_0)^3 + \cdots$, which shows that the second term in the rhs of Eq. (13.1) accounts for group velocity dispersion (GVD) at second and higher-orders. The nonlinear terms in Eq. (13.1) account for the optical Kerr effect, plasma induced defocusing and multiphoton absorption (MPA):

$$N(\mathcal{E}) = T^2 N_{\mathrm{Kerr}}(\mathcal{E}) + N_{\mathrm{Plasma}}(\mathcal{E}) + TN_{\mathrm{MPA}}(\mathcal{E}) \tag{13.2}$$

The Kerr term (13.3)

$$N_{\mathrm{Kerr}}(\mathcal{E}) = \mathrm{i}k_0 n_2 \left\{(1-\alpha)|\mathcal{E}(x,y,z,t)|^2 + \alpha\int_{-\infty}^{t} \mathcal{R}(t-\tau)|\mathcal{E}(x,y,z,\tau)|^2 d\tau\right\}, \tag{13.3}$$

is split into an instantaneous component due to the electronic response in the polarization and a delayed component, of fraction α, due to stimulated molecular Raman scattering [34]. The function $\mathcal{R}(t)$ mimics the molecular response with a characteristic time τ_{dK} and frequency Ω:

$$\mathcal{R}(t) = \mathcal{R}_0 \exp(-t/\tau_{dK}) \sin \Omega t \tag{13.4}$$

where $\mathcal{R}_0 = (1 + \Omega^2\tau_{dK}^2)/\Omega\tau_{dK}^2$. In air at 800 nm, $\tau_{dK} = 70$ fs and $\Omega =$ 16 THz [34, 35].

The plasma term (13.5)

$$N_{\mathrm{Plasma}}(\mathcal{E}) = -\frac{\sigma}{2}(1 + \mathrm{i}\omega_0\tau_c)\rho(x,y,z,t), \tag{13.5}$$

accounts for plasma absorption (real part) and plasma defocusing (imaginary part). The cross section σ for inverse Bremsstrahlung follows the Drude model [36] and reads:

$$\sigma = \frac{k\omega_0\tau_c}{n_0^2\rho_c(1+\omega_0^2\tau_c^2)} \tag{13.6}$$

where τ_c is the electron collision time. In air, $\tau_c = 350$ fs and $\sigma = 5.1 \times 10^{-18}$ cm^2. Therefore $\tau_c \gg \omega_0^{-1}$, and in this limit, the defocusing term can be expressed as a function of the critical plasma density since $\sigma\omega_0\tau_c\rho \simeq k_0\rho/n_0\rho_c$. The MPA term in Eq. (13.2)

$$N_{\mathrm{MPA}}(\mathcal{E}) = -\frac{W(|\mathcal{E}|^2)U_i}{2|\mathcal{E}|^2}(\rho_{at}-\rho)\mathcal{E}, \tag{13.7}$$

accounts for energy absorption due to multiphoton ionization and depends on the ionization rate $W(|\mathcal{E}|^2)$ of the medium, its ionization potential U_i and its neutral atom density ρ_{at}. In the case of multiphoton ionization, simplifications are obtained from the scaling $W(|\mathcal{E}|^2) = \sigma_K|\mathcal{E}|^{2K}$, which leads to $N_{\mathrm{MPA}}(\mathcal{E}) = -\frac{\beta_K}{2}|\mathcal{E}|^{2K-2}(1-\rho/\rho_{at})$, where the cross section for multiphoton absorption reads $\beta_K = K\hbar\omega_0\rho_{at}\sigma_K$.

The operators U and $T \equiv 1 + \mathrm{i}\omega_0^{-1}\partial/\partial t$, account for space-time focusing and self-steepening of the pulse (also called optical shock generation) [29,37, 38]. They generally describe the deviations from the slowly varying envelope approximation in time [28–30].

13.2.2 Plasma Generation by Optical Field Ionization

The evolution equation for the electron density reads

$$\frac{\partial\rho}{\partial t} = W(I)(\rho_{at}-\rho) + (\sigma/U_i)\rho I. \tag{13.8}$$

In Eq. (13.8), the photoionization rate $W(I)$ describes the probability of ionization of an atom with potential U_i (the forbidden band in a solid). The general formulation of Keldysh [39] and Perelomov et al. (PTT) [40] describes the ionization rate $W(I)$ in the multiphoton regime, valid for $I \leq 10^{13}$ W/cm^2, as well as in the tunnel regime when $I \geq 10^{14}$ W/cm^2 [41]. It also covers the intermediate regime which actually is the regime in which infrared fs filamentation takes place [42]. Fig. 13.1 shows the ionization rates for oxygen and nitrogen computed at 800 nm from the full Keldysh–PPT formulation (solid curves). About two orders of magnitude above these rates, the fine curves show their counterpart obtained with a recently determined prefactor for diatomic molecules [43]. The multiphoton ionization rates are plotted in dash-dotted lines and the ADK rates are shown with dashed curves. Note that the ionization rates for nitrogen are several orders of magnitude smaller than for oxygen.

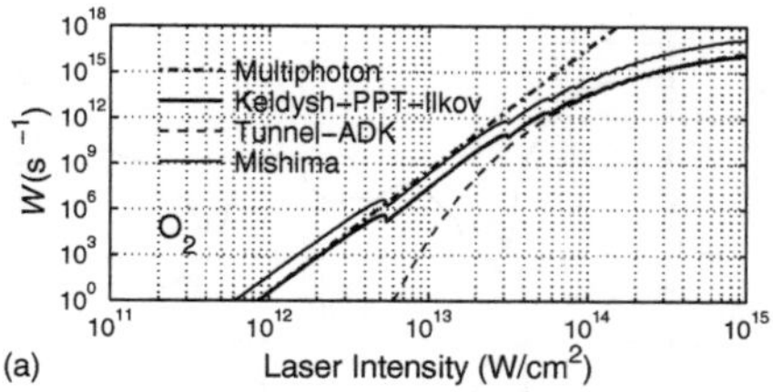

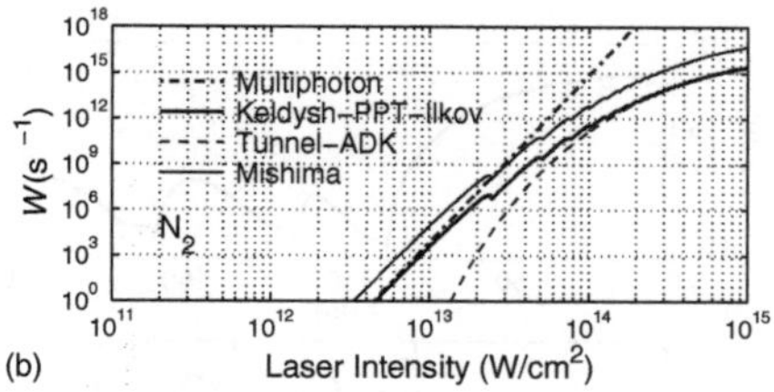

Fig. 13.1. Ionization rates for (**a**) oxygen and (**b**) nitrogen vs. laser intensity. The *bold solid curves* are computed according to the Keldysh [39], PPT [40], Ilkov et al. [42] formulations. The *dashed-dotted lines* show the multiphoton ionization rates and the *dashed curves* show the tunnel rates (ADK) according to [41]. The *fine solid curves show* the rates determined by Mishima et al. [43] with a specific prefactor for diatomic molecules

13.3 Typical Results of Numerical Simulations

13.3.1 Fluence, Beam Width, Intensity and Electron Density in a Single Filament

Figure 13.2 gives typical results of numerical simulations obtained with a simulation code which has been shown to reproduce experimental results accurately in several experimental situations [11, 22, 23, 30, 33, 44–47]. Figs. 13.2a,b show the beam width as a function of the propagation distance for an ultraviolet ($\lambda_0 = 248$ nm) and an infrared ($\lambda_0 = 800$ nm) pulse ($R(z)$ is the half width at half maximum (HWHM) of the fluence distribution $F(r, z) \equiv \int_{-\infty}^{+\infty} I(r, z, t)dt$). The results exhibit a first part (purely Kerr compression) where the beam is self-focused and collapses on itself at the end of this stage. The local intensity of the electric field strongly increases and becomes sufficient to ionize the molecules of air. Ionization of the molecules of oxygen and of nitrogen requires the simultaneous absorption of 3-4 ultraviolet or 8-11 infra-red photons. A non-negligible ionization of air requires a threshold intensity and therefore appears very abruptly, in the center of the beam, as indicated by the dashed curves in Figs. 13.2c,d. The generation of a plasma of free electrons and ions by multiphoton ionization acts like a saturating mechanism. Actually, two effects are efficient simultaneously in the saturation process: plasma defocusing only acts on the trailing part of the pulse but multiphoton absorption, the energy losses associated with plasma generation, act on the high intensity part of the beam. These effects limit the peak intensity of the laser in the filament to a value in the vicinity of a few 10^{13} W/cm^2 for infrared (800 nm) pulses, with an associated electron density of a few 10^{16} cm^{-3}, and less for UV pulses. These saturation levels depend on the ionization rates which are used in the computation. Here, the Mishima et al. [43] formulation has been used. Higher intensities up to 10^{14} W/cm^2 and electron densities up to 10^{17} cm^{-3} can be obtained with different ionization rates [30, 45]. While a complex space-time dynamics occurs,

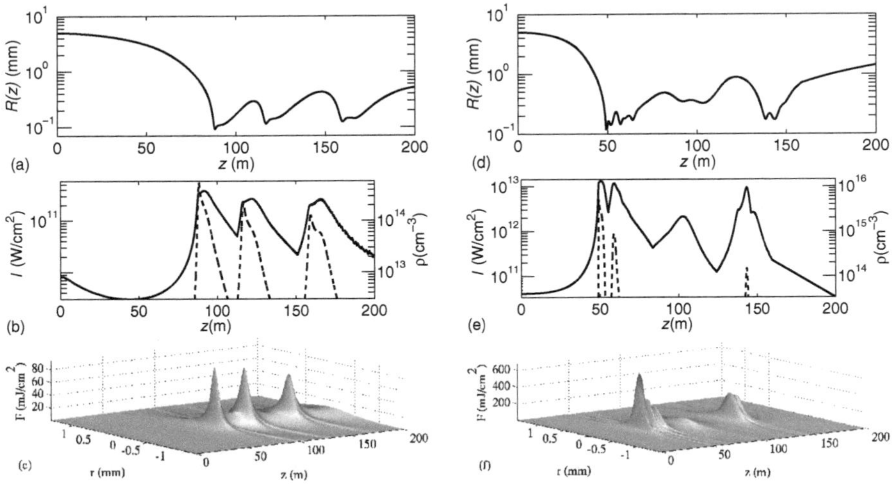

Fig. 13.2. Propagation of a collimated beam (diameter of 1 cm) in air for an ultraviolet (*first column*: $\lambda = 248$ nm, $\tau_{FWHM} = 100$ fs, 1 mJ) laser pulse and an infrared (*second column*: $\lambda = 800$ nm, $\tau_{FWHM} = 100$ fs, 3 mJ). The initial power of the pulse is slightly larger than the critical power for self-focusing $P_{cr}^{\mathrm{UV}} = 0.12$ GW, $P_{cr}^{\mathrm{IR}} = 3.2$ GW. The beam radius $R(z)$, the maximum intensity (*solid curve, left axis*), the density of free electrons on-axis (*dashed curve, right axis*) and the fluence distribution are plotted as a function of the propagation distance

the numerical simulations show a competition between focusing and defocusing effects taking place in the transverse diffraction plane, which leads to the formation of a filament with an average diameter of a few hundreds of μm. Each pinching of the beam by the optical Kerr effect causes multiphoton absorption and ionization (mainly oxygen is ionized), which involves defocusing and an increase in the beam width. This process persists in principle as long as the power of the beam exceeds the critical power for self-focusing. The intense core of the beam contains a fraction of a few percent of the beam energy. Most of the beam energy is contained in the feet which feed the core. The consequence of this property is the ability of the energy reservoir to regenerate the filament core when it happens to be extinguished, even by a central stopper [4–7, 9]. The filament therefore persists over several tenths of Rayleigh lengths, a process that does not need any external guiding such as a fiber and that has been called for this reason a self-guided propagation. During this self-guided propagation, the energy losses are weak and the pulse undergoes important structural modifications. Indeed, the pulse tends to adjust its size in the vicinity of an intensity close to the ionization threshold. The number and the frequency of the rebounds of ionization, intensity and fluence (Fig. 13.2e,f) depend on the initial conditions, in particular on the convergence of the beam and its size.

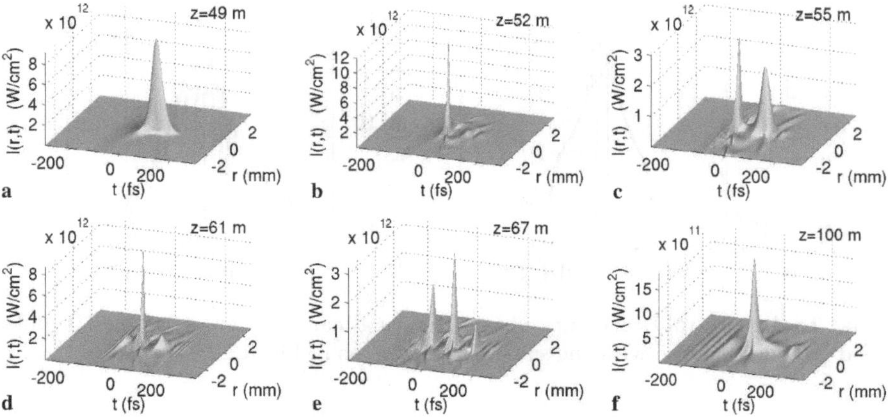

Fig. 13.3. Evolution of the temporal profile of the pulse during the nonlinear propagation. The initial pulse (**a**) undergoes contraction in the transverse diffraction plane (**b**) plasma defocusing of its trailing part (**c**) which is subsequently refocused by the optical Kerr effect (apparent splitting beyond the nonlinear focus) (**d**, **e**, **f**) During the propagation in the form of a filament, successive focusing-defocusing cycles reinforce and accentuate the shortening and the stiffening of the pulse beyond the nonlinear focus

13.3.2 Evolution of the Temporal Profiles

Figure 13.3 shows the evolution of the profile of the infrared pulse in space and time in the case $P \simeq P_{cr}$. At the beginning of the filament, the pulse is split into two shorter pulses. This temporal splitting partly reflects a spatial effect due to the plasma defocusing of the trail of the pulse and its subsequent refocusing by the optical Kerr effect. This effect has been called Dynamic Spatial Replenishment [48–52] and constitutes the main reason of the asymmetry in time. To a lesser extent, the splitting in time also follows from the arrest of collapse by GVD (more efficient in the UV) and from the energy depletion of the center of the beam due to multiphoton absorption. The sequence of figures 13.3 shows the dynamic evolution of the pulse with the appearance of shorter subpulses through these recurrent splitting in time. In this example, the temporal profile of the pulse is multipeaked. In general, it can be rather complex during the filamentation process. However, clean isolated pulses can be generated at specific locations and under well controlled conditions, their duration can be as short as a single cycle.

13.3.3 Evolution of the Power Spectra

As we have just seen, filamentation leads to extremely short structures of a few fs duration. Correspondingly a large spectrum is generated. Figure 13.4 shows a typical example of the spectral broadening which occurs during

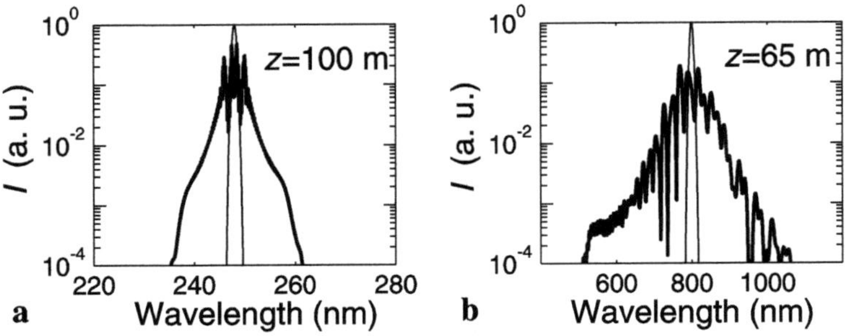

Fig. 13.4. Spectral Broadening obtained during filamentation of (**a**) the UV pulse and (**b**) the IR pulse with the same parameters as in Fig. 13.2

filamentation. The spectral range of the pulse if multiplied by a factor 10 (or more) leading, in the case of the IR pulse (Fig. 13.4b), to the appearance of a continuum of emission covering the visible and extending towards the infra-red domain. The UV pulse (Fig. 13.4a) undergoes a smaller, yet also important spectral broadening. This strong spectral broadening is due to self phase modulation, which corresponds to the generation of new frequencies in the spectrum of the laser pulse due to the temporal variation of the refraction index as $n = n_0 + n_2 I(r, z, t)$. The simplest model accounting for the evolution of the temporal phase links the instantaneous frequencies to the pulse intensity as:

$$\omega(t) = -\frac{\partial \phi}{\partial t} \sim \omega_0 - \frac{n_2 \omega_0}{c} z \frac{\partial I(t)}{\partial t} \qquad (13.9)$$

The generation of new frequencies thus depends on the slope of the pulse, the propagation distance in the Kerr medium and its nonlinear index. In a purely Kerr medium, the front part of the pulse generates redder frequencies, the back part bluer frequencies. Inspection of Eq. (13.9) explains why a broad continuum is generated during filamentation. There is a long interaction distance z of a pulse with very steep edges dI/dz.

To summarize, simulations predict a clamping of the peak intensity to a few 10^{13} W/cm^2. During filamentation, extremely short sub-pulses are generated with duration of a few fs. A large spectral broadening is simultaneously obtained. Finally, the electron density of the plasma generated by infrared filamentation reaches a few 10^{16} cm^{-3}. A competition between the optical Kerr effect and multiphoton absorption associated with ionization and plasma defocusing leads to a strongly dynamic propagation in the form of a filament.

13.3.4 Pulse Self-shortening

Very recently, the possibility to generate isolated nearly single cycle pulse by filamentation in noble gases has been demonstrated both theoretically [27] and experimentally [25], a result that might impact several domains related to high-field physics. Fig. 13.5 shows the computed evolution of a 1 mJ, 25 fs infrared laser pulse focused in the middle ($z = 95$ cm) of a gas cell filled with argon at 0.8 atm. The complex structure in space and time reflects the highly nonlinear filamented regime which leads to successive splitting in time and eventually forms a 3 fs isolated structure at the end of the filament, with a peak intensity of about 5×10^{13} W/cm^2. This pulse self-shortening process can even be simplified by using a gas density gradient, achieved by a control of the pressure along the propagation distance. In this case, numerical simulations show that three parameters (maximum gas density, length of the density gradient and position with respect to the focus of the lens) are available to switch on and off the filamentation process, and finally ensure the obtention of a single cycle pulse in vacuum. Fig. 13.6a shows the evolution of the pulse duration as a function of the propagation distance when the gas pressure increases up to 0.5 atm at 85 cm and decreases back to zero, the FWHM length of the pressure gradient being 18 cm [26]. The dynamics in the filament is limited to the self-focusing stage followed by the plasma defocusing stage in the center and trailing part of the pulse, which typically produces the fishbone structure shown in Fig 13.6(b) in the r, t space. Finally, the decrease of the pressure is equivalent to switch off all nonlinear effects, which leads to the pancake pulse structure shown in Fig. 13.6c, i.e. to a 2 fs pulse with peak intensity of a few 10^{13} W/cm^2 at the end of the filamentation stage.

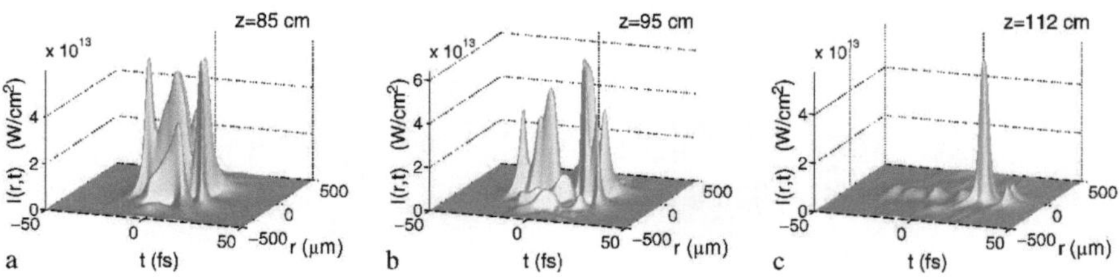

Fig. 13.5. Numerical simulations show the self-shortening of a 1 mJ, 25 fs infrared laser pulse in a gas cell filled with argon at 0.8 atm (according to Couairon et al. [27])

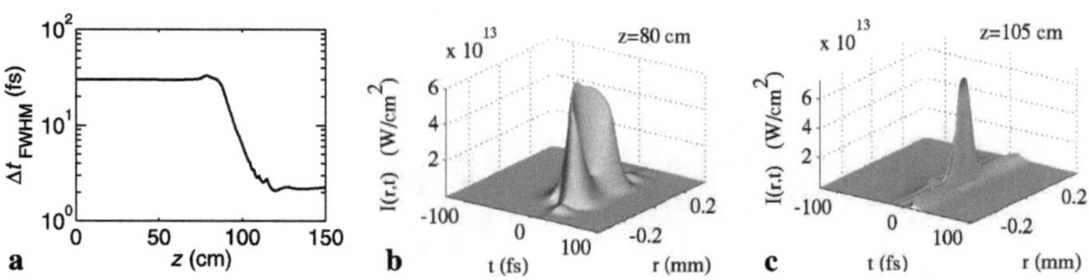

Fig. 13.6. (**a**) Pulse duration (FWHM) computed on the intensity integrated spatially over 100 μm during the filamentation of a 25 fs laser pulse in an argon pressure gradient (according to Couairon et al. [26]). (**b–c**) Intensity distributions computed at the beginning (**b**) and at the end (**c**) of the pressure gradient

13.4 Multifilamentation

The situation described so far relates to a pulse having an incident power around P_{cr}, where $P_{cr} = 3$ GW for air at 800 nm is the threshold power for filamentation. If $P \gg P_{cr}$, multifilamentation occurs triggered by short range modulational instability [53]. Irregularities in the incident beam profile, even modest, are rapidly reinforced and lead to a breakdown of the beam in several hot spots which act as nuclei for several filaments. Mlejnek et al. [4] performed the first realistic (3+1) dimensional simulations about multiple filamentation and called *optical turbulence* this propagation regime, which describes the self-guided propagation rather well on long distances from an initially collimated beam (see figure 13.7). The origin of multiple filamentation patterns traces back to the works of Bespalov and Talanov [53] on the modulational instability of powerful laser beams. Noise in the input beam should be amplified and lead to a seemingly random nucleation of hot spots or filaments. Yet, modulational instability also leads to a selection of a specific transverse length scale which should appear as a main feature of the filamentation patterns. Recently, alternative explanations for multiple filamentation were proposed by Fibich and coworkers [54–56]. They showed that deterministic vectorial effects could prevail over the amplification of noise in the process of multiple filamentation. A small ellipticity of the input beam should also lead to well determined multiple filamentation patterns [8, 57]. Building on this idea, it was shown that multiple filaments can be organized

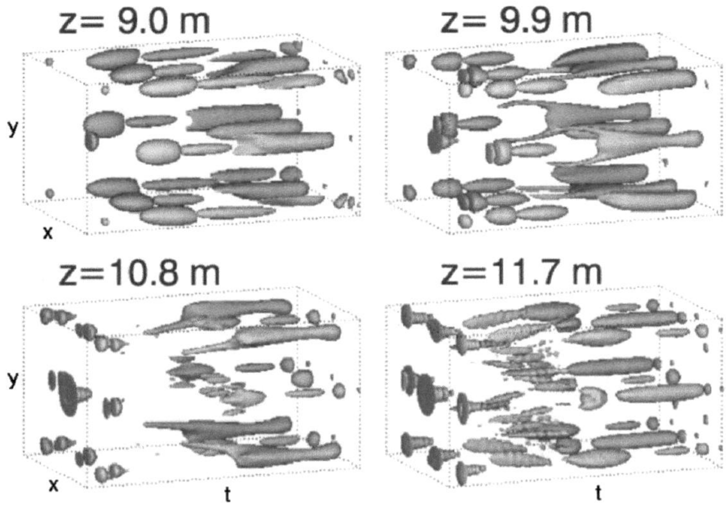

Fig. 13.7. Numerical simulations show the appearance of multiple filaments whose distribution changes with the propagation distance. The input power was $P_{in} = 35P_{cr}$. The transverse (x, y) box-size is $\sim 3 \times 3$ mm and box length in the temporal (t) direction represents ~ 200 fs. (according to Mlejnek et al. [4])

by various control mechanism acting either on the intensity or on the phase of the input field [33, 58]. For instance, large intensity gradients induced by a non-circular diaphragm are shown to lead to beam break-up and multiple filaments regularly located at the periphery of the input beam (see Fig. 13.8). The multiple filaments further coalesce into a single filament on axis, which constitute a very simple process to enlarge the propagation distance of a multifilamenting pulsed beam and the energy content of a single filament.

Using large negative initial chirps so as to delay the beginning of the filamentation process, it was shown numerically by means of (3+1)D simulations that the effect of azimuthal perturbations on input beam traversing a circular diaphragm also leads to long and intense light filaments that mutually interact via the background energy reservoir (see Fig. 13.8b). Similar organized multifilamentation patterns, although with a symmetry breaking, were obtained numerically when the measured beam intensity was introduced in the code as an initial condition. The pattern in figure 13.9(a) was obtained during an horizontal filamentation campaign [11, 12] which led to the observation of weakly ionizing filaments over kilometric distances, connected by a network of moderately intense energy clearly visible on Fig. 13.9a. Figure 13.9b shows good agreement between this measurement and the numerically obtained pattern from (3+1) D simulations starting with the beam measured at the output of the laser. This demonstrates that the organized features of the filament are governed more by the input pulse than by shot to shot fluctuations or air turbulence. The energy exchange between the background energy reservoir and the filaments constitutes the process sustaining the propagation, extinction and nucleation of filaments over long distances [11–13].

It should be noted that for long propagation distances, it is necessary to take into account the effect of air turbulence. Up to now, this problem was dealt with only very partially [59, 60].

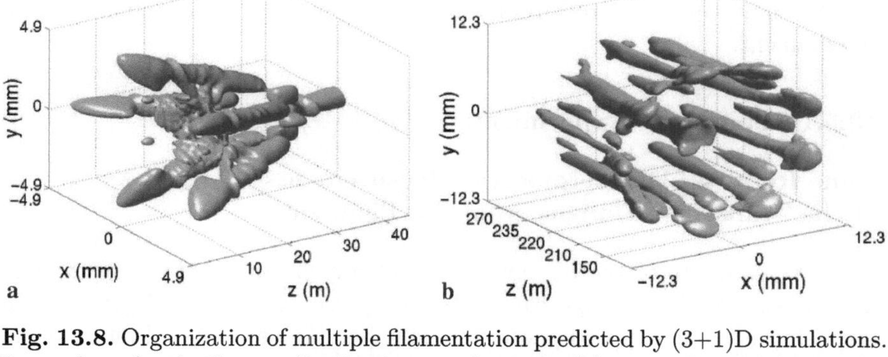

Fig. 13.8. Organization of multiple filamentation predicted by (3+1)D simulations. Iso-surfaces for the fluence distribution are shown for (**a**) a 10 mJ, 130 fs laser pulse whose beam is initially reshaped by a five-foil mask. (**b**) a 150 mJ, 500 fs, 800 nm pulse whose initial beam includes azimuthal perturbations of order 10. Note the non-uniform spacing along the z-axis. (according to Méchain et al., Refs. [11, 33])

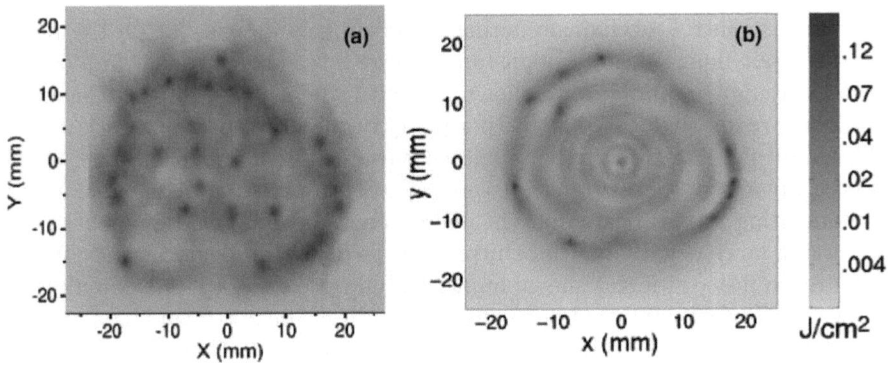

Fig. 13.9. Comparison between the multiple filamentation patterns obtained in (**a**) experiments and (**b**) simulations after propagation over 68 m of a 190 mJ, 800 nm laser pulse stretched to 1.2 ps by a negative chirp (according to Méchain et al., Refs. [11,12])

To summarize, multiple filamentation patterns arise for powers $P \gg P_{cr}$ and result from the growth of modulationally unstable perturbations in the beam. Their nucleation and coalescence are seemingly disordered, yet they can be organized by means of several control processes. They are sustained by the background energy reservoir which can refill a specific filament or reform another light string if it is destroyed by an obscurant on its path. This leads to their observation at considerable distances reaching several kilometers.

13.5 Experiments

The various results of simulations have been tested in laboratory and outdoor experiments. One of the difficulties is the high intensity of the filament core which requires special precautions in carrying out the measurements inside the filament.

13.5.1 Laboratory Experiments

Time Resolved Diffractometry, Measurements of the Transverse Dimension of the Filament

A coarse estimate of the diameter of the filament can be done by examining the size of the damage caused by the pulse in a blade of glass or the micro-burns on an exposed photographic paper. The attenuated image of the impact of the filament on a diffuser can also be recorded. The principle of a more precise measurement is shown in figure 13.10 [61,62]. The laser pulse is divided into a principal pulse, which is used to form a filament, and a much

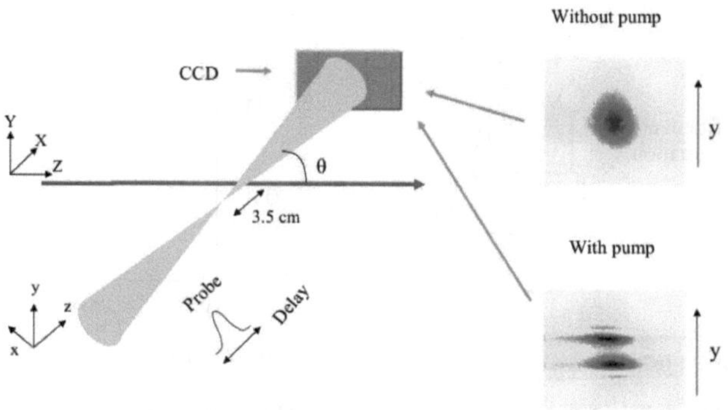

Fig. 13.10. Experimental diagram for time-resolved optical diffractometry, allowing the measurement of the filament size and the duration of the generated plasma. θ represents the angle between the propagation vectorsof the pump and probe beams. The probe beam is focused 3.5 cm before the filament created by the pump. Note the fringes in the far field, when there is spatial and temporal overlap between the probe beam and the plasma column generated by the filament. The full width of the far-field image in the y-direction is 1 mm (from S. Tzortzakis et al., [61,62])

less intense pulse, which is used as a probe. The probe beam crosses the filament and thus sees a rectilinear phase object. The phase front of the probe pulse undergoes a deformation which is analyzed in the far field (right part in figure 13.10). A rectilinear phase object gives place to a system of fringes, whose variation depends on the changes in the refraction index. By moving the probe pulse perpendicularly to the propagation axis of the filament, the dimension of the object which modifies the refraction index (filament) is obtained, about 80 μm in the initial stage of the filamentation. This measurement is in agreement with numerical results [30,63,64]. Yang et al. [65] measured the phase contrast and obtained similar results. Several groups announced the presence of filaments with a millimeter size after a propagation over several tens of meters [10,66,67]. In these filaments, the intensity of the pulse is lower than the ionization threshold, but the optical Kerr effect still plays its role.

Peak Intensity of a Filament

Measurements of the peak intensity in a filament are far from being easy. Estimates based on the measurements of the energy and the dimensions of the filament give a set of dispersed values ranging between 10^{12} and 10^{14} W/cm^2. A more precise measurement was performed by Lange et al. [68,69]. It consists in introducing a filament, after its propagation over a distance of 50 cm in air, inside a cell containing a noble gas, like argon (see 13.11). For intensities of about 10^{13}–10^{14} W/cm^2, the atoms generate a great number of odd high-order harmonics [70]. A simple and well established relation links

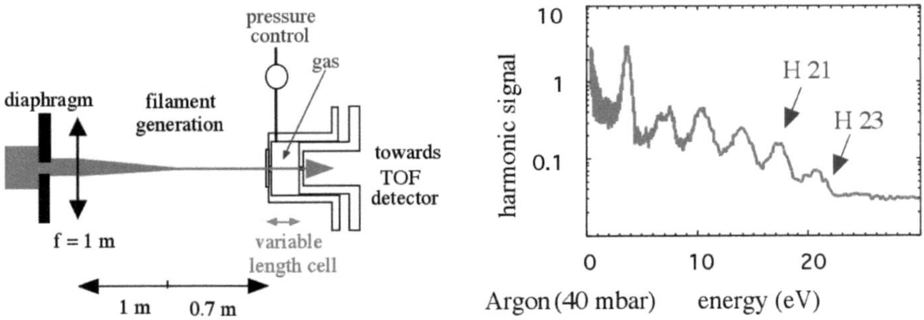

Fig. 13.11. Principle for the measurement of the peak intensity in a filament. The odd harmonics generated in a cell filled with argon at low pressure are detected in a second cell filled with Xenon. The energy of the free electrons in the xenon gas is measured by a detector of time of flight, which informs about the energy of VUV photons responsible for ionization. The figure on the right-hand side shows the odd harmonics ranging between the 11th and 23rd orders (according to Lange et al. [69])

the order of the highest harmonic to the intensity of the laser [71, 72]:

$$\hbar\omega = U_i + 3.2U_p, \qquad (13.10)$$

where U_i is the ionization potential of the atom and U_p is the ponderomotive energy (the energy of the free electrons oscillating in the laser field), which is directly proportional to the peak intensity:

$$U_p = \frac{1}{2}m_e\langle v^2\rangle = \frac{e^2E^2}{4m_e\omega^2} = 9.33 \times 10^{-14} I\lambda^2. \qquad (13.11)$$

Here, I is expressed in W/cm^2, λ in μm and U_p is in eV. Once generated, the various harmonics are introduced into a second cell where a detector measures the energy of the UV photons produced by the filament. Figure 13.11 shows that harmonics are generated up to the 23rd order in argon, which corresponds to a peak intensity in the filament of 5×10^{13} W/cm^2, in agreement with the evaluations obtained from numerical simulations.

Pulse Duration

The pulse duration can be usually measured by auto or cross correlation techniques or by reconstruction of the field evolution via techniques such as SPIDER or FROG. However, the temporal shape of the pulse in the form of a filament undergoes a strong reorganization leading to high intensities, which make these techniques unsuitable. To measure the duration of the pulse in a filament, Lange used the method described on figure 13.12 [68, 73]. Two filaments cross in a silica blade. The cross diffraction signal emitted in the

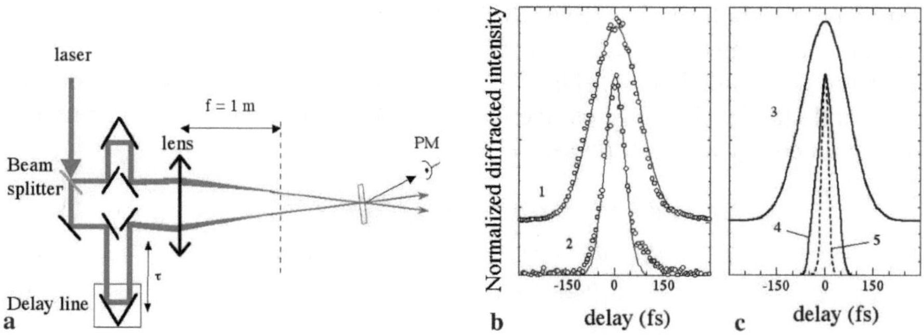

Fig. 13.12. Principle of the measurement of the duration of the self-guided (filamented) pulse in air (according to Lange [68]). Two filaments are formed in air. They cross in a silica blade, which gives rise to a four wave mixing signal. (**a**) Experimental setup used to measure the duration of self-guided pulses (**b**). The measurement of the signal as a function of the delay between the two filaments gives the cross correlation trace of the filaments. 1: trace of two non filamented pulses; 2: trace corresponding to two filaments. (**c**) Numerical results taking into account the temporal resolution of the detector (*full curve*)

direction $2k_1 - k_2$, (where k_1 and k_2 are the wavevectors of the two filaments) is measured according to the delay between the two filaments. The obtained cross-correlation trace indicates a reduction of the pulse duration in the filaments by a factor 4 (see figure 13.12b,c). The resolution of the correlator, about 20 fs, did not make it possible to resolve the substructures predicted by numerical simulations. Several groups, however, have reported a pulse shortening. Mikalauskas et al. [67] showed by a measurement of third order autocorrelation that a pulse at 532 nm, with an initial duration of 900 fs, was shortened by a factor 6 after a propagation in the form of a filament over 16 m in air. Tzortzakis et al. [74] have shown that ultraviolet picosecond pulses in the form of filaments are structured and strongly shortened. Couairon et al. [46], by measuring cross-correlation traces between a pulse at the end of a filament and a non filamented pulse from the same laser ($\lambda = 800$ nm, duration 120 fs), report a temporal compression ratio of 10 for the self-guided pulse. Hauri et al. showed a pulse shortening down to nearly one optical cycle by filamentation of a 1 mJ, 42 fs, 800 nm laser pulse in a low pressure argon gas cell [25].

Plasma Lifetime

The plasma lifetime is obtained from time-resolved diffraction measurements [61] and from electric measurements of the plasma conductivity [75–77]. The first method makes it possible to obtain the initial density and its evolution on a picosecond scale. The second gives the variation of the density on a time scale larger than the nanosecond. The principle of diffraction measurements is indicated in Fig. 13.10. The contrast of the fringes observed in far field is measured as a function of the delay between the probe pulse and the self-guided pulse. The evolution equations for the electron density (ρ_e) and the densities of positive (p) or negative (n) ions (ρ_p, ρ_n) generated instantaneously at $t = 0$ reads:

$$d\rho_e/dt = \alpha\rho_e - \eta\rho_e - \beta_{ep}\rho_e\rho_p \tag{13.12}$$

$$d\rho_p/dt = \alpha\rho_e - \beta_{ep}\rho_e\rho_p - \beta_{np}\rho_n\rho_p \tag{13.13}$$

$$d\rho_n/dt = -\eta\rho_e - \beta_{np}\rho_n\rho_p \tag{13.14}$$

The coefficient α corresponds to the multiplication of electrons by avalanche in the presence of an electric field (either the laser field, or an external applied field). The quantity $\eta = 2.45 \times 10^7$ s^{-1} is the coefficient of electron attachment to the neutral oxygen molecules. The coefficients β_{ep} and β_{np} correspond to electron-ion and ion-ion recombination processes. The solution to the equation reads:

$$\rho_e(t) = \frac{\rho_e(0)\exp[-(\eta-\alpha)t]}{1-\rho_e(0)\beta_{ep}(\exp[-(\eta-\alpha)t]-1)/(\eta-\alpha)} \tag{13.15}$$

When $\eta \gg \alpha$, the electron density varies according to the law:

$$\rho_e(t) = \frac{\rho_e(0)\exp[-\eta t]}{1+bt} \tag{13.16}$$

It is initially dominated by the decay rate $b = \rho_e(0)\beta_{ep}$.

Experiments indeed show a fast initial decrease, in the form $\rho_e(t) = \rho_e(0)/(1 + bt)$, with a coefficient $b = 4.7 \times 10^8$ s^{-1} close to the value found in the literature [61]. This nonexponential decrease indicates that the capture of electrons on the parents ions dominates the initial evolution of the plasma during the first nanoseconds. The later evolution, when $\rho_e < 10^{15}$ cm^{-3}, is due to the attachment of electrons on the oxygen molecules. The electron density decreases exponentially with a characteristic time $\tau = 1/\eta \sim 130$ ns [76].

Plasma Density

The values published for the mean plasma density in a filament vary over several orders of magnitude between 10^{14} and 10^{17} cm^{-3} [65,75,77,79]. This great dispersion is explained by the fact that certain measurements were performed in the multifilamentation regime. In this case, the average value integrated on the whole section of the beam is obtained, whereas the measurements performed in a single filament give the density inside a filament. The values corresponding to only one filament lie between 10^{16} and 10^{17} cm^{-3}. Precise measurements of the evolution of the plasma density, using an all optical method of time resolved diffraction allows us to infer with precision the initial density $\rho_e(0)$ from Eq. (13.16), which reaches 10^{17} cm^{-3} in the vicinity of the focus. Measurements of the electric conduction give $\rho_e(0) \sim 10^{16}$ cm^{-3}.

Plasma Resistivity

Ladouceur et al. [76] measure an average resistance varying in time between 3.6×10^5 Ωm^{-1} (initially) and 6.4×10^7 Ωm^{-1} after 150 ns. The resistance is directly proportional to the length of the plasma column in the filament until 6m. The evolution of the resistance as a function of time is explained by the capture of the electrons on the oxygen molecules. Tzortzakis et al. [75] find 2×10^6 Ωm^{-1} during the first ns, at an inter-electrode distance ranging between 2 and 10 cm. In these two cases, the diameter of the filament is about 80 μm, which leads to a value of the resistivity (resistance per conducting unit of area) about $1\Omega/cm$. These values are in agreement with a plasma density of 10^{16} cm^{-3}.

Length of Plasma Column

The maximum length of the plasma string generated by filamentation can be roughly evaluated by the ratio of the pulse energy to the energy losses per length unit [78]. Figure 13.13b shows that for a given pulse energy, the maximum filamentation length is expected for a specific pulse duration. Several techniques have been developed to measure the length of the plasma column generated by filamentation. In addition to the measurement of the conductivity [61, 77], the measurements of the backscattered fluorescence induced

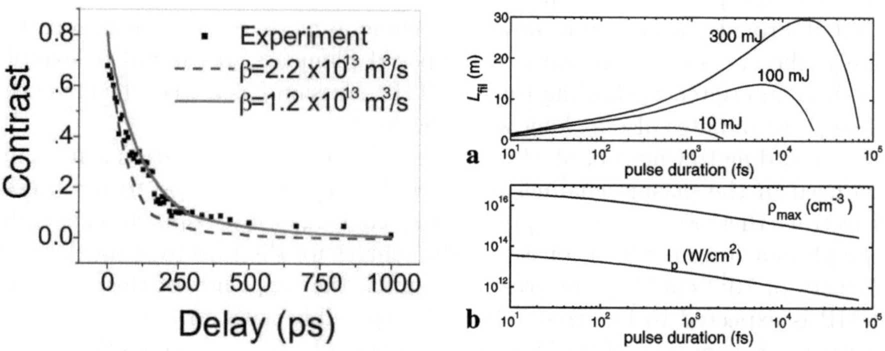

Fig. 13.13. *Left column*: Contrast of the probe wave front central fringe as a function of the delay between probe and self-guided pulse. The curves are obtained from the analytic formula (13.15) [61]. *Right column*: (**a**) Filamentation length in air for energies of 10, 100 and 300 mJ, and (**b**) maximum intensity and electron density as a function of the pulse duration [78]

by multiphoton ionization of excited nitrogen molecules constitute a non-intrusive method to detect the plasma channel, which is useful to detect long propagation distances in the atmosphere but does not give access to the dynamics of plasma generation. Using this technique associated with a lidar technique [80], plasma channels covering about 200 m were detected from 40 mJ laser pulses [81]. A linear antenna or a circular antenna around the filament has also been used to detect the electromagnetic pulse (EMP) radiated from the filament and compared with good agreement to the measurements of the backscattered fluorescence [82]. The latter method can detect the dynamics of plasma generation but is useless over long propagation distances. In contrast with the other detection methods, the luminescence can be detected remotely with a telescope.

Electromagnetic Pulse Emitted from the Plasma

It has been recently observed that plasma filaments formed by ultrashort laser pulses propagating in gases emit a broadband EMP of terahertz radiation in the direction transverse to the propagation axis of the filament [17, 76, 83]. The measurement principle from Ref. [83] is shown in Fig. 13.14. The sub-THz emission is detected by a heterodyne receiver using a local oscillator functioning at 941 GHz or 1181 GHz. A signal is detected at these two frequencies, in the form of a narrow cone of emission, perpendicular to the propagation axis z of the filament. The intensity of the sub-THz emission from the filament is constant over a distance z of about one meter, in good agreement with the length of the plasma channel, as detected independently by direct electric conductivity of air. These measurements were confirmed by calorimetric measurements, using a cooled bolometer [84]. Up to now, however, the calibration of the emitted Thz power has not been determined. One interesting aspect of THz emission in the context of filamentation studies is that it provides a new, non invasive technique to detect air ionization. Finally, the emission from two closely spaced filaments was found to exhibit positive interference, showing that the THz emission has a spatial coherence. The pulse radiates also at long wavelengths [81].

According to Cheng et al. [18], a longitudinal charge separation is initially induced in the plasma channel by the radiation pressure due to the pulse. Electrons then oscillate at the plasma frequency $\omega_p^2 = \rho_e e^2/\epsilon_0 m$, and generate the plasma current which constitute the sources for the EMP. For an electron density of 10^{16} cm^{-3} in the plasma channel, the frequency of this sub-THz EMP is expected to be around 100 GHz (possibly with an initial radiation lasting a few ps at a THz frequency). Although its main prediction could be experimentally verified, this model has been disputed [85–87]. Sprangle et al. [19] show that the source of the radiative EMP is the ponderomotive force which contains Fourier components with superluminal phase velocities, a mechanism similar to the Cerenkov radiation. A sine-qua non condition for EMP radiation is therefore the presence of a spatial modulation, such as that

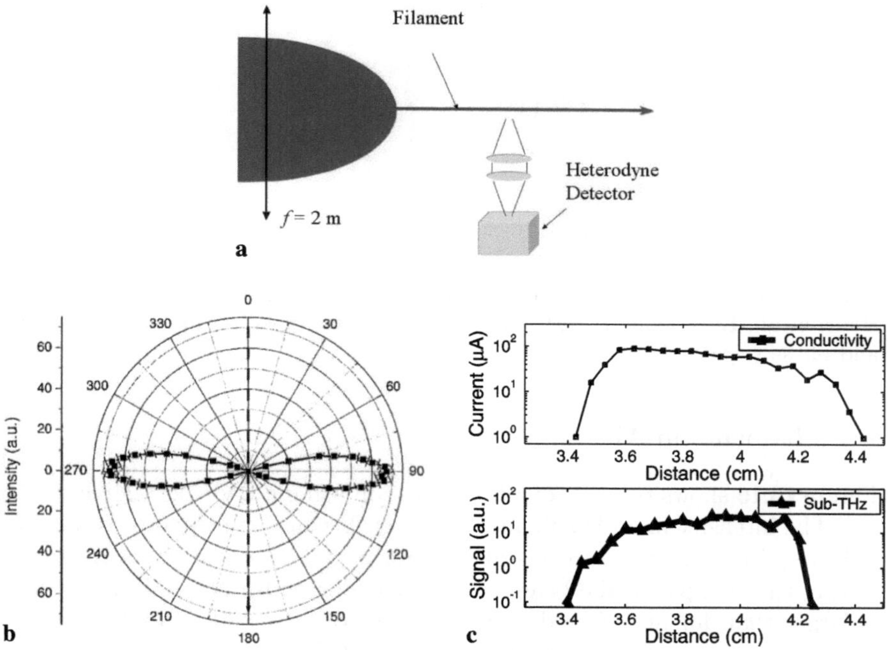

Fig. 13.14. (**a**) General diagram for the detection of the sub-THz radiation from a plasma channel generated by a filament. (**b**) The radiation diagram shows two lobes with an intensity maximum. The lobes are perpendicular to the axis of the filament [83]. (**c**) Comparison of filamentation lengths measured by conductivity and by sub-THz detection

arising from the focusing-defocusing cycles in the filament, which produces the superluminal components in the ponderomotive force.

Experimental Results on Pulse Self-shortening by Filamentation

It has been long recognized that filamentation of ultrashort laser pulses is associated with a self-shortening process due to the erosion of the trailing part of the pulse by the plasma generated by the leading part [88] and the steepening of the front part. The first experimental report of self-shortening induced by filamentation traces back to Ref. [73]. Self-shortening from 200 to 40 fs was reported in air [67], from 130 fs to 10 fs by concatenation of cross-polarized filaments [46]. Recent experiments in argon gas cells demonstrated that this self-shortening process lead to nearly single cycle pulses with a flat phase front and an excellent beam quality [25]. Fig. 13.15a shows the temporal profile recorded by a spider technique after successive filamentation in two argon cells of a 1 mJ, 42 fs infrared laser pulse. The resulting 5.7 fs pulse possesses a nearly flat temporal phase as shown in Fig. 13.15b. In addition, the locking of the carrier envelope offset is not destroyed by the filamentation process.

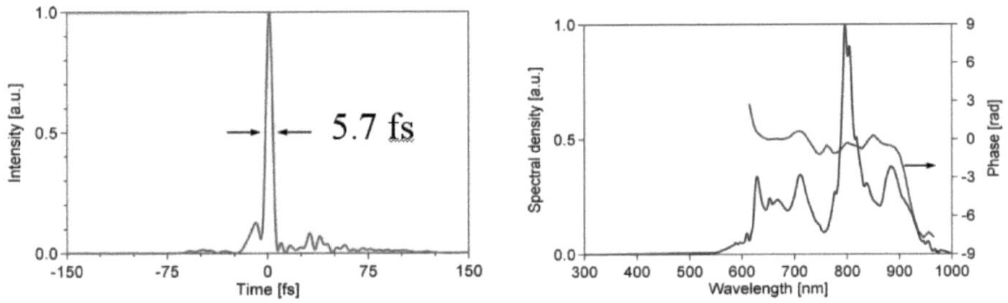

Fig. 13.15. Pulse self shortening by filamentation in argon gas cells (according to Hauri et al., Refs. [25, 27]). (**a**) Temporal profile. (**b**) Spectral intensity (*left axis*) and phase (*right axis*)

UV Filaments in Air

Figure 13.16 shows the case of a filamentation using a UV laser (248 nm) [74, 89]. The intensity in an UV filament is lower, because multiphoton ionization only requires the simultaneous absorption of 3-4 photons; it occurs earlier during the collapse of the beam on its axis. The width of resulting filament is larger than for an infrared filament. A good agreement is found between the measured size of the beam and the computed values. Generation of third and fifth harmonics in a UV filament has been reported recently [90].

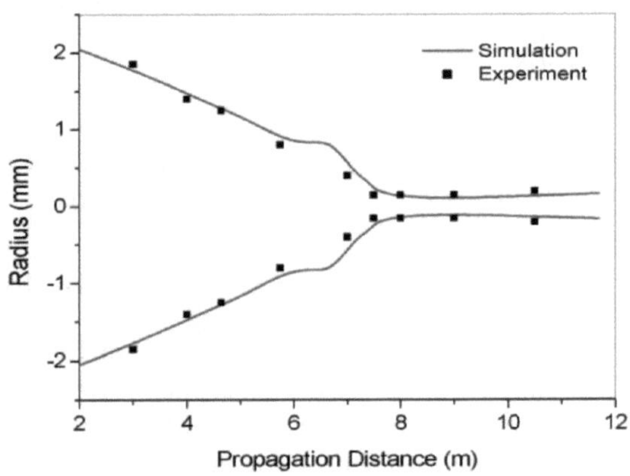

Fig. 13.16. Comparison between the diameter of a UV filament (347 nm) and simulations (according to S. Tzortzakis et al., ref [74])

13.5.2 Long Range (Outdoor) Experiments

Several groups gave persistent reports on multifilamentary structures propagating over several hundreds of meters [2, 10, 65, 66]. The formation of filaments at kilometric distances in the sky was also reported from the detection

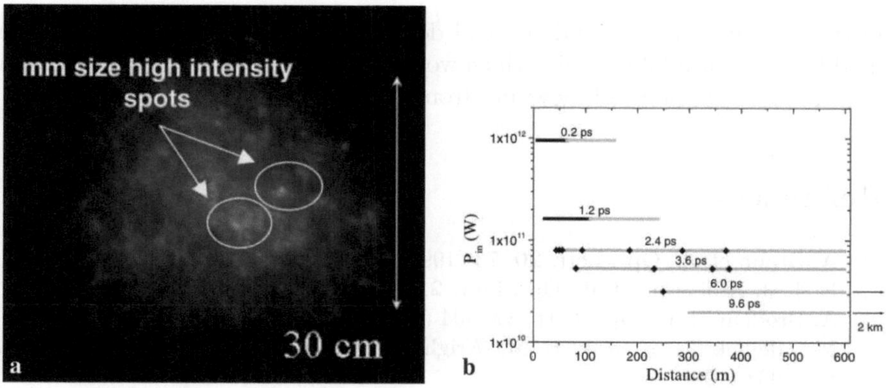

Fig. 13.17. (**a**) Beam cross section intercepted by a white screen at a propagation distance of 2350 m. The initial negative chirp corresponded to a duration of 9.6 ps. (**b**) Filamentation length as a function of the chirp of the input laser pulse. The duration of the non-chirped pulse is 100 fs. Black points and lines refer to locations where air ionization could be detected, grey lines to distances where bright light channels were observed (according to Méchain et al., Ref. [11, 12])

of the supercontinuum by a telescope [13, 16]. During an horizontal propagation campaign led with the Teramobile laser [11, 12, 91], the presence of multiple filaments up to 600 m was clearly demonstrated from step by step measurements of the plasma density using three different techniques (conductivity measurements, detection of the luminescence and detection of the sub-THz EMP). The presence of light channels of millimetric size was demonstrated up to 2.2 km. As shown on Fig. 13.17, the domain where multiple filaments were observed with intensities sufficient to ionize air molecules extends on more than 450 m. The intensity of the filaments lies between 10^{10} and 10^{13} W/cm^2.

13.6 Conclusion

In conclusion, important results on femtosecond filamentation in air, and more generally in Kerr media, have been obtained during the last decade. The physics associated with filamentation is extremely rich, and still needs a more complete understanding in order to clearly demonstrate the potential of filaments for applications. From the theoretical point of view, the models can be confronted to experiments over laboratory scales mainly. It is still a formidable computational task to simulate numerically a realistic propagation and filamentation over kilometric distances. From the experimental point of view, remarkable progress have been done recently in the three dimensional reconstruction of the dynamics of filaments in Kerr media [92–94]. From this technique, it is expected that a complete and simultaneous characterization in the spatial, temporal and spectral domains of the filamenting laser pulse might be soon available in air, which would lead to a breakthrough both from the fundamental point of view and from the point of view of applications.

References

1. A. Braun et al., Opt. Lett. **20**, 73 (1995).
2. E. T. J. Nibbering et al., Opt. Lett. **21**, 62 (1996).
3. A. Brodeur et al., Opt. Lett. **22**, 304 (1997).
4. M. Mlejnek, M. Kolesik, E. M. Wright, and J. V. Moloney, Phys. Rev. Lett. **83**, 2938 (1999).
5. F. Courvoisier et al., Appl. Phys. Lett. **83**, 213 (2003).
6. M. Kolesik and J. V. Moloney, Opt. Lett. **29**, 590 (2004).
7. A. Dubietis, E. Gaižauskas, G. Tamošauskas, and P. D. Trapani, Phys. Rev. Lett. **92**, 253903 (2004).
8. A. Dubietis, G. Tamošauskas, G. Fibich, and B. Ilan, Phys. Rev. Lett. **29**, 1126 (2004).
9. W. Liu et al., Appl. Phys. B **80**, 857 (2005).
10. B. L. Fontaine et al., Phys. Plasmas **6**, 1615 (1999).
11. G. Méchain et al., Appl. Phys. B **79**, 379 (2004).
12. G. Méchain et al., Optics Commun. **247**, 171 (2005).
13. M. Rodriguez et al., Phys. Rev. E **69**, 036607 (2004).
14. F. Théberge, W. Liu, Q. Luo, and S. L. Chin, Appl. Phys. B **80**, 221 (2005).
15. J. Kasparian et al., Opt. Lett. **25**, 1397 (2000).
16. L. Wöste et al., Laser und Optoelektronik **29**, 51 (1997).
17. A. Proulx, A. Talebpour, S. Petit, and S. L. Chin, Optics Commun. **174**, 305 (2000).
18. C.-C. Cheng, E. M. Wright, and J. V. Moloney, Phys. Rev. Lett. **87**, 213001 (2001).
19. P. Sprangle, J. R. Peñano, B. Hafizi, and C. A. Kapetanakos, Phys. Rev. E **69**, 066415 (2004).
20. W. Hoyer et al., Phys. Rev. Lett. **94**, 115004 (2005).
21. S. Tzortzakis et al., Phys. Rev. Lett. **87**, 213902 1 (2001).
22. L. Sudrie et al., Phys. Rev. Lett. **89**, 186601 (2002).
23. A. Couairon et al., Phys. Rev. B **71**, 125435 (2005).
24. A. Dubietis, G. Tamosauskas, I. Diomin, and A. Varanavičius, Opt. Lett. **28**, 1269 (2003).
25. C. P. Hauri et al., Appl. Phys. B **79**, 673 (2004).
26. A. Couairon et al., Opt. Lett. **30**, 2657 (2005), in press.
27. A. Couairon et al., J. Mod. Optics (2005), in press.
28. T. Brabec and F. Krausz, Phys. Rev. Lett. **78**, 3282 (1997).
29. A. L. Gaeta, Phys. Rev. Lett. **84**, 3582 (2000).
30. A. Couairon et al., J. Opt. Soc. Am. B **19**, 1117 (2002).
31. A. Couairon, Phys. Rev. A **68**, 015801 (2003).
32. P. Sprangle, J. R. Peñano, and B. Hafizi, Phys. Rev. E **66**, 046418 (2002).
33. G. Méchain et al., Phys. Rev. Lett. **93**, 035003, 1 (2004).
34. J.-F. Ripoche et al., Opt. Commun. **135**, 310 (1997).

35. E. T. J. Nibbering et al., J. Opt. Soc. Am. B **14**, 650 (1997).
36. E. Yablonovitch and N. Bloembergen, Phys. Rev. Lett. **29**, 907 (1972).
37. A. A. Zozulya, S. A. Diddams, and T. S. Clement, Phys. Rev. A **58**, 3303 (1998).
38. A. A. Zozulya, S. A. Diddams, A. G. V. Engen, and T. S. Clement, Phys. Rev. Lett. **82**, 1430 (1999).
39. L. V. Keldysh, Sov. Phys. JETP **20**, 1307 (1965).
40. A. M. Perelomov, V. S. Popov, and M. V. Terent'ev, Sov. Phys. JETP **23**, 924 (1966).
41. M. V. Ammosov, N. B. Delone, and V. P. Krainov, Sov. Phys. JETP **64**, 1191 (1986).
42. F. A. Ilkov, J. E. Decker, and S. L. Chin, J. Phys. B: At. Mol. Opt. Phys. **25**, 4005 (1992).
43. K. Mishima et al., Phys. Rev. A **66**, 033401, 1 (2002).
44. S. Tzortzakis et al., Phys. Rev. Lett. **86**, 5470 (2001).
45. A. Couairon and L. Bergé, Phys. Rev. Lett. **88**, 135003 1 (2002).
46. A. Couairon et al., Optics Commun. **225**, 177 (2003).
47. A. Couairon, Eur. Phys. J. D **27**, 159 (2003).
48. M. Mlejnek, E. M. Wright, and J. V. Moloney, Opt. Lett. **23**, 382 (1998).
49. M. Mlejnek, E. M. Wright, and J. V. Moloney, Optics express **4**, 223 (1999).
50. M. Mlejnek, E. M. Wright, and J. V. Moloney, IEEE J. Quant. Elec. **35**, 1771 (1999).
51. M. Mlejnek, M. Kolesik, E. M. Wright, and J. V. Moloney, Computer Simulation **56**, 563 (2001).
52. J. V. Moloney, M. Kolesik, M. Mlejnek, and E. M. Wright, Chaos **10**, 559 (2000).
53. V. I. Bespalov and V. I. Talanov, ZhETF Pis'ma **3**, 471 (1966), [JETP Lett. 3 (1966) 307].
54. G. Fibich and B. Ilan, Opt. Lett. **26**, 840 (2001).
55. G. Fibich and B. Ilan, Physica D **157**, 112 (2001).
56. G. Fibich and B. Ilan, Phys. Rev. Lett. **89**, 013901 1 (2002).
57. G. Fibich, S. Eisenmann, B. Ilan, and A. Zigler, Opt. Lett. **29**, 1772 (2004).
58. H. Schroeder, J. Liu, and S. L. Chin, Opt. Express **12**, 4768 (2004).
59. V. P. Kandidov et al., Quantum Electronics **29**, 911 (1999).
60. V. P. Kandidov, O. G. Kosareva, E. I. Moshaev, and M. P. Tamarov, Atmos. Oceanic Opt. **13**, 394 (2000).
61. S. Tzortzakis, M. Franco, B. Prade, and A. Mysyrowicz, Opt. Commun. **181**, 123 (2000).
62. S. Tzortzakis, Thèse de doctorat, École Polytechnique, École Polytechnique, Palaiseau, 2001.
63. A. Chiron et al., Eur. Phys. J. D **6**, 383 (1999).
64. H. R. Lange et al., Opt. Lett. **23**, 120 (1998).
65. H. Yang et al., Phys. Rev. E **66**, 016406 (2002).
66. S. L. Chin et al., Appl. Phys. B **74**, 67 (2002).
67. D. Mikalauskas, A. Dubietis, and R. Danielus, Appl. Phys. B **75**, 899 (2002).
68. H. R. Lange, Thèse de doctorat, École Polytechnique, École Polytechnique, Palaiseau, France, 1998.
69. H. R. Lange et al., Phys. Rev. Lett. **81**, 1611 (1998).
70. P. Salières, A. l'Huillier, P. Antoine, and M. Lewenstein, Adv. At. Mol. Opt. Phys. **41**, 83 (1999).

71. P. B. Corkum, Phys. Rev. Lett **71**, 1994 (1993).
72. K. C. Kulander, K. J. Schafer, and J. L. Krause, in *Proceedings of the workshop on super intense laser atom physics (SILAP III)*, edited by B. Piraux (Plenum Press, New York, 1993), Vol. 316, p. 95.
73. H. R. Lange et al., in *CLEO/IQEC conference* (OSA, San Francisco, 1998).
74. S. Tzortzakis et al., Opt. Commun. **197**, 131 (2001).
75. S. Tzortzakis et al., Phys. Rev. E **60**, R3505 (1999).
76. H. D. Ladouceur et al., Opt. Commun. **189**, 107 (2001).
77. H. Schillinger and R. Sauerbrey, Appl. Phys. B **68**, 753 (1999).
78. A. Couairon, Appl. Phys. B **76**, 789 (2003).
79. C. Y. Chien et al., Opt. Lett. **25**, 578 (2000).
80. A. Iwasaki et al., Appl. Phys B **76**, 231 (2003).
81. S. A. Hosseini et al., Appl. Phys. B **77**, 697 (2003).
82. S. A. Hosseini, B. Ferland, and S. L. Chin, Appl. Phys. B **76**, 583 (2003).
83. S. Tzortzakis et al., Opt. Lett. **27**, 1944 (2002).
84. G. Méchain et al., Appl. Phys. B **77**, 707 (2003).
85. G. Shvets, I. Kaganovich, and E. Startsev, Phys. Rev. Lett. **89**, 139301 (2002).
86. V. T. Tikhonchuk, Phys. Rev. Lett. **89**, 209301 (2002).
87. C.-C. Cheng, E. M. Wright, and J. V. Moloney, Phys. Rev. Lett. **89**, 139302 (2002).
88. A. Sergeev et al., Phys. Rev. A **46**, 7830 (1992).
89. S. Tzortzakis et al., Opt. Lett. **25**, 1270 (2000).
90. N. Kortsalioudakis et al., Applied Physics B **80**, 211 (2005).
91. www.teramobile.org, .
92. A. Matijošius et al., Nonlinear Analysis: Modelling and Control **9**, 259 (2004).
93. A. Matijosius et al., Opt. Lett. **29**, 1123 (2004).
94. M. A. C. Potenza et al., Opt. Commun. **229**, 38 (2004).

14 Pulse Self-Compression in the Nonlinear Propagation of Intense Femtosecond Laser Pulse in Normally Dispersive Solids

Ruxin Li, Xiaowei Chen, Jun Liu, Yuxin Leng, Jiansheng Liu, Yi Zhu, Xiaochun Ge, Haihe Lu, Lihuang Lin, and Zhizhan Xu

State Key Laboratory of High Field Laser Physics and Shanghai Institute of Optics and Fine Mechanics, Chinese Academy of Sciences, P.O. Box 800–211, Shanghai 201800, P. R. China

Summary. The self-compression phenomena of intense femtosecond pulses in normally dispersive solids were investigated experimentally. Both un-chirped and negatively chirped laser pulses were used as input pulses. It is demonstrated that intense femtosecond laser pulses can be compressed by the nonlinear propagation in the transparent solids, and the temporal and spectral characteristics of output pulses were found to be significantly affected by the input laser intensity, with higher intensity corresponding to shorter compressed pulses. By the propagation in a 3 mm thick BK7 glass plate with the laser power of GW level, a self-compression from 50 fs to 20 fs was achieved, with a compression factor of about 2.5. However, the output laser pulse was observed to split into two peaks when the input laser intensity is high enough to generate super-continuum and conical emission. When the input laser pulse is negatively chirped, the spectra of the pulse are reshaped and narrowed due to strong self-action effects, and the temporal pulse duration is found to be self-shortening. With the increase in the input pulse intensity, the resulted self-compressed pulses became even shorter than the input laser pulse, and also shorter than $sech^2$ transform-limited pulse according to the corresponding spectra.

14.1 Introduction

The nonlinear propagation of intense femtosecond laser pulses in air, solids and liquids has recently been subjected to intense investigation [1–5]. Filamentation, self-channelling, conical emission, supercontinuum generation (SG), strong space-time coupling and pulse temporal self-compression are observed when the power of input laser pulses is higher than the critical power for self-focusing. The underlying physics in these processes are the interplay of Kerr nonlinearity, diffraction, multi-photon ionization (MPI), ionization-induced-refraction, group velocity dispersion (GVD) and self phase modulation (SPM). These new phenomena associated with the nonlinear propagation have led to several important applications such as femtosecond laser remote sensing and lighting control. Pulse temporal self-compression has been less intensely investigated. Although its origin is basically a strong space-time coupling during the propagation, the detailed physics has not been fully understood and the valid parameter space has not yet been revealed. The self-compression scheme can be used in producing ultra-short laser pulses.

Pulse compression by dispersion compensation following a spectral broadening due to SPM during the nonlinear propagation has been intensively investigated since the invention of the first mode-locked laser. The so-called extra-cavity compression technique plays a fundamental role in generating state of the art ultra-short laser pulses. Pulse compression using single-mode optical fibers [6] or gas-filled hollow fibers [7–9] followed by dispersion control unit has led to the generation of extremely short laser pulses of monocycle level. However, the output pulse energy is limited to the range of nanojoules to submillijoules due to the damage or ionization of the nonlinear media.

Hauri and co-workers reported [11] recently the generation of 0.38 mJ, 5.7 fs pulses via the cascaded self-guided propagation of original 0.84 mJ, 43 fs pulses in noble gases and the dispersion compensation with chirped mirrors. This method can greatly reduce the experimental difficulty as compared with gas-filled hollow fiber technology.

The compression based on the spectral broadening during the propagation of intense laser pulses in bulk media such as glass is another attractive technology [12, 13]. This technique has, in principle, no limit in energy up-scaling and is potentially applicable for the compression of much higher energy pulses.

However, in the above-mentioned schemes the dispersion control unit following the spectral broadening requires elaborate design [10] and sometimes leads to a considerable loss in pulse energy [8, 9]. Moreover, the broadened spectrum induced by SPM is often accompanied with complex nonlinear chirps that are difficult to compensate for, which will result in large temporal pedestals in the compressed pulses.

Pulse self-compression without using an independent dispersion control unit is attractive as a pulse compression technique due to its simplicity in implementation, its underlying physics is however more complicated. Early in the study of the nonlinear propagation of the femtosecond laser pulses in transparent media, it was found that the self-focusing effect of the intense femtosecond pulses could lead to pulse compression spatially and temporarily in normal-dispersion regime [14–16].

In 2000, I. G. Koprinkov et al. [17] reported their observation of the ultrashort pulse self-compression (SC) in gases for the first time, which could be interpreted by the theory of gas-induced soliton [18]. Most recently, N. L. Wagner et al. [19] demonstrated self-compression from 30 fs to 13 fs by the propagation of pulses inside a hollow waveguide filled with low-pressure argon gas. Rather than the effect of self-focusing (SF), they attributed their results mainly to the role of MPI excited by the intense laser pulses and the guiding in the hollow waveguide. Hauri and co-workers also showed [11] via calculation self-compression can lead to the production of 1.3 cycle pulse during the self-guided propagation in gas filled tube.

Although the role of GVD and the nonlinear parameters in solids differ by orders of magnitude from those in gases, it was predicted that SC is also possible in bulk media as a result of self-channelling of intense femtosecond pulses [20], even to a few cycles [21, 22]. However, the investigation of SC in solids are mainly restricted to theoretical calculations and primarily at relatively low power of the order of MW, which is just above the critical power for SF in solids. And the pulse duration and beam diameter considered in those works are quite different from the requirement of compressing intense laser pulses.

Although SPM is generally associated with spectral broadening, it is not always the case if the input pulse is initially chirped. The sign of the initial chirp of the pulse decides whether the SPM will compress or broaden the spectrum. Spectral broadening occurs in the case that the initial pulse is positively chirped or un-chirped. With negatively chirped pulses, SPM can result in spectral compression in normally dispersive nonlinear media. Spectral narrowing in optical fiber was theoretically studied with negatively chirped pulses in 1993 [23]. It is shown that a minimum pulse duration will be obtained at almost the same time of the maximum spectral compression. Soon M. Oberthaler et al. [24] experimentally demonstrated the spectral narrowing in fibers and considered that narrow-bandwidth picosecond sources could be obtained by transforming femtosecond laser pulse in this way. Some years later, B. R. Washburn et al. [25] reported that a longer transform-limited pulse of about 600 fs was obtained from a 110 fs pulse source that was negatively chirped to 665 fs. The pulse duration accompanied with spectral narrowing is always broadened compared to original transform-limited pulses.

In this work, we use intense femtosecond laser pulses with a duration of $\sim$50 fs and peak power up to GW, and launch them into a normally dispersive solid (BK7 glass) in loose focusing condition. It is demonstrated that these intense femtosecond pulses can be compressed solely by the nonlinear propagation in solids. The exciting SC phenomena are investigated in details under a variety of experimental conditions. The experimental results show that the temporal duration of the compressed pulses is significantly affected by the input pulse intensity, with higher intensity corresponding to shorter compressed pulses. But the resulted pulses split into two peaks as the input laser intensity is high enough to generate supercontinuum and conical emission. This work shows a good agreement with the previous theoretical predictions, and will provide a significant experimental reference for the study on the nonlinear propagation of intense femtosecond pulses. On the other hand, the SC in solids shows a new compression method scalable for higher energy femtosecond pulses. We also investigate the propagation of intense femtosecond pulses in the BK7 glass with initially negative chirp. We find that the spectral narrowing and temporal compression can be obtained simultaneously in normally dispersive medium. The spectrum of the output pulse is reshaped and narrowed due to strong self-action effects, and the temporal pulse duration is self-shortening, instead of broadening at the same time. With the increase in the input pulse intensity, the resulted self-compressed pulses became even shorter than the original laser pulse source, and also shorter than sech^2 transform-limited pulse according to the corresponding spectrum.

In Sect. 14.2 we describe the experimental setup, in Sect. 14.3 we present the results and discussions about pulse self-compression in normally dispersive solids of laser pulses without chirp and with initially negatively chirp. And in Sect. 14.4 we summarize this work.

14.2 Experimental Setup

The experimental setup is shown in Fig. 14.1. The laser source used in our experiment is a commercially available chirped pulse amplification Ti:sapphire laser system (Spectral-Physics Spitfire_50 fs) running at 1-KHz repetition rate, producing 0.5 mJ/pulse, and about 50 fs in duration with a central wavelength at 800 nm. Typically, the beam quality parameter M^2 is about 1.3 with 7 mm in beam diameter (at $1/e^2$ of the peak intensity) and the bandwidth (FWHM) of the pulse is about 22 nm. The laser beam first passes through an attenuator consisting of a half-wave plate (HWP) and a polarizer, which can continuously adjust the laser energy, and then is focused by an $f = 1.0$ m lens. The focused beam is collimated by an $f = 0.5$ m sliver-coated concave mirror, and a piece of BK7 glass used as the nonlinear medium is inserted between the focusing lens and the collimation concave mirror. The collimated beam is then separated by a beam splitter (B.S.). One weak part is sent to a spectral phase interferometry for direct electric-field reconstruction (SPIDER) (APE, Co. Ltd.). And the other part is sent

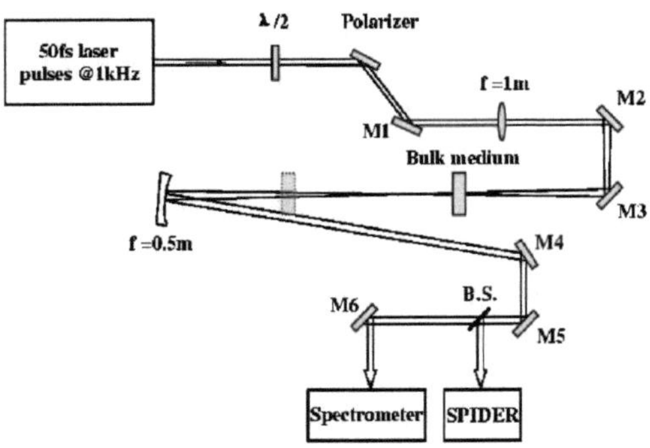

Fig. 14.1. Schematic drawing of the experimental setup

to a grating spectrometer (SpectraPro-300i, Acton Research Corporation) to monitor the spectrum. By using a SPIDER, the real-time temporal intensity profile and spectral phase of the pulses transmitted through the BK7 glass can be measured simultaneously.

14.3 Experimental Results and Discussions

14.3.1 Self-Compression in Solids of Un-chirped Laser Pulses

In the experiment, we used a piece of 3-mm-thick BK7 glass plate as the nonlinear medium, and placed it in the optical path with the front surface locating at 860 mm away from the focusing lens, where the input laser beam is convergent. At this position, the beam radius at $1/e^2$ of the peak intensity on the entrance surface of the glass plate is about 400 μm, which is significantly larger than the beam size in the previous studies [20–22]. In this section we mainly analyze the temporal and spectral behavior of the laser beam transmitted through the glass as a function of the input laser intensity on the glass surface. Since the glass position is fixed, the intensity can be varied by adjusting of the HWP.

The pulse duration, spectral bandwidth (FWHM) and the corresponding transform-limited (TL) pulse duration of the transmitted pulses are shown in Fig. 14.2 as a function of the input pulse energy. These data were taken with the initial pulse duration of ~50 fs, and the spectral bandwidth and TL pulse duration were estimated with the Gauss fit of the experimental spectrum. Note that the input laser peak power we used is higher than the threshold power for SF (P_{cr}) by three orders of magnitude. For BK7 glass,

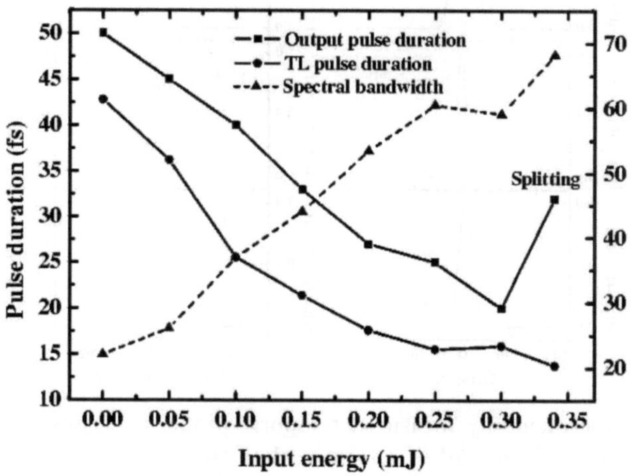

Fig. 14.2. Output pulse duration, spectral bandwidth (FWHM) and the corresponding transform-limited (TL) pulse duration versus input pulse energy

the value of the nonlinear index coefficient is $n_2 = 3.45 \times 10^{-16}\,\mathrm{cm^2/W}$, corresponding to $P_{cr} = \lambda^2/2\pi n_0 n_2 \approx 1.8\,\mathrm{MW}$ [26]. At lower energy (0.05 mJ, $P_{in} \approx 1\,\mathrm{GW}$, $I_{in} \approx 3.97 \times 10^{11}\,\mathrm{W/cm^2}$), the temporal shape of the transmitted pulse are nearly identical to those of the incident pulse, but the pulse is self-compressed to 44 fs, and the spectrum is somewhat broadened with the appearance of a small blue pedestal. The preferred generation of the blue frequency components should be a result of space-time focusing and self-steepening [21, 27]. At higher energy the pulse duration shows further shortening, and collapses rapidly with the increase in the input laser energy. The pulse duration reaches the minimum of 20 fs at the energy of 0.3 mJ, which is near the TL pulse duration corresponding to the spectral bandwidth. At the same time, the spectral bandwidth is broadened distinctly from the initial bandwidth of 22 nm to 59 nm with the increase in the input pulse energy. As the input energy is increased to 0.34 mJ, corresponding to $I_{in} \approx 2.7\times10^{12}\,\mathrm{W/cm^2}$ on the entrance surface of the glass, the pulse envelope breaks into two peaks. Under this condition, supercontinuum and conical emission were observed, but the transmitted energy was not greatly reduced, and the main beam quality was still good. Moreover, the bandwidth (FWHM) of the spectrum of the on-axis beam is broadened to 68 nm.

In Fig. 14.3, we compare the experimental spectrum, temporal profile and spectral phase of the transmitted pulses at the energy of 0.3 mJ (b–b1) and 0.34 mJ (c–c1) with those of the initial pulses (a–a1). As seen in

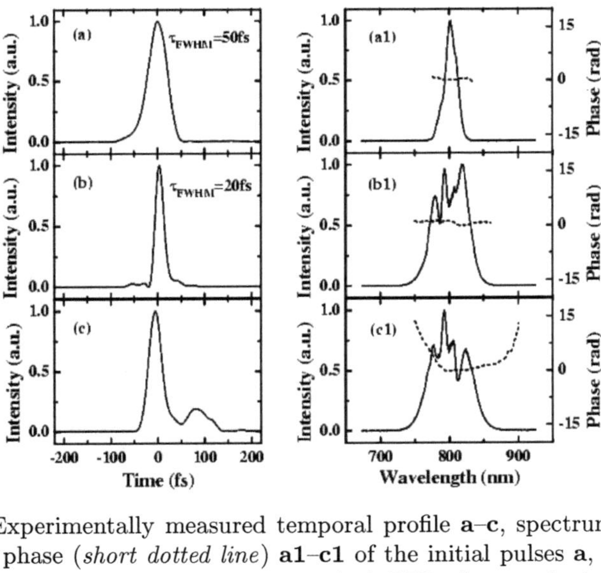

Fig. 14.3. Experimentally measured temporal profile **a**–**c**, spectrum (*solid line*) and spectral phase (*short dotted line*) **a1**–**c1** of the initial pulses **a**, **a1** and those of the pulses transmitted through a 3-mm-thick BK7 glass as the input energy is increased. The input energy for **b**, **b1** is 0.3 mJ, and for **c**, **c1** is 0.34 mJ

Fig. 14.3b, the compressed pulse is relative clean without obvious pedestals, but displays a steep front profile with small oscillatory tails on both sides. The steepening of the leading edge should be a shock behavior induced by the self-defocusing action of the plasma [27], while the oscillatory tails on the edge of the pulse can be a feature of the high-order dispersion. Figure 14.3c shows the two-peak structure of the transmitted pulses at the energy of 0.34 mJ, in agreement with the previous observations qualitatively [28,29]. The intensity spectra under these two conditions show a similar oscillatory structure. It is a typical characteristic of the SPM and consistent with the spectrum shown in [7]. It is worthy to note that, in comparison with the spectrum with blue-shifted pedestal at the energy of 0.05 mJ, the increase in the input energy does not lead to increased broadening of the pedestal, but results in the broadening towards both blue and red side. Besides, in contrast to the flat distribution of phase in the case of 0.3 mJ (short dotted line in Fig. 14.3b1), the spectral phase at the energy of 0.34 mJ shows a distinct positive group delay dispersion (short dotted line in Fig. 14.3c1).

The experimental phenomena presented above can be understood as follows: when the input intensity is low, the main effects during the propagation include the slight SPM, space-time focusing and self-steepening, which result in the appearance of the spectral blue tail [21, 27]. With the increase in the input intensity, the effect of the self-focusing becomes stronger. The strong self-focusing moves the off-axis energy towards the peak of the pulse, and then the pulse is compressed spatially and temporally [15, 20]. At the same time, the pulse intensity becomes so high during the self-focusing process in the nonlinear medium that it leads to the formation of low-density electron plasma by MPI. Since the plasma formation causes defocusing and nonlinear absorption of the back edge, it pushes the peak intensity to the front edge of the pulse. As a result, the pulse envelope shows a profile with a steep leading edge [21], and corresponds to a additional redshifted spectral broadening at higher input energy. At further high input intensity, the sharp pulse peak decays before it outputs the glass, whereas the back of the pulse refocuses [14]. Two-peak pulse is thus formed.

The pulse self-compression was also observed in the condition of divergent input beam, i.e., the bulk medium is placed somewhere after the lens focus (the little rectangle in dotted line in Fig. 14.1). Similar with the condition of convergent input beam, the pulse duration of the output pulses is also related to the input pulse intensity, with higher intensity corresponding to shorter resulted pulses. However, the compression factor is less than two no matter how we change the input pulse intensity. Figure 14.4 shows the temporal and spectral characteristic profiles of the resulted pulses when the BK7 glass plate is placed in different positions in the optical path.

Although the compression factor is a little smaller when the bulk medium is placed after the lens focus than before, we found in the experiments that

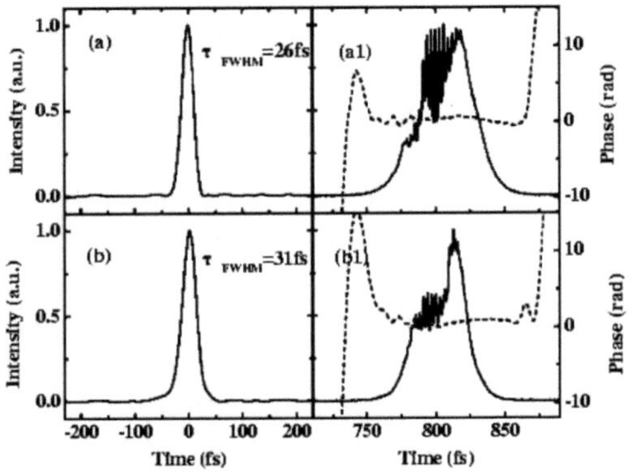

Fig. 14.4. Experimentally measured temporal profile **a**, **b**, spectrum (*solid line*) and spectral phase (*short dotted line*) **a1**, **b1** of the output pulses in the condition of divergent input laser beam, with the entrance surface of the glass plate 160 mm **a**, **a1** and 195 mm **b**, **b1** after the lens focus

the spatial mode was significantly improved of the laser beam transmitted through the nonlinear medium. Figure 14.5 compares the spatial intensity distribution of the output pulses measured by a CCD instrument (Spiricon, LBA-300PC). The spatial mode improvement of the laser pulses is probably due to the spatial self-focusing effect of the high-power femtosecond pulses in the bulk medium, under the condition of divergent propagation beam which pushes beam energy of the center high-power (above P_{cr} of the BK7 glass) part to the pulses peak and suppresses the fringe low-power section. Therefore, we choose to put the BK7 glass plate after the lens focus in the following experiments.

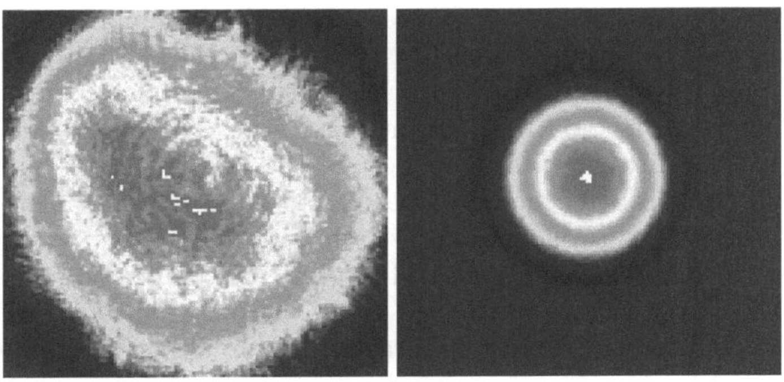

Fig. 14.5. Comparison of the cross-section intensity distribution of the initial laser beam and the output beam transmitted the bulk medium which is placed after the lens focus

14.3.2 Self-Compression in Solids of Negatively Chirped Laser Pulses

We also investigate the SC by using negatively chirped pulse. In the experiment, the negatively chirped pulse is obtained by increasing the distance of the grating pair in the compressor of the chirped-pulse-amplification system, which broaden the output pulses from the laser system from ~50 fs to ~75 fs. We still chose the 3-mm-thick BK7 glass plate as the nonlinear medium and placed it 145 mm after the lens focus. And the peak power and the peak intensity on the surface of the BK7 glass is at the order of GW and 10^{11} W/cm^2, respectively. Therefore the peak power we used in the experiments is about three orders higher than that used in the previous study [15], in which pulse splitting was observed with negatively chirped pulses and narrower beam (70 μm FWHM) in a much thicker (2.54 cm) bulk medium.

The pulse duration and spectral bandwidth of transmitted pulses after the BK7 glass are shown in Fig. 14.6 as a function of the input pulse energy. When the energy increases from 0.05 mJ to 0.3 mJ, corresponding to the intensity on the front surface of the glass from 1.33×10^{11} W/cm^2 to 7.96×10^{11} W/cm^2, the pulse duration is compressed from the initial 75 fs to 27 fs consistently. On the other hand, the spectral bandwidth (FWHM) is narrowed at first, and reaches the minimum about 13 nm (FWHM) when the input pulse energy equals 0.1 mJ, then begins to broaden with the increase in the input pulse energy, as shown by the dashed line in Fig. 14.6. When the input pulse energy further increases to 0.35 mJ, corresponding to the intensity on the front surface of the glass plate about 8.97×10^{11} W/cm^2, rainbow-like conical emission starts to occur.

Figure 14.7 shows the evolution of the spectral and temporal profile of transmitted pulses after the BK7 glass with the increase in the input pulse energy. The spectrum is narrowed when the input pulse energy is low due to SPM with negatively chirped pulse. And when the input pulse energy

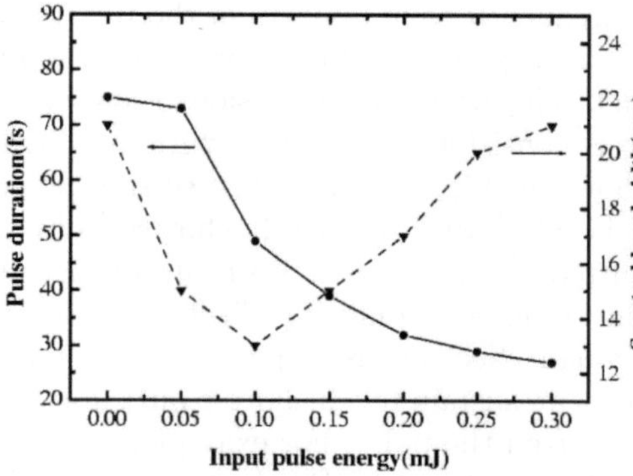

Fig. 14.6. Pulse duration, spectral bandwidth (FWHM) of the transmitted pulse with air and a piece of 3-mm-thick BK7 glass versus input pulse energy

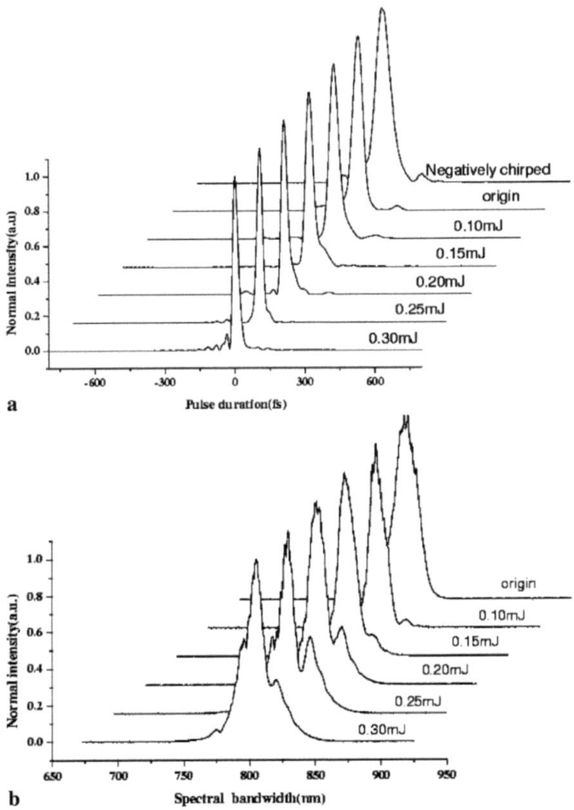

Fig. 14.7. Experimental measured evolution of the temporal profile **a**, intensity spectrum **b** of the initial negatively chirped pulses and those of the output pulses transmitted through a piece of 3 mm-thick BK7 glass with the increase in the input pulse energy

is 0.10 mJ, a small frequency wing appears in the red frequency region. The wing in the blue frequency region also occurs when the input pulse energy increases to 0.15 mJ. The wings on both sides grow up and the spectrum at the center part is consistently narrowed with the increase of the input pulse energy, as shown in Fig. 14.7b. This spectrum narrowing of the center part is due to the SPM with negatively chirped pulses. The blue-shifted wing of the transmitted pulse spectrum is a result of space-time focusing and self-steepening. The combined nonlinear effects of self-focusing, SPM, self-steepening, and Raman effect etc reshape the spectrum from Gaussian to other function which can sustain much shorter pulse for the same bandwidth (FWHM) and self-shorten the pulse. For example, with 0.15 mJ input pulse energy, the spectral bandwidth (FWHM) is narrowed to 15 nm with an near Lorentzian function spectral profile (the Lorentz fit of the measured intensity spectrum is shown by dashed line in Fig. 14.8a1) and the pulse duration is self-compressed to 39 fs at the same time, which is shorter than the Sech^2 transform-limited pulse duration for 15 nm spectral bandwidth (FWHM), as shown in Fig. 14.8. The pulse duration is also self-compressed consistently as the input pulse energy increasing, as shown in Fig. 14.7a. It can be seen that the steepening of the leading edge of the pulse occurs when the input pulse energy increases to 0.15 mJ, which results in the blue-shifted wing in the pulse spectrum. And the self-steepening becomes more obvious with the increase of the input pulse energy, corresponding to the growth of the blue-shifted wing in pulse spectrum. A 29 fs and a 27 fs pulse that are shorter than Sech^2 transform-limited for 20 nm and 21 nm spectral bandwidth (FWHM) are obtained with 0.25 mJ and 0.30 mJ input, respectively. Note that the spectral bandwidth and pulse duration vary little when the input pulse energy increases from 0.25 mJ to 0.3 mJ. It means that the resulted compressed pulses are stable at a relatively large input energy region. The steepening of the leading edge and blue-frequency broadening as shown in Fig. 14.7a should be a shock behavior induced by the self-defocusing action due to higher order negative susceptibilities $\chi^{(n)}$ $(n > 3)$ and ionization [17, 19, 30]. We also measured the beam intensity profile with a CCD instrument (Spiricon, LBA-300PC) at last. The transmitted beam is compressed due to self-focusing in the glass and the spatial model of the transmitted beam after the glass improved greatly compared to the original beam, as shown by the inset in Fig. 14.8a.

We retrieve all the compressed pulse profile by Fourier transforming the spectrum and phase. The spectrum is independently measured by the grating spectrometer and the spectral phase is measured by SPIDER. Figure 14.8a shows the retrieve pulse and the measured pulse with 0.15 mJ input pulse energy. We find that the retrieved temporal profile (dashed line) coincides with the measured pulse profile (solid line) very well. It is proved that the pulse profile measured by SPIDER in our experiment is reliable.

Actually, the spectral and temporal compression occurs simultaneously with the BK7 glass plate at a wide range of positions. We had tried to move the 3-mm-thick BK7 glass 180 mm after the lens focus. When the input pulse is negatively chirped to 65 fs and the pulse energy is 0.25 mJ, the spectral bandwidth (FWHM) is narrowed from 22 nm to 16 nm, and the pulse duration is compressed to 41 fs. When the input pulse energy is increased to 0.3 mJ, the spectral bandwidth (FWHM) broadens to about 21 nm and the pulse duration is still compressed to 34 fs. The evolution trend of the spectrum and temporal profile are similar.

The above experiments demonstrate that spectrum narrowing and pulse self-compression of high-intensity femtosecond pulses is a general phenomenon that can be observed with negatively chirped femtosecond pulses in a variety of normally dispersive media. From a simple physical standpoint, we understand the spectral and temporal simultaneous compression of negatively

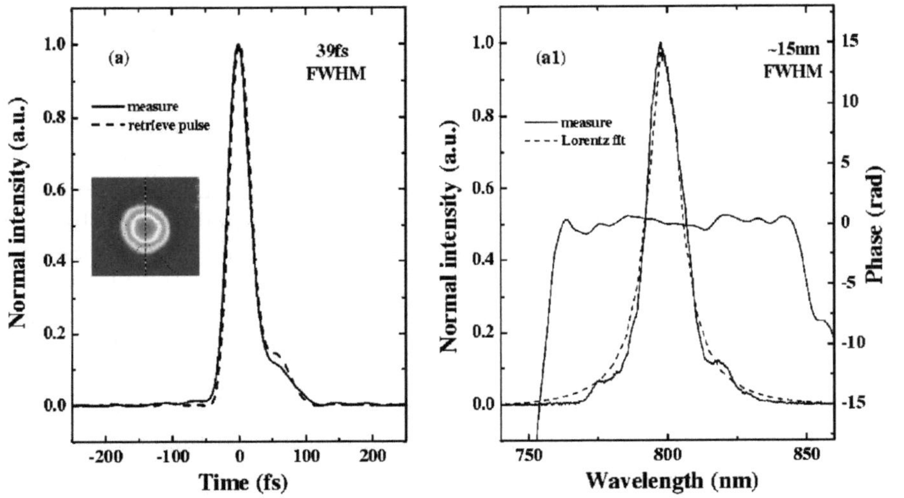

Fig. 14.8. Comparison of the experimental measured (*solid line*) temporal profile **a**, spectrum and spectral phase **a1** of the output pulses transmitted through a piece of 3-mm-thick BK7 glass with the input pulses negatively chirped and the input pulse energy of 0.15 mJ. The *inset* in **a** is the cross-section intensity distribution of the output pulse beam, the *dashed line* in **a** is the retrieved temporal profile corresponding to the measured spectrum and spectral phase, and the *dashed line* in **a1** is the Lorentz fit of the measured intensity spectrum

chirped pulses as follows: when the input intensity is low, SPM narrowed the spectrum. Combined with the media dispersive and space-time focusing, the transmitted pulse is shortened. As the intensity of the input pulse increase, the combined nonlinear effects of self-focusing, SPM, self-steepening, Raman effect, etc. reshape the spectrum from Gaussian to other function which can sustain much shorter pulse for the same bandwidth (FWHM). Strong self-focusing effect moves the off-axis energy toward the peak of the pulse and compressed it in both spatial and temporal domain [15, 20].

14.4 Conclusions

In conclusion, the self-compression of intense femtosecond laser pulses has been experimentally demonstrated in normally dispersive solids. By the propagation in a thin BK7 glass plate, a self-compression from 50 fs to 20 fs was achieved, corresponding to a compression factor of ~2.5. Our experimental results verified that the role of the high-order dispersion and high-order nonlinearity induced by the multiphoton ionization is non-negligible in the nonlinear propagation process. The scheme to achieve ultra-short pulses by SC in solid medium is relatively simple to operate and promising for compressing laser pulses with higher energy. With an initially negatively chirped laser pulse, we found that spectral narrowing is followed by pulse shortening instead of pulse broadening. Compressed pulses shorter than $Sech^2$ transform-limited of the spectral bandwidth (FWHM) are obtained. We show a method for pulse compression by reshaping the spectral profile, not only by broadening the spectrum. The spectrum is reshaped without energy loss and the temporal form is also reshaped using this method. The above discussed pulse self-compression method offers a new possibility to compress intense laser pulse down to a few cycles with cascaded compressor geometry which is under development.

Acknowledgments

This work is supported partially by Natural Science Foundation of China (Grant Nos. 69925513 and 19974058), Chinese Ministry of Science and Technology through contract G1999075204, Chinese Academy of Sciences through contracts KGCX2-SW-10 and KGCX2-SW-114, and the Major Basic Research Project of Shanghai Commission of Science and Technology.

References

1. A. Braun, G. Korn, X. Liu, D. Du, J. Squier, and G. Mourou, "Self-channeling of high-peak-power femtosecond laser pulses in air," Opt. Lett. **20**, 73–75, 1995.
2. A.A. Zozulya, "Propagation dynamics of intense femtosecond pulse: multiple splittings, coalescence, and continuum generation," Phys. Rev. Lett. **82**, 1430–1433, 1999.
3. A. Brodeur and S.L. Chin, "Ultrafast white-light continuum generation and self-focusing in transparent condensed media," J. Opt. Soc. Am. B **16**, 637–650, 1999.
4. I. Alexeev, A. Ting, D.F. Gordon, E. Briscope, J.R. Penano, R.F. Hubbard, and P. Sprangle, "Longitudinal compression of short laser pulses in air," Appl. Phys. Lett. **84**, 4080–4082, 2004.
5. E.T.J. Nibbering, P.F. Curley, G. Grillon, B.S. Prade, M.A. Franco, F. Salin, and A. Mysyrowicz, "Conical emission from self-guided femtosecond pulse in air," Opt. Lett. **21**, 62–64, 1996.
6. A. Balltuska, Z. Wei, M.S. Pshenichikov, D.A. Wiersma, "Optical pulse compression to 5 fs at a 1-MHz repetition rate" Opt. Lett. **22**, 102–104, 1997.
7. M. Nesoli, S. De Silvestri, and O. Svelto, "Generation of high energy 10fs pulses by a new pulse compression techinique," Appl. Phys. Lett. **68**, 2793–2795, 1996.
8. B. Shenkel, J. Biegert, and U. Keller, "Generation of 3.8-fs pulses from adaptive compression of a cascaded hollow fiber supercontinuum" Opt. Lett. **28**, 1987–1989, 2003.
9. K. Yamane, Z. Zhang, K. Oka, R. Morita, and M. Yamashita, "Optical pulse compression to 3.4 fs in the monocycle region by feedback phase compensation," Opt. Lett. **28**, 2258–2260, 2003.

10. Z. Cheng, F. Krausz, and Ch. Spielmann, "Compression of 2 mJ kilohertz laser pulses to 17.5 fs by pairing double-prism compressor: analysis and performance," Opt. Comm. **201**, 145, 2002.
11. C.P. Hauri, W. Kornelis, F.W. Helbing, A. Couairon, A. Mysyrowicz, J. Biegert, and U. Keller, "Generation of intense, carrier-envelope phase-locked few-cycle laser pulses through filamentation," Appl. Phys. B **79**, 673–677, 2004.
12. C. Rolland and P.B. Corkum, "Compression of high-power optical pulses," J. Opt. Soc. Am. B **5**, 641–647, 1988.
13. E. Mével, O. Tcherbakoff, F. Salin, and E. Constant, "Extracavity compression technique for high-energy femtosecond pulses," J. Opt. Soc. Am. B **20**, 105–108, 2003.
14. A.T. Ryan and G.P. Agrawal, "Pulse compression and spatial phase modulation in normally dispersive nonlinear Kerr media,"Opt. Lett. **20**, 306–308, 1995.
15. S.A. Diddams, H.K. Eaton, A.A. Zozulya, and T.S. Clement, "Characterizing the nonlinear propagation of femtosecond pulses in bulk media," IEEE J. Select. Topics Quantum Electron. **4**, 306–316, 1998.
16. N. Aközbek, C.M. Bowden, A. Talebpour and S.L. Chin "Femtosecond pulse propagation in air: Varitional analysis" Phys. Rev. E **61**, 4540–4545, 2000.
17. I.G. Koprinkov, A. Suda, P. Wang, and K. Midorikawa, "Self-compression of high-intensity femtosecond optical pulses and spatiotemporal soliton generation," Phys. Rev. Lett. **84**, 3847–3850, 2000.
18. L. Bergé and A. Couairon, "Gas-induced solitons," Phys. Rev. Lett. **86**, 1003–1006, 2001.
19. N.L. Wagner, E.A. Gibson, T. Popmintchev, I.P. Christov, M.M. Murnane, and H.C. Kapteyn, "Self-compression of ultrashort pulses through ionization-induced spatiotemporal reshaping," Phys. Rev. Lett. **93**, 173902–1, 2004.
20. S. Henz and J. Herrmann, "Self-channeling and pulse shortening of femtosecond pulses in multiphoton-ionized dispersive dielectric solids," Phys. Rev. A **59**, 2528–2531, 1999.
21. Hélène Ward and Luc Bergé, "Temporal shaping of femtosecond solitary pulses in photoionized media," Phys. Rev. Lett. **90**, 053901, 2003.
22. Z. Wu, H. Jiang, Q. Sun, H. Yang, and Q. Gong, "Filamentation and temporal reshaping of a femtosecond pulse in fused silica," Phys. Rev. A **68**, 063820–1-8, 2003.
23. S.A. Planas, N.L. Pires Mansur, C.H. Brito Cruz, and H.L. Fragnito, "Spectral narrowing in the propagation of chirped pulses in single-mode fibers," Opt. Lett. **18**, 699–701, 1993.
24. M. Oberthaler and R.A. Höpfel, "Special narrowing of ultrashort laser pulses by self-phase modulation in optical fiber," Appl. Phys. Lett. **63**, 1017–1019, 1993.
25. B.R. Washburn, J.A. Buck, and S.E. Ralph, "Transform-limited spectral compression due to self-phase modulation in fibers," Opt. Lett. **25**, 445–447, 2000.
26. J.K. Ranka, R.W. Schirmer, and A.L. Gaeta, "Observation of pulses splitting in nonlinear dispersive media," Phys. Rev. Lett. **77**, 3783–3786, 1996.
27. A.L. Gaeta, "Catastrophic collapse of ultrashort pulses" Phys. Rev. Lett. **84**, 3582–3585, 2000.

28. J.K. Ranka and A.L. Gaeta, "Breakdown of the slowlyvarying envelope approximation in the self-focusing of ultrashortpulses," Opt. Lett. **23**, 534–536, 1998.
29. S. Tzortzakis, L.Sudrie, M. Franco, B. Prade, A. Mysyrowicz, A. Couairon, and L. Bergé, "Self-guided propagation of ultrashort IR laser pulses in fused silics," Phys. Rev. Lett. **87**, 213902–1-4, 2001.
30. A. Couairon, "Dynamics of femtosecond filamentation from saturation of self-focusing laser pulses," Phys. Rev. A **68**, 015801-1, 2003.

15 Ultraintense Tabletop Laser System and Plasma Applications

S. Martellucci[1], M. Francucci[1], and P. Ciuffa[2]

[1] University of Rome "Tor Vergata", Department of "Ingegneria dell'Impresa", Via del Politecnico n. 1, 00133 Rome, Italy
smart@uniroma2.it or francucci@ing.uniroma2.it

[2] Elettronica s.p.a., Systems Technology and Processing Department, 00131 Rome, Italy
patrizio.ciuffa@elt.it

Summary. The advent of the ultraintense Tabletop Pulsed Laser Systems (TPLSs) has opened new and thus for unexpected frontiers allowing for the development and the progress of new research areas, in particular in scientific and industrial fields. Usually, these laser systems are used in order to generate plasma on solid or gaseous targets, so-called laser-induced plasma (LIP). The generated plasma behaves like a source of visible, UV and X radiation that can be used for many applications. In particular, X-rays emitted from LIP are used in X spectroscopy, microlithography, microscopy, imaging, radiographies (for example of biological samples). Conversion efficiency studies from laser radiation to X-rays for different targets are central for the energy balance of the source as an important performance parameter and reason of industrial attractiveness. More in generally, TPLSs are used for radiation–matter interaction studies, for fundamental plasma parameter determination, for astrophysical applications, for inertial confinement fusion, for studies in high energy physics or in the compact particle accelerator field, for quantum electrodynamics studies, defense systems, etc. TPLS design and realization are difficult tasks that require interdisciplinary cooperative efforts among researchers from different disciplines: physics, chemistry, engineering, material science, etc. This chapter describes a TPLS that is in operation at the Tor Vergata University laboratories, which is based on multistage pulsed Nd:YAG/Glass laser source (1064 nm/15 ns/10 J/TEM_{00} emission mode/pulse repetition rate = 1 shot per minute). Finally, some examples of the application of this TPLS are given.

15.1 Introduction

In terms of international competition Japan, USA and Europe recognized the potential of microsystems in the late 1980s. Microsystems will increase the competitiveness of products by lowering the manufacturing costs, by increasing the performance, or both. Promising areas are, of course, mass markets like automotive and consumer products. This potential explains the amount of effort invested in this field by large Japanese, American and European companies. The cost of microsystems depends mostly on the technology (number of steps), rather than on the actual complexity of the fabricated device. This is a new chance for innovative small and medium-sized companies. Completely new products can be imagined, that will lead to niche markets, where the European small and medium-sized enterprises are competitive. Europe is strengthening its position in this field by investing in research and development and engaging in interdisciplinary activities with industrial executives and R&D engineering groups. In this context, TPLSs are relevant particularly attention to Extreme Ultraviolet (EUV) microlithography from laser-induced plasma for microelectronics applications. This chapter emphasizes the practice rather than the theory and treats application fields of laser induced plasma, in particular concerning X spectroscopy, radiobiology, plasma backscattering radiation study and conversion efficiency estimation from laser light to plasma X radiation.

15.2 The Ultrashort and Ultrapower Lasers and Their Evolution in Time

Many research activities in the world are devoted to studying of plasma state physics and are carried out by using a simpler and less expensive experimental apparatus with smaller dimensions than to traditional radiation sources (e.g. synchrotron), called a Tabletop Pulsed Laser System (TPLS). It can be installed on a laboratory optical table and used to produce plasmas in a wide interval of charge states, densities and temperatures. Currently, thanks to a strong reduction in the laser pulse duration ($\tau \cong n \times 10\,\text{fs}$), TPLSs can reach enormous intensities in the focal plane, for example 10^{20}–$10^{21}\,\text{W/cm}^2$, corresponding to a laser pulse power of $10^{15}\,\text{W}$ (that is 1 PW). Thus such TPLSs are considered ultrashort, ultraintense and ultrahigh power laser sources.

Starting with the invention of the laser in 1960, laser peak power has progressed through a series of improvements of approximately three orders of magnitude each time. This continuous growth in laser peak power has been mainly the result of laser pulse duration reduction and, secondarily, of laser pulse energy increasing. In fact the first pulsed lasers, working in free running regime (that is phase between various longitudinal modes changes from mode to mode and position of pulse maximum traveling inside laser optical cavity), emitted pulses with duration of $10\,\mu\text{s}$ and peak power of around 1 kW ($10^3\,\text{W}$).

Till 1962, thanks to the modulation of the laser cavity quality factor (Q-switching technique), pulsed lasers were able to emit same energy of free running regime, but with pulse duration, τ, approximately 1000 times smaller (about ns) and therefore MW ($10^6\,\text{W}$) peak power range. Then in 1964, the advent of the mode locked working laser regime (based on the blocking of relative phase between various longitudinal modes inside laser optical cavity, with soliton formation and propagation) has reduced laser pulse duration, τ, of another factor of 1000, reaching ps values, therefore reaching laser pulse peak power of the order of GW ($10^9\,\text{W}$). To such GW/cm^2 laser intensity, refractive index n of material invested from laser beam becomes dependent on intensity, I, at first order through the following linear law:

$$n = n(I) = n_0 + n_1 I \,, \tag{15.1}$$

where n_1 is a constant depending on the considered non-linear optical material. Consequently, for laser pulses with gaussian radial distribution of intensity, a central laser beam zone sees a refractive index greater than edges. Therefore, optical elements act as convergent lenses (Kerr effect, a third order non-linear optical phenomenon and crossed means called Kerr medium). So, wave front deformation is produced and self focusing of laser beam radiation is observed, as shown in following figure:

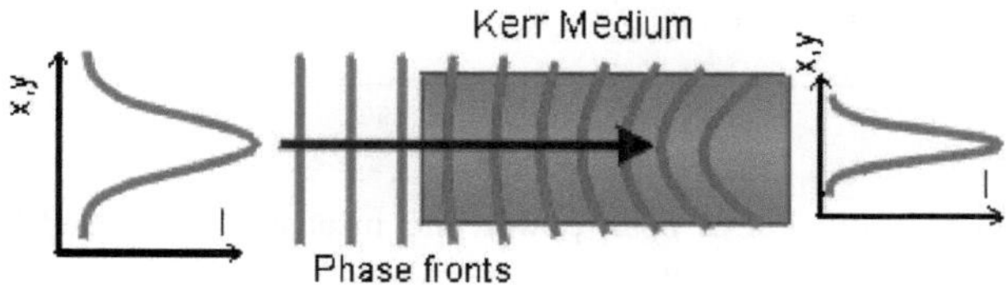

Fig. 15.1. Kerr effect scheme

From 1970 to 1985, due to Kerr effect, the increase of laser intensity beyond GW/cm^2 was a major difficulty (The Nd:YAG/Glass laser discussed here, without an increase of the laser beam dimensions through a beam expander has a glass refractive index that limits pulse intensity to approximately 10^9 W/cm^2). Without approaching different techniques, enlargement of laser beam dimensions with an increasing of laser source sizes, realization costs and a decay of pulse repetition rate was the unique research way.

In 1985, the situation changed drastically with the introduction of Chirped Pulse Amplification (CPA) technique [1,2], with intensity amplification of ultrashort laser pulses to 10^{21} W/cm^2. In the CPA technique, ultrashort laser pulse (tens of fs) emitted through an oscillator (such as Ti:Sa solid state oscillator) is not amplified directly, but it is stretched in temporal duration, τ, (tens of ps), then amplified at lower intensity level and finally compressed to tens or hundreds of fs with extremely higher intensities. Therefore, the CPA technique allows a compromise between two needs apparently in contrast: 1) maximize energy fluence for increasing energy extraction efficiency; 2) maximize laser beam intensity with negligible undesired non linear optical effects. In addition, CPA technique has the advantage of permitting the realization of compact TPLSs able to emit pulses with peak power until PW (10^{15} W).

Dimension reduction between TPLS (realized with Q-switching or CPA techniques) and synchrotron is similar to what has taken place in electronics since 1960 when, with advent of integrated circuits, sizes of electrical circuits were reduced remarkably from millimetric to micrometric scale. Finally, the following figure shows time evolution of peak power and intensity of focused laser radiation concerning pulsed lasers and relative techniques of beam processing.

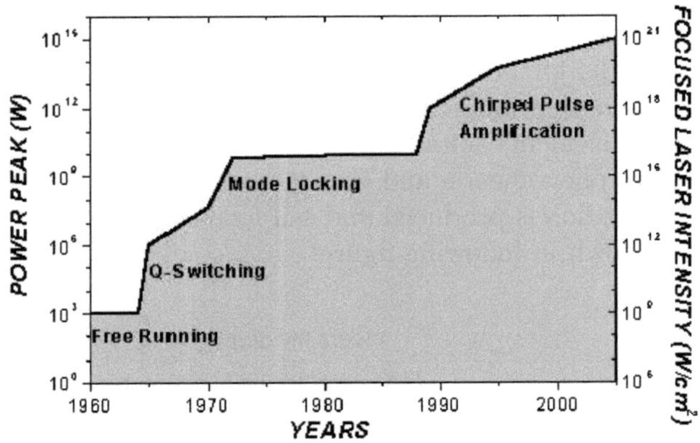

Fig. 15.2. Time evolution of peak power and intensity of focused laser radiation for pulsed lasers V_s used techniques

15.2.1 IR–soft X-ray conversion efficiency model [3]

X-ray energy (E_{x}) emitted from laser-induced plasma can be determined by measuring voltage signal through an opportune detector (as PIN photodiode protected with opportune filters) connected with an oscilloscope, then by determination of plasma X radiation energy for a unit of solid angle (indicated with $\mathrm{d}E_{\mathrm{x}}/\mathrm{d}W_x$) thanks to an opportune analytic relation and finally by integrating it on half solid angle 2π srad. Such a measurement is fundamental for models concerning conversion efficiency between laser radiation energy (in our case infrared light) and X-ray energy emitted from laser induced plasma. To such a purpose, the orthogonal cartesian reference system in polar coordinates reported in the next figure is considered, where the y axis corresponds to propagation direction of radiation and where fundamental angles are $\theta' = \pi/2 - \theta$ and $\varphi' = \pi/2 - \varphi$.

In such reference system, analytic expression to calculate X-ray energy emitted from plasma for every laser shot is the following:

$$E_{\mathrm{x}} = \int_{\Omega_x} \frac{\partial E_{\mathrm{x}}}{\partial \Omega_x} \mathrm{d}\Omega_x = \int_{\Omega_x} \frac{\partial E_{\mathrm{x}}}{\partial \Omega_x} \frac{r \cos\varphi' \mathrm{d}\theta' \cdot r \mathrm{d}\varphi'}{r^2} = \int_{\Omega_x} \frac{\partial E_{\mathrm{x}}}{\partial \Omega_x} \cos\varphi' \mathrm{d}\theta' \mathrm{d}\varphi' \, . \tag{15.2}$$

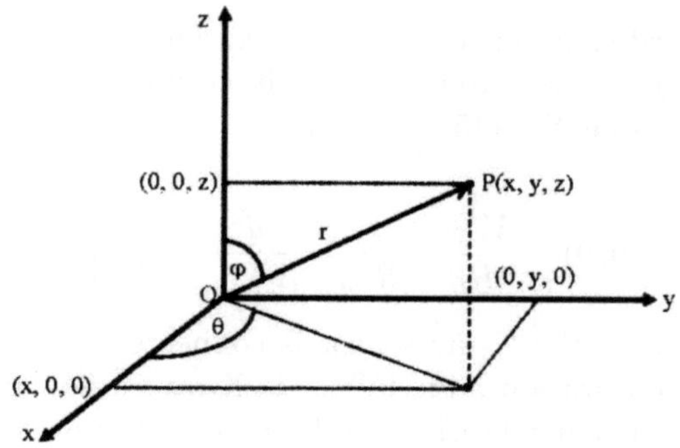

Fig. 15.3. Orthogonal cartesian reference system in polar coordinates

Assuming angular distribution of plasma X-rays with $(\cos\theta')^{0.6}$ and $(\cos\varphi')^{0.6}$, respect to peak value $\partial E_{\mathrm{x}}(y, 0°, 0°)/\partial\Omega_x (\equiv e_\Omega(d,0,0))$, we have:

$$E_{\mathrm{x}} = e_\Omega(r = y = d, 0, 0) \cdot \int_{+\frac{\pi}{2}}^{-\frac{\pi}{2}} (\cos\theta')^{0.6}\,\mathrm{d}\theta' \cdot \int_{+\frac{\pi}{2}}^{-\frac{\pi}{2}} (\cos\varphi')^{1.6}\,\mathrm{d}\varphi' \,. \quad (15.3)$$

Resolving numerically the two integrals that appear in this last equation, one obtains searched result:

$$E_{\mathrm{x}} \cong e_\Omega(r = y = d, 0, 0) \cdot 2.3 \cdot 1.708 = e_\Omega(r = y = d, 0, 0) \cdot 3.93 \quad [\mathrm{J}]\,, \quad (15.4)$$

where $r = y = d$ is the distance between target and X-ray detector along normal direction to target.

Fortunately, $e_\Omega(d, 0, 0)$ can be measured through an X-ray sensor, along the normal direction to target (maximum emission direction) and spaced d from target, and is calculated by means of the following relation:

$$e_\Omega(d,0,0) = \frac{\partial E_{\mathrm{x}}(x = y = d, \theta' = 0°, \varphi' = 0°)}{\partial\Omega_x} = \frac{\partial Q_x}{\partial\Omega_x \Re_{\mathrm{xrd}} T_f}\,, \quad (15.5)$$

where ∂Q_x is infinitesimal electric charge produced on X-ray sensor, $\Re_{\mathrm{xrd}}$ is X-ray sensor responsivity (in C/J) and T_f is transmission coefficient of all filters put between plasma and X-ray sensor. By X-ray detector connected to a load, the amount $\partial Q_x/\partial\Omega_x$ can be estimated through the relation:

$$\frac{\partial Q_x}{\partial\Omega_x} \cong \frac{\frac{V_{\mathrm{peak}}}{R_L} \cdot \tau_x}{\frac{A^{\mathrm{XRD}}_{\mathrm{exposed}}}{d^2}}\,, \quad (15.6)$$

where the produced charge is calculated as an integral of the triangular current pulse, $i(t)$, generated through the voltage pulse, $v(t)$, visualized on an opportune oscilloscope. So, (15.5) becomes:

$$e_{\Omega}(d,0,0) \cong \frac{V_{\text{peak}}}{R_L} \cdot \frac{\tau_x}{\Re_{\text{xrd}}} \cdot \frac{d^2}{A^{\text{XRD}}_{\text{exposed}}} \cdot \frac{1}{T_f}\left[\frac{\text{J}}{\text{srad}}\right] , \qquad (15.7)$$

R_L is the load where the X-ray sensor is connected, τ_x represents half the maximum X-pulse duration and $A^{\text{XRD}}_{\text{exposed}}$ is X-ray sensor area exposed to X radiation. Now, replacing (15.7) in (15.4), one obtains the following final formula for plasma X-ray energy determination:

$$E_{\text{x}} \cong 3.93 \cdot e_{\Omega}(d,0,0) \cong 3.93 \cdot \frac{V_{\text{peak}}}{R_L} \cdot \frac{\tau_x}{\Re_{\text{xrd}}} \cdot \frac{d^2}{A^{\text{XRD}}_{\text{exposed}}} \cdot \frac{1}{T_f}[\text{J}]$$
$$\Rightarrow E_{\text{x}} \propto V_{\text{peak}} \, . \qquad (15.8)$$

Now, it is possible to determine the conversion efficiency from measured laser radiation energy (E_{laser}) to X-ray energy (E_{x}) through relation:

$$\eta\,(\text{laser} \to \text{X-rays}) = \frac{E_{\text{x}}}{E_{\text{laser}}} \Rightarrow \eta\,(\text{IR} \to \text{X-rays}) = \frac{E_{\text{x}}}{E_{\text{IR}}} \, . \qquad (15.9)$$

Conversion efficiency studies from laser radiation to X-rays emitted from laser induced plasma have been conducted in Tor Vergata University using six different solid targets (see relative section) and high intensity ultrashort tabletop infrared multistage Nd:YAG/Glass laser system described later; used X-ray sensor has been a silicon PIN photodiode with $d = 0.35$ m, $A^{\text{XRD}}_{\text{exposed}} = 100\,\text{mm}^2$, $\Re_L = 50\,\Omega$, $R_{\text{xrd}} \cong 0.25$ C/J, $T_{\text{f,FWHM}} = 0.008 = 0.8\%$ (due to an Al filter 40 µm thick put in front of PIN photodiode), V_{peak} and $\tau_{\text{x,FWHM}}$ measured directly on oscilloscope (typically $\tau_{\text{x,FWHM}} \cong 15\,\text{ns}$).

15.3 Chemical Deep UV-Soft X-ray Revealers

When plasma is generated through focused laser radiation of intensity $I > 10^7\,\text{W/cm}^2$ on a target, dense and warm matter behaves like a non-monochromatic dot source of visible, UV and X radiation that can be used for various physical applications such as, UV and X-ray spectroscopy and microradiographies. When one works with radiation emitted from laser-induced plasma, its detection and recording are very important operations. More exactly, to detect plasma radiation (like UV and X-rays), it is possible to use detectors based on vacuum and PIN photodiodes while to record this radiation one usually uses particular revealers. Here we discuss only revealers that record radiation spectral-power spatial distribution. To make that, one can use different chemical deep UV and soft X-ray revealers, like Q-plates, emulsion revealers and DEF or RAR films (generally called photographic films), which we are using at Tor Vergata University and will describe in the next two sections.

15.3.1 Q-plates Emulsion Revealers [3]

"Ilford" Q-Plates have specific application for revelation of heavy ions in the mass spectrography, where emulsions cannot be used normally because particles would be stopped by the top gelatin layer. For this reason, such a layer is absent in the "Ilford" Q-Plates. The emulsion is spun into a thick film on a glass support with a technique that allows silver haloid concentration on the emulsion surface, conferring a better sensibility. extreme surface fragility makes it unsuitable for conventional photography. Therefore, Q-Plates are useful in applications where heavy atomic particles or radiation with $\lambda < 200\,\mathrm{nm}$ must be revealed.

Photographic film response to X-rays is of practical interest for diagnostic of warm plasmas produced by laser. In fact, such films are used to reveal X-rays diffused through crystals or reticules, in the spectrometers, or to carry out calibrated measures on large X-ray fluence areas. In addition, spatial resolution of photographic films is of the order of 1–10 μm, with respect to 10–20 μm of CCD cameras, but they require long development times and have a nonlinear behavior versus X-ray fluence. Thus, photographic film calibration is necessary. In our experiment, three "Ilford" Q-Plates have been exposed directly to plasma X radiation, with different fluence in various zones of Q-Plates, by using ladder filter thickness. The experiment used "Ilford" Q-Plates with dimensions $(32\times20)\,\mathrm{mm}^2$, working on a red light lamp and only touching Q-Plates edges. Then, measuring optical density with a densitometer, we obtained the calibration curve of Q-Plates. In the next figure, the experimental set-up is depicted.

"Ilford" Q-Plates have been spaced 40 cm from the plasma, inserting them on an opportune aluminum support; rectangular aluminum filters have total thickness equal to 1.5 μm or 7.5 μm or 10 μm or 20 μm or 30 μm, with gradually reduced width (ladder) inserted in a small paper envelope.

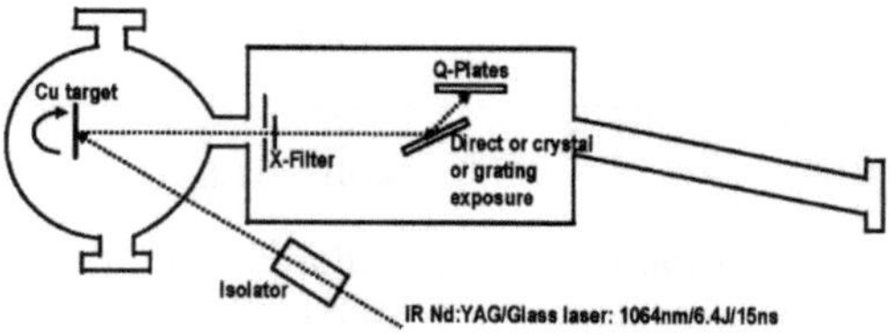

Fig. 15.4. Set-up for experiments with "Ilford" Q-plates using the Tor Vergata high intensity tabletop pulsed Nd:YAG/Glass laser source

For Q-Plate development, three small basins have been used that contain the following distilled water solutions: a) developer-type Kodak Dental X-Ray Developer 5060686, 25% volume concentration, working time with agitation = 4 minutes; b) washing-through 5% volume concentration acetic acid solution, working time with agitation = 1 minute; c) fixer-type Kodak Dental X-Ray Fixer 5060694, 25% volume concentration, working time with agitation = 4 minutes. After the Q-Plate fixing phase is finished, laboratory light can be switched on and, then, it is necessary final operation of Q-Plate washing through normal water for 8 minutes. In the following figures, we report some results obtained during our experiments carried out with "Ilford" Q-Plates.

From the last figure, it is possible immediately to observe that, comparing the results obtained with second and third "Ilford" Q-Plate (that is by using total number of incident X-photons six times greater with around the same intensity), optical density growth of Q-Plates has not been observed. Thus, their optical fixing law does not seem to be integral or optical saturation phenomena happen at working wavelength.

Considering the second "Ilford" Q-Plate, one can deduce the Optical Density-Intensity diagram, where intensity is photon number per area unit divided laser shot number; as example, for the second Q-Plate, intensity (I) is expressed through $\frac{\sharp\,\text{photons}}{5\ \text{laser shots}\cdot\mu\text{m}^2}$. For such a purpose, it is important to determine parameter I; consequently, one must know the X-ray energy per solid angle unit, e_{Ω}, emitted from laser induced plasma. In the case of our

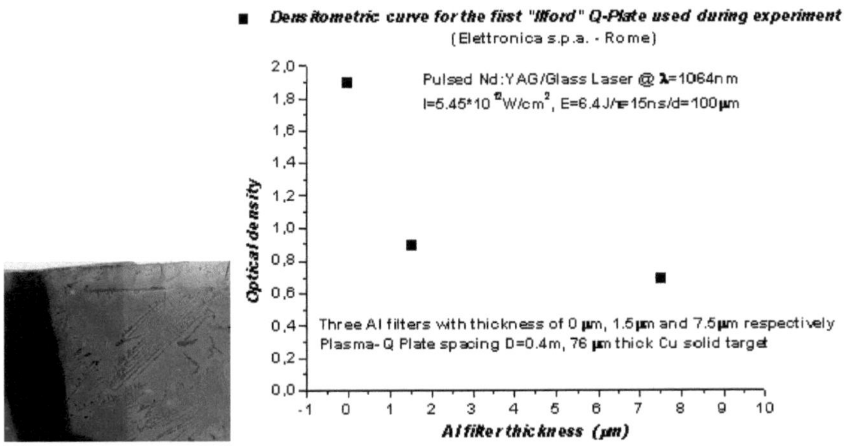

Fig. 15.5. *On the left*: vision of the first "Ilford" Q-Plate after its development. *On the right*: densitometric diagram of the first "Ilford" Q-Plate obtained thanks to Elettronica s.p.a. of Rome. Five laser shots have been necessary during the experiment

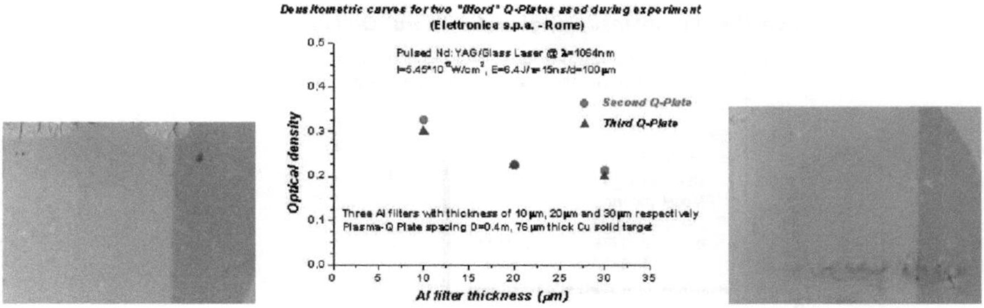

Fig. 15.6. *On the left*: vision of the second "Ilford" Q-Plate after its development. *In the center*: densitometric diagram of the second and third "Ilford" Q-Plate obtained thanks to Elettronica s.p.a. of Rome. *On the right*: vision of the third "Ilford" Q-Plate after its development. Five and 30 laser shots have been necessary during experiments with the second and third "Ilford" Q-Plate, respectively

experiment, X-ray energy is equal to 1.24 keV, corresponding to Cu L-shell energy; so, from (15.7) and figure parameters, one obtains:

$$e_\Omega = 0.724\frac{\text{mJ}}{\text{srad}} \Rightarrow e_s|_{d=0.4m} = 4.53\frac{\text{mJ}}{\text{m}^2} \Rightarrow I = 22.8\frac{\text{photons}}{\text{laser shots} \cdot \mu\text{m}^2}\,. \tag{15.10}$$

Then, by multiplying the intensity I for an aluminum filter transmissivity at 1.24 keV considering different Al thicknesses and for additional factors represented through the total number of laser shots (that is five for our experiment with the second "Ilford" Q-Plate), it is possible to calculate the total X-photon number per area unit (in our case expressed in μm^2) that has produced optical density measured for Q-Plate.

Finally, the Optical Density-Intensity diagram obtained at Tor Vergata University for the "Ilford" Q-Plates is shown in the following figure.

15.3.2 DEF and RAR Film

Another experiment was made to fix a slit 50 µm large on photographic film type Kodak DEF-5. These films (to direct exposure) are suitable like great area revealers for X-rays with an energy greater than a water window one (that is $hf > 0.56$ keV), in particular because they have a thin gelatinous screen that stops soft X-rays with low energy. Moreover, their emulsion is made through thin flexible film that can be cut. In addition, these films can be calibrated [4, 5] to show sensibility better than 1 photon/μm^2 and their spatial resolution arrives until to 1 μm^2 (that is 1 µm × 1 µm).

The following figure shows a calibration example of Kodak DEF film versus incident photon number per area unit for three different photon energies.

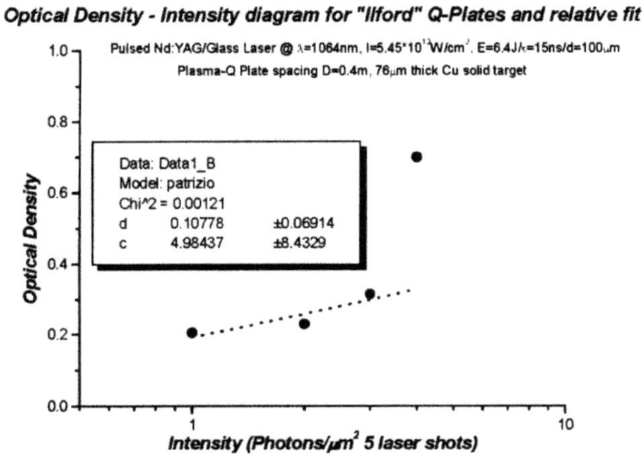

Fig. 15.7. The Tor Vergata calibration curve for the "Ilford" Q-Plates valid when external parameters are those reported on diagram (in particular when laser shot number is five and the X-ray energy emitted from laser induced plasma is 1.24 keV, corresponding to Cu L-shell energy)

In our experiment, DEF-5 Kodak film has been put in an opportune film support to contact with 50 μm large slit. Such support permits implantation of eventual input filters; in our case, we have used an aluminum filter 6 μm thick to stop visible and UV radiation. Then, this structure has been fixed on a support and exposed to soft X radiation emitted from the Tor Vergata Nd:YAG/Glass laser source-induced Mg plasma (L-shell at 0.93 nm), with plasma-DEF film spacing of 15 cm. Subsequently, DEF-5 Kodak film has been

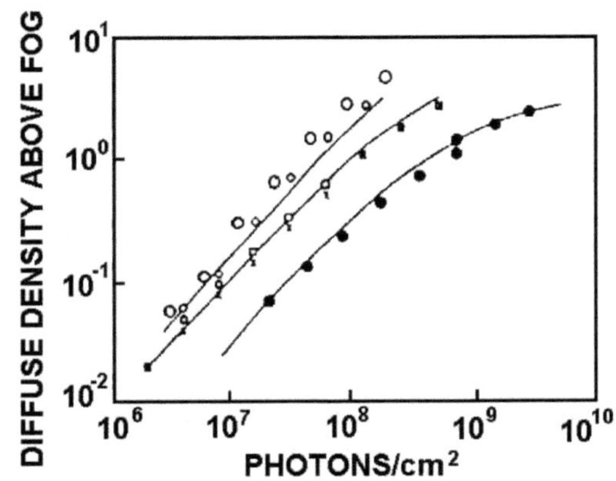

Fig. 15.8. Calibration curve for Kodak DEF film versus incident photon number per area unit for three different photon energies. For example, *full black circles* are associated to photons with energy equal to 0.93 keV, corresponding to Cu L-shell energy

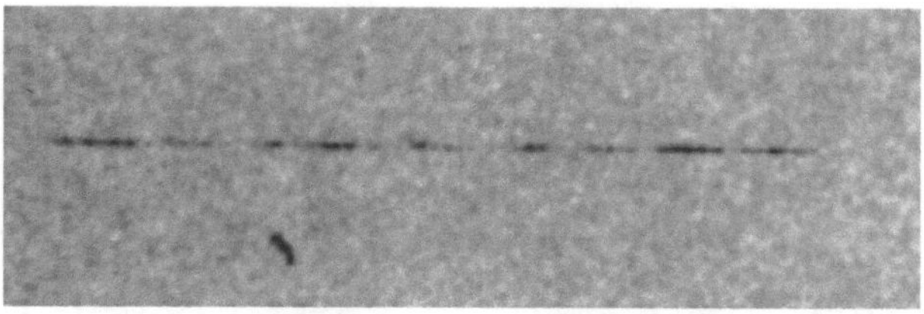

Fig. 15.9. Development of Kodak DEF-5 film through conventional method after exposure to soft X-rays emitted from the Tor Vergata Nd:YAG/Glass laser source-induced Mg plasma (L-shell at 0.93 nm). Non-uniformity is due to debris presence in the slit

developed through conventional methods and the following figure illustrates the obtained result.

Moreover, some examples of X microradiographies concerning biological samples, obtained by using soft X-rays produced by Tor Vergata Nd:YAG/Glass laser source induced Cu plasma and fixed on DEF-5 Kodak or 50 ASA photographic film are reported. In particular, in the next figures we show two soft X-ray microradiographies of biological cellules and structures concerning leaves of watered plant doped through water-copper sulfate (H_2O-$5CuSO_4$) solution, that causes Cu concentration growth along lymphatic conduits.

As in the first operation, in the film support, we have put 50 ASA photographic film rear (sensitive to soft X-rays) and the small leaves front, fixed on edge. Then, film support has been put tens of centimeters far from Cu plasma, along normal direction of plasma maximum emission, and one or two laser shots have been used. Moreover, observing the obtained results, one notes that leave regions constituted by chlorophyll conduits are more absorbent than others, where are present only protein layers, and consequently appear clearer on film. For biological cellules, DEF-5 Kodak film has been used; they have been fixed through glue on X-ray transparent plastic film that has been put inside of film support. Finally, outside of it, an aluminum filter 6 mm thick has been placed to stop visible and UV radiation. Then, support has been fixed to around 15 cm from Tor Vergata Nd:YAG/Glass laser-induced Cu plasma, along normal direction of maximum emission.

In conclusion, it is necessary to underline that in our soft X-ray spectroscopy or microscopy experiments with laser-induced plasma, we also use Kodak RAR 2492 film as X-ray fixer, which has similar characteristics to DEF-5 Kodak film. Some examples of X spectra and microradiographies, obtained through Kodak RAR 2492 film, are reported later.

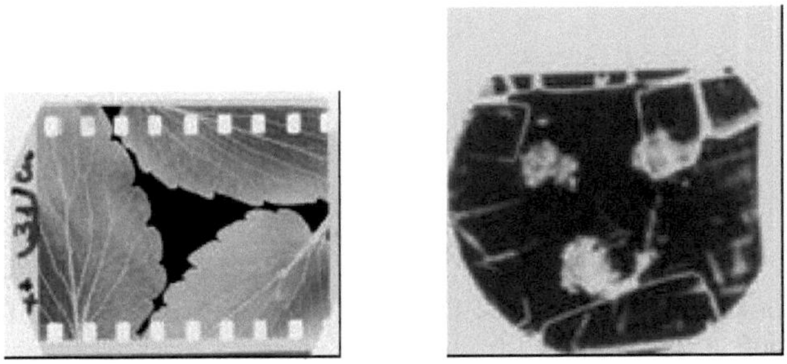

Fig. 15.10. First *on the left*: soft X-ray microradiography of leaves, doped through water-copper sulfate (H_2O-$5CuSO_4$) solution, obtained on 50 ASA photographic film. Second *on the left*: soft X-ray microradiography of biological cellules obtained on DEF-5 Kodak film. The Tor Vergata Nd:YAG/Glass laser source was used to generate plasma on Cu target

15.4 Tor Vergata Nd:YAG/Glass Laser Facility

Laser–plasma interaction is presently a subject of large and growing interest [6]. It attracts not only the attention of plasma physics community, but also of other communities like those of laser physics, spectroscopy, multiphoton physics, microlithography and so on. The growing interest is due to the availability in many basic research laboratory of laser systems with unique properties (in particular TPLS) [7] and powerful computing means, and to the perspective of several new possible applications [8, 9], besides traditional area of inertial confinement fusion. At the international level, the laser-plasma field is now dividing into many branches concerning astrophysics [10, 11], inertial confinement fusion [12, 13], quantum electrodynamics and high energy physics [14–16], atomic physics [17], solid state physics [18], X-ray spectroscopy [19–25], radiographies [26], and imaging [27].

A high temperature artificial plasma is created when an intense laser pulse is focused onto the surface of target. To produce plasma, laser pulse intensity should reach or exceed the threshold for plasma formation, which is typically about 10^7 Watt/cm^2. Intense laser light is initially absorbed in a thin layer of material (of the order of skin depth) which is heated, melted and vaporized; then, an expanding plasma cloud is generated at target surface. With a laser pulse width longer than 10 ps, plasma expansion timescale is shorter than the pulse duration, and hydrodynamics effects become important. A gradient in the plasma density develops and laser energy can be absorbed by this plasma, as it propagates towards the target until it reaches the critical surface, where plasma frequency equals laser frequency. Plasma frequency characterizes local charge oscillations in the plasma and is proportional to square root of electron density. Moreover, laser induced plasmas are good radiators and visible, UV and X radiation can be emitted from hot plasmas due to line transitions, radiative recombinations and bremsstrahlung radiation [28, 29].

Radiative properties represent classical example of non-interfering probe. Many important plasma information can be obtained directly through its emission spectrum as ionization balance, density, temperature, chemical composition, opacity, etc.; therefore, radiative properties are powerful diagnostic method of plasma state. Such consideration can alone explain the growing interest addressed to radiative properties of hot and dense matter. So, X-ray spectroscopy of multicharged ions appears as one of the most powerful techniques available for investigation of high temperature plasmas. For example, X-ray emission spectra of He-like ions have been used as diagnostics for several types of hot plasmas (magnetically confined plasmas, laser induced plasmas, astrophysical plasmas).

These artificial plasmas can be produced by experimental apparatuses that use laser sources (laser induced plasma) and can be used in numerous applications. These apparatuses are the TPLSs, with dimensions comparable with those of an optical table and use high intensity ultrashort tabletop pulsed laser systems as a light source, that are focused on solid or liquid or gaseous target to generate plasma.

Here a brief description for one of these laser systems is reported. The considered apparatus is a Tor Vergata ultraintense multistage pulsed Nd:YAG/Glass laser facility ($\lambda = 1064\,\text{nm}$, pulse duration $\tau = 15\,\text{ns}$, max pulse energy $E_{\text{max}} = 10\,\text{J}$, emission mode = TEM_{00} parallel to optical table, Pulse Repetition Rate (PRR) = 1 shot/minute), based on Q-switching technique and working on University of Rome "Tor Vergata"-Engineering Faculty–laboratories of the research group of Prof. Sergio Martellucci.

This laser system is used to generate hot plasmas (electron temperature T_e = 100–300 eV) on solid targets with different fundamental parameters as optical thickness (opacity), electron temperature and density, charge state, etc. and, consequently, to make X spectroscopy, microscopy, microradiographies and imaging by using soft X-rays emitted from laser induced plasma with energies comprised between 0.5 keV and 2 keV.

In the following figures, a Tor Vergata Nd:YAG/Glass laser apparatus is schematically described and represented.

It essentially consists of a double-stage vacuum chamber, a high power multistage tabletop pulsed solid state Nd:YAG/Glass laser source (1064 nm/10 J/15 ns), spectrometer to detect X-rays emitted from laser induced plasma and PIN (or Vacuum) photodiode to measure their energy.

Plasma is produced in the vacuum chamber using fundamental output of nanosecond Nd:YAG/Glass laser as excitation source and with base pressure of 10^{-2} mbar. The laser source can deliver pulses with power P $\cong$ 1 GW (10^9 W) and its pulse repetition rate is limited to 1 shot/minute to minimize thermal lens effect. A single lens with about 20 cm focal length focuses

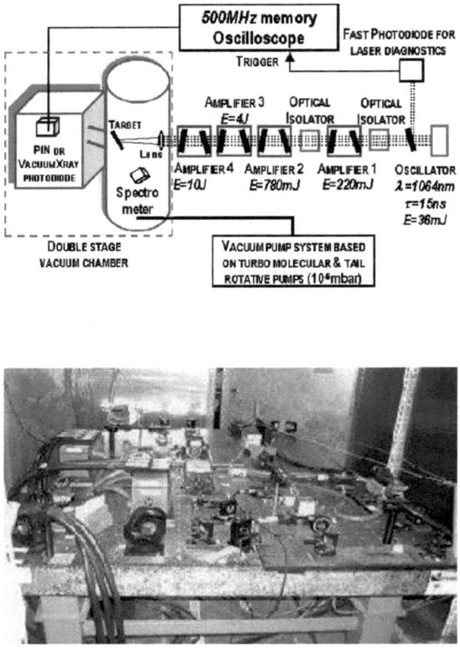

Fig. 15.11. Tor Vergata high intensity TPLS based on infrared solid state Nd:YAG/Glass laser: flow diagram and photo

laser beam on target at 45° incidence angle. Moreover, estimated laser beam diameter at focal point is around 100 μm.

Temporal profile of laser pulse is monitored for each laser shot with a fast photodiode, that provides also trigger signal to permit oscilloscope synchronization with X-ray signal measured by Quantrad PIN X-ray photodiode model 100-PIN-125N. Active surface of photodiode is shielded from UV and visible part of plasma radiation by means of thick Al layer (usually tens of μm), variable with considered X-ray region. Moreover, X-ray signal measured by PIN (or Vacuum) photodiode is visualized and recorded through fast digital oscilloscope and X-ray pulses show the same temporal profile of laser pulses with similar FWHM (around 15 ns). Finally, it is necessary to underline that X radiation, emitted from laser-induced plasma, can be detected through spherically bent mica or quartz crystal spectrometer [30, 31].

15.5 Applications of the Tor Vergata Nd:YAG/Glass Laser Facility and Experimental Results

In the following sections, several experimental results concerning laser-plasma physics are reported. These results are been obtained by using Tor Vergata ultraintense multistage tabletop pulsed Nd:YAG/Glass laser system (max laser intensity on target $I = 8.49 \times 10^{12}$ W/cm^2). The considered spectral region of produced soft X-rays is typically 0.5 keV–1.65 keV (correspond to 7.5 Å–24.8 Å in wavelength terms). Initially, some examples of X spectra concerning He-like Mg XI and near Ne-like Cu XX plasma ions are reported (Sects. 15.5.1 and 15.5.2 respectively). Then we report on an example of soft X-ray microradiography concerning a biological sample (Saponaria officinalis leaves) doped with different concentrations of cadmium acetate, and a study of plasma backscattering radiation, when used target is copper (Sects. 15.5.3 and 15.5.4 respectively). Finally, a study of conversion efficiency from the Tor Vergata Nd:YAG/Glass laser infrared radiation to soft X-rays emitted from laser induced plasma in the spectral region comprised between 1.3 keV and 1.55 keV in energetic terms (or between 8 Å and 9.56 Å in wavelength terms) is reported for six different solid targets (Sect. 15.5.5).

15.5.1 X-ray Spectroscopy

An important application of X-rays emitted from laser-induced plasma is the X spectroscopy, in particular, in the soft X-ray field as it happens with the Tor Vergata Nd:YAG/Glass laser source. The following figure reproduces the emission spectrum of He-like Mg XI plasma ions in the spectral region (9.1–9.35) Å obtained with the above-mentioned laser source [19].

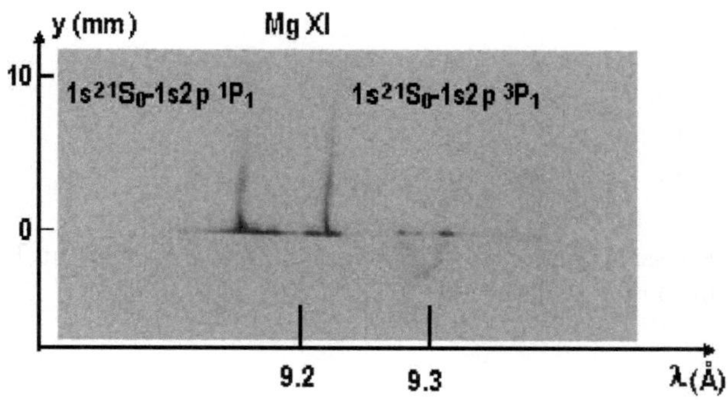

Fig. 15.12. Space resolved image in the direction of plasma expansion for He-like Mg XI ion emission spectrum in the spectral region (9.1–9.35) Å

This spectrum has been obtained with spectral resolution $\lambda/\Delta\lambda = 10\,000$ and with spatial resolution of 20 μm along the direction of plasma expansion using a spectrometer based on spherically bent crystal [32–34], which has worked in a one-dimensional (1D) scheme. Recorded spectral range contains the resonance line w (1s2p $^1P_1$–1s^2 $^1S_0$) and the recombination line y (1s2p $^3P_1$–1s^2 $^1S_0$) of He-like Mg XI ions and satellite structures due to 1s2lnl′–1s^{2}nl′ radiative transition stabilization of doubly excited states in the Li-like ion [35]. Satellite lines are placed in the edge of resonance line towards greater wavelengths with respect to resonance line ones because, during stabilization, additional nl′ electron (called spectator electron) perturbs other orbital by shielding nuclear charge, resulting in a transition wavelength somewhat longer than that of parent transition. While doubly excited states 1s2lnl′ are predominately formed via resonant process of dielectronic recombination process, 1s2l levels in the He-like ions are mainly populated by electron impact excitation [36, 37].

For $n = 2$ satellites are quite far from the resonance line, while for $n > 3$ the splitting becomes very small and many satellite lines overlap with resonance line. Therefore, all resonance lines contain contribution arising from their correspondent satellite lines. In the most common case of hot plasmas that are not too dense, contribution of such satellite lines is very low and can be neglected.

However, it should be noted that satellite and resonance line intensities show a different dependence on ion charge (Z) and plasma parameters, as electron density and plasma temperature. There are conditions under which satellite line intensities are comparable with resonance line intensity; our experimental conditions show under which conditions satellite lines give strong contribution to whole emission spectrum. In the next figure it is evident that at sufficient large distance from target surface radiation emitted from laser induced plasma is dominated by w and y lines, while satellite structure contribution appears negligible.

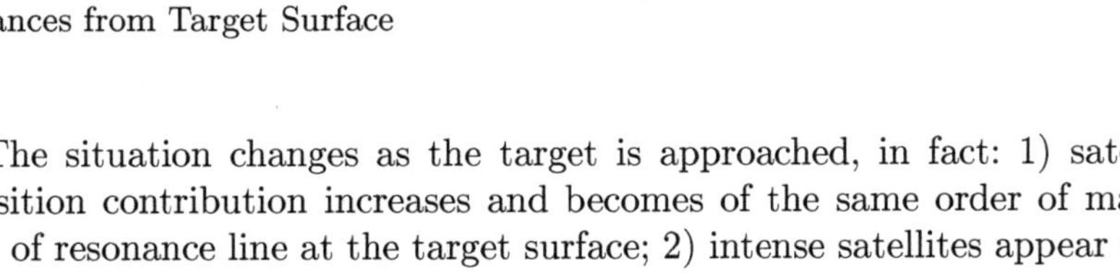

Fig. 15.13. Emission Spectrum of He-like Mg XI Plasma Ions at Three Different Distances from Target Surface

The situation changes as the target is approached, in fact: 1) satellite transition contribution increases and becomes of the same order of magnitude of resonance line at the target surface; 2) intense satellites appear close to the intercombination line; 3) the resonance line becomes more asymmetric with new spectral lines observable in its long-wavelength wing.

Neither Stark nor opacity broadening can account for experimental line width of the resonance line. According to the model developed by Dr. Faenov and Dr. Rosmej, the resonance line structure arises from contribution of Rydberg satellite transitions of type 1s2lnl′–1s^{2}nl′ [38] (satellite lines).

15.5.2 Identification and Precise Measurements of the Wavelengths of High-n Transitions of Near Ne-like Cu Plasma Ions

Precise measurements of the Rydberg series in highly charged ions provide rigorous tests of atomic structure calculations and X-ray transitions from high Z atoms with charge states around Neon-like isosequence have been object of considerable interest [39, 40]. X-ray lasing has been demonstrated in Ne-like ions [41] and the need to understand kinetics of this system has caused development of very precise collisional-radiative modeling tools (CR model) and request of high resolution experimental data to be used for benchmarking multi-electron atomic structure calculations.

Neon-like ions are also of interest as diagnostic tracers of plasma conditions: in equilibrium, their great stability across a range of temperatures makes them unambiguous emitters in X-ray spectra. X-ray spectrum of a Ne-like ion is characterized by closed shell configuration and seven series of transitions will appear in the L-shell emission spectrum of Ne-like Cu ion. These transitions arise from $2s^22p^6$–$2s^22p^5nl'$ and $2s^22p^6$–$2s2p^6nl'$ excitations. By varying energy in the laser pulses and by changing spot size of laser focus, we achieve different temperature conditions in moderate density plasmas. Thus, changes are observed in the plasma emission spectra obtained with different laser irradiation conditions. These spectrum changes allow to discriminate between different plasma charge states when many lines are emitted near to each other. In other words, acting for example on laser intensity, it is possible to alter number of lines concerning O-like Cu^{21+}, F-like Cu^{20+} and Ne-like Cu^{19+} ions present in Cu plasma with different charge states.

Three examples of measured spectra obtained by Tor Vergata Nd:YAG/ Glass laser source induced Cu plasma are shown in the next figure. Laser energy used to obtain these spectra has been 8 J and 2.5 J, while line identifications have been made by using the RELAC atomic structure code. Among all transitions belonging to L-shell, 2s-nd transitions give prevailing contribution. In the next figure (top on the left), in addition to Ne-like transitions in Cu plasma ions, several high-n Rydberg lines from more highly charged O-like and F-like Cu ions are present. Their contribution becomes completely negligible (see next figure, on the right-top and on the bottom), when a lower value of laser energy (2.5 J) has been used or when focalization lens has been defocused, that is when lower value of laser intensity on target has been used.

These spectra permit one to conclude that laser intensity (I) is the fundamental parameter that determines plasma conditions and, consequently, observed plasma emission spectra. Moreover, thanks to the previous figure, it is possible to identify Ne-like Cu^{19+}, F-like Cu^{20+} and O-like Cu^{21+} ion lines; in particular, Ne-like Cu^{19+} ion lines are those 7D, 7G, 6C, 6F, 6D, F-like Cu^{20+} ion lines are those ρ, π, μ, κ, ι, θ, η, ζ, ε, δ, α and O-like Cu^{21+} ion lines are those ν, γ, β.

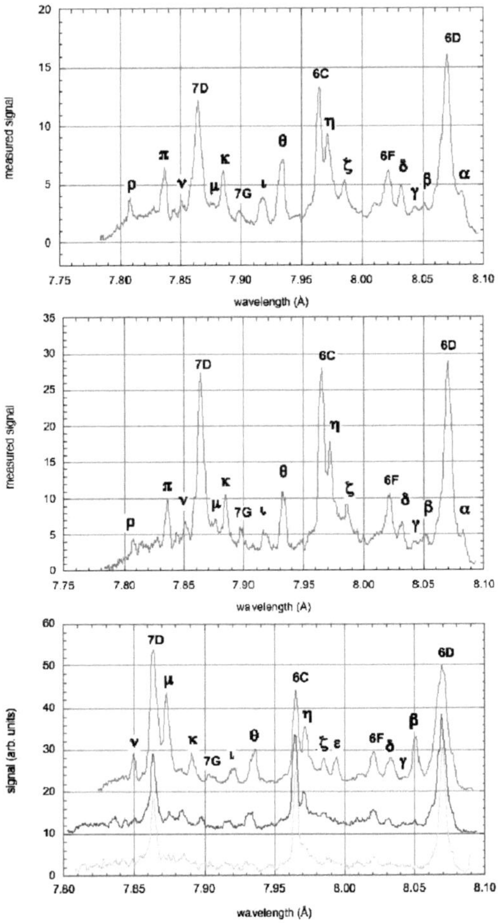

Fig. 15.14. *Top on the left*: near Ne-like Cu XX ion emission spectrum in the spectral region (7.75–8.1)Å obtained with the Tor Vergata Nd:YAG/Glass laser source, with laser energy of 8 J and with lens in best focus position. *Top on the right*: near Ne-like Cu XX ion emission spectrum in the spectral region (7.75–8.1)Å obtained with the Tor Vergata Nd:YAG/Glass laser source, with laser energy of 2.5 J and with lens in best focus position. *On the bottom*: near Ne-like Cu XX ion emission spectrum in the spectral region (7.75–8.1)Å obtained with the Tor Vergata Nd:YAG/Glass laser source, with laser energy of 8 J and with focalization lens in three different positions: best focus (*top spectrum*), defocus of 5 mm (*central spectrum*), defocus of 10 mm (*bottom spectrum*). Finally, for all spectra, identification of high-n transitions (*spectral lines*) is reported

Finally, through comparison with theoretical spectra obtained by using steady state collisional-radiative model (steady state CR model), it has been possible to determine fundamental parameters of Cu plasma produced by the

Tor Vergata Nd:YAG/Glass laser source. So, Cu plasma conditions, whose X spectra are reported in the previous figure, are the following in the hypotheses of isothermal and spherical plasma:

1. $T_i = T_e$ where T_i = ion temperature and T_e = electron temperature;
2. $100\,\text{eV} \leq T_e \leq 300\,\text{eV}$, depending on laser intensity on target;
3. $5\times10^{20}\,\text{cm}^{-3} \leq n_e \leq 10^{21}\,\text{cm}^{-3}$, depending on laser intensity on target (n_e = electron density);
4. $L = 150\,\mu\text{m}$, where L is plasma diameter;
5. opaque plasma (or plasma with optical thickness), that is one must consider opacity effect for reproducing theoretically the spectral line shapes and intensities observed experimentally;
6. absence of hot electrons, that is of electrons with higher energy (of the order of keV) than average one;
7. plasma charge state depending on laser intensity on target.

15.5.3 X-ray Microscopy and Radiobiology [26]

Soft X-rays emitted from laser induced plasma are widely used for biological sample imaging and different techniques have been developed to make X-ray imaging, as soft X-ray contact microscopy based on X-ray absorption in the water window ($\lambda = 2.2$–$4.3\,\text{nm}$) and concerning living cells and as biological sample microradiography (like leaves or small animals), that uses soft X-rays with energy comprised between 1 keV and 1.5 keV ($\lambda = 8$–$12\,\text{Å}$ in wavelength terms).

By passing X-rays through small size biological sample (usually $20 \times 20\,\mu\text{m}^2$), one can obtain its image with magnification around 1000 and with very high spatial resolution (around 30 nm) using CCD camera. Projection microradiographies of larger biological samples (of the order of a few mm) can also be obtained by using spectrometer based on spherically bent mica or quartz crystal, with spatial resolution of some μm and with selecting possibility a precise wavelength thanks to the use of crystal as monochromator.

Due to penetration depth of X-rays inside the biological sample, X-ray imaging can provide more and better structural information with higher spatial resolution (30 nm) than conventional techniques, such as optical and electron microscopy. Therefore, X-ray imaging is a useful tool in biology that permits one to obtain images even of whole living cells. Moreover, by using X-rays, one can execute measurements of trace elements (in particular heavy metals, like Ca, Fe, Ni, Cd, etc.) and, therefore, make maps of accumulation sites in several biological samples by different techniques, as X-ray absorption, X-ray fluorescence microprobes and fluorescence microtomography. Consequently, semi-quantitative estimations of element concentration down

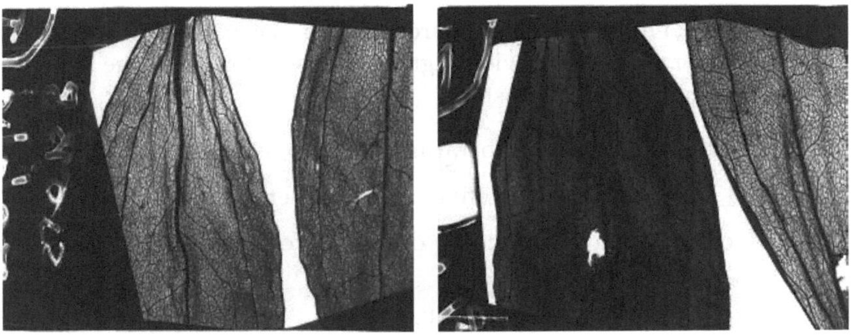

Fig. 15.15. Examples of soft X-ray microradiographies of Saponaria officinalis leaves obtained by using X radiation emitted from Tor Vergata Nd:YAG/Glass laser induced Cd plasma

to a few pg/cm^2 with sub-micron spatial resolution in a relatively large area (mm^2) can be obtained. In particular, it is possible to make sample imaging to localize specific elements already present in the sample or even artificially absorbed through doping procedure. In this way, one can evidence inside the biological structures where a particular substance is present.

In the following figure, microradiographies of Saponaria officinalis leaves doped with different concentrations of cadmium (Cd) are reported. These radiographies are been obtained by using soft X-rays with energy of about 1 keV emitted from the Tor Vergata Nd:YAG/Glass laser induced Cd plasma and with the aim of distinguishing different levels of cadmium doping inside Saponaria leaves, based on X-ray absorption by leaves in non monochromatic X-ray microradiography.

Potted Saponaria officinalis plants were grown for two months in the open air. Then, some plants were treated for 3 or 10 days with different concentrations of cadmium acetate and values of dopant concentrations were 2 mM and 15 mM. Other Saponaria officinalis plants, instead, were not doped with cadmium (control plants), were grown under the same conditions of doped plants and were maintained with pure water. Both the treated and control samples were dehydrated through the same procedure and time duration. Moreover, the doped and control Saponaria officinalis leaves were mounted together on holder, equipped with the same polypropylene filter, and exposed simultaneously to avoid errors due to intensity fluctuations between a laser shot and another one.

Saponaria leaves with similar morphology and the same mass per surface unit within a few percent (3.8 mg/cm^2 in our case), after being dehydrated, were chosen. Then, Saponaria leaves were exposed to soft X-rays in the range 1–1.5 keV (that is 0.8–1.2 nm, which includes K-edge of cadmium absorption and, therefore, this radiation is mainly absorbed in the Saponaria leave sites where cadmium is present) emitted from laser induced plasma, using Cd as solid target in a vacuum chamber. Samples were exposed to soft X-rays produced by four laser shots, plasma–sample spacing was 15 cm, estimated X-ray fluence was about 30 mJ/cm/shot on the sample and the laser source used to generate plasma was the Tor Vergata high intensity Nd:YAG/Glass TPLS with an energy per laser pulse, effectively used during experiment, of 5 J.

On the left, the last figure is shows a comparison between the control leaf (on the left side) and 2 mM cadmium doped Saponaria leaf (on the right side), while in the center a comparison between the control leaf (on the right side) and 15 mM cadmium doped Saponaria leaf (on the left side) is reported. From these comparisons, it is possible to conclude that Saponaria leave microradiographies are darker when cadmium concentration on leaves increases. Therefore, the control (no doped with cadmium) sample microradiographies appear brighter than doped sample ones. In particular, Saponaria leave microradiographies doped with 15 mM of cadmium are darkest, due to great absorption of X-rays emitted from laser induced Cd plasma.

In conclusion, the most important result of this experiment, which was executed in the Tor Vergata laboratories, was that our laser-plasma system, used to make soft X-ray microradiography, was able to distinguish between biological samples with different Cd doping concentrations through different darkening of microradiography, due to different absorption on sample of soft X-rays generated by laser induced plasma.

15.5.4 Plasma Backscattering Radiation Study

The production of laser-induced plasma is an important and interesting phenomenon. In particular, when one works with a solid target, intense laser light is initially absorbed in a thin layer of material (of the order of skin depth), which is heated, melted and vaporized; then, an expanding plasma cloud is generated at target surface. As already said previously, working with laser pulse widths longer than 10 ps, the timescale of plasma expansion is shorter than pulse duration and hydrodynamics effects become important. A gradient in the plasma density develops and laser energy can be absorbed by this plasma as it propagates towards target until it reaches critical surface where electron density, n_e, is equal to critical density, n_c, and where plasma frequency, $f_p(n_e)$, equals laser frequency f_0.

So, in correspondence of the critical surface of solid target, plasma behaves as mirror and is able to reflect laser radiation mainly on specular direction respect to incidence one.

Backscattering radiation coming from laser-induced plasma is very dangerous for the laser source and for the optical elements because it can damage them (in particular of first amplifiers and oscillator rods), compromising laser performances. So, plasma backscattering radiation study is important when one works with laser-induced plasma, in order to minimize this problem.

In the case of the Tor Vergata Nd:YAG/Glass laser source-induced plasma, we used MOPA (Master Oscillator Power Amplifier) laser architecture with optical isolators. In particular, varying the energy of the Tor Vergata Nd:YAG/Glass laser source, we measured the energy, $E^{(s)}_{\text{refl}}$, of component with S vertical polarization of plasma backscattering radiation and its reflectivity, $R^{(s)}$, along the incident direction, when copper is used as solid target. The results of this experiment are reported in the following table and graphs.

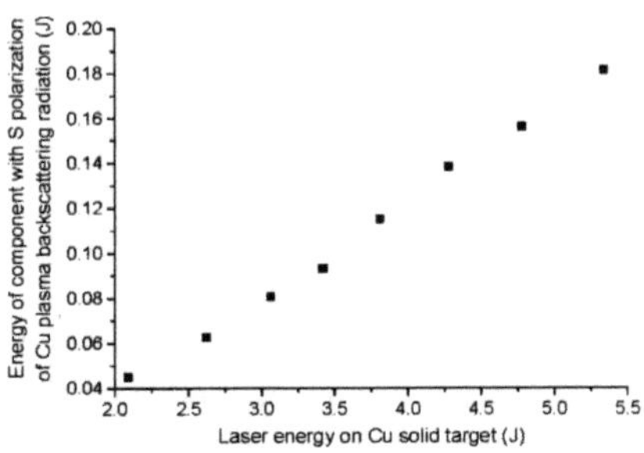

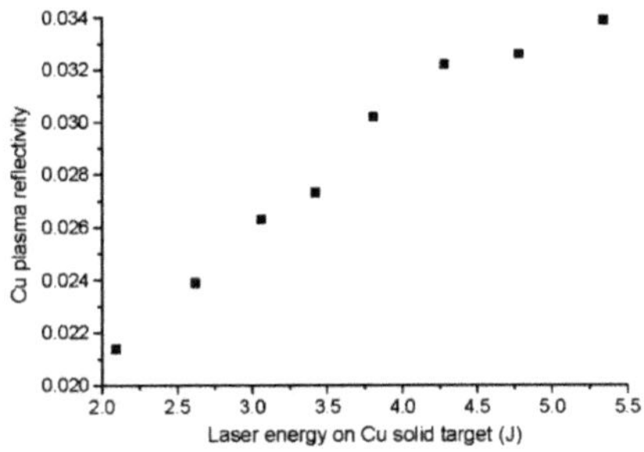

Fig. 15.16. *On the left*: energy of component with S polarization of plasma backscattering radiation along incident direction when copper is used as solid target. *On the right*: corresponding Cu plasma reflectivity relative to S polarization backscattering radiation

From the table and graphs that reported above, it is possible to conclude that the energy of the component with S polarization of Cu plasma backscattering radiation and its reflectivity along the incident direction increases when

Table 15.1. Energy, $E^{(s)}_{\text{refl}}$, of the component with S polarization of plasma backscattering radiation and its reflectivity, $R^{(s)}$, along incident direction, with 56° polarizer plate, $\lambda/4$ plate for producing incident circular polarization and linear reflected polarization of laser beam and copper as solid target.

E_{laser}(J)	$E^{(s)}_{\text{refl}}$(J)	$R^{(s)}_{\text{copper}} = E^{(s)}_{\text{refl}}/E_{\text{laser}}$
2.09	0.0448	0.0214
2.62	0.0627	0.0239
3.06	0.0806	0.0263
3.42	0.0932	0.0273
3.81	0.115	0.0302
4.28	0.138	0.0322
4.78	0.156	0.0326
5.34	0.181	0.0339

the laser energy grows on a solid target. These results were obtained with a Cu target, but they are valid in general for solid targets.

Finally, always using table data, one can conclude that in the case of the Tor Vergata Nd:YAG/Glass laser source the energy of Cu plasma backscattering radiation assumes values comprised between 45 mJ and 180 mJ around, with reflectivity comprised between 2.1% and 3.4% around.

15.5.5 IR–soft X-ray Conversion Efficiency Experimental Results

Results concerning conversion efficiency, η, from infrared (IR) laser radiation to soft X-rays emitted from the Tor Vergata Nd:YAG/Glass laser source-induced plasma, using various solid targets, is reported to find a target with the best IR-soft X-ray conversion efficiency between those considered. The studied spectral region to estimate η is that of soft X-rays around 1.5 keV. More exactly, this region is comprised between 1.3 keV and 1.55 keV in energetic terms (or between 8 Å and 9.56 Å in wavelength terms). Such a spectral region has been selected through an aluminum filter with a total thickness of 40 μm, put in front of PIN photodiode used for detection of soft X-rays emitted from plasma. During the experiment, we used the following solid targets: 1) MAGNESIUM (Mg, atomic number $Z = 12$); 2) TITANIUM (Ti, atomic number $Z = 22$); 3) IRON (Fe, atomic number $Z = 26$); 4) COPPER (Cu, atomic number $Z = 29$); 5) ZINC (Zn, atomic number $Z = 30$); 6) YTTRIUM (Y, atomic number $Z = 39$).

For all executed measurements, the laser energy (E_{laser}) varies between 3.25 J and 9.91 J, while the focalization lens is always maintained in the best focus position to have a laser spot with diameter of around 100 mm on the target and, consequently, laser beam intensity between 2.76×10^{12} W/cm^2 and 8.41×10^{12} W/cm^2. The obtained results are shown in the following figures.

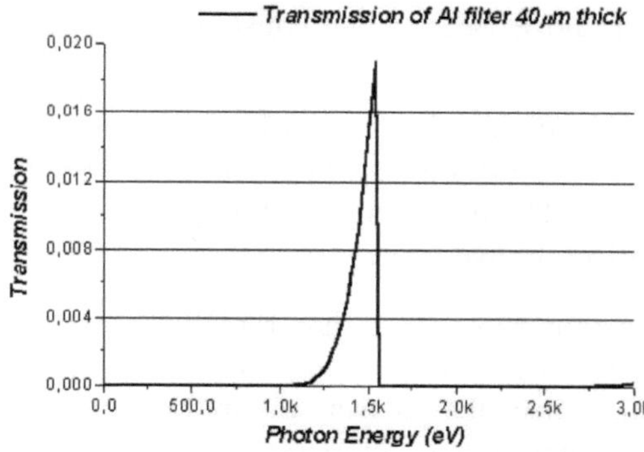

Fig. 15.17. Transmission coefficient of aluminum filter 40 μm thick versus incident photon energy in the considered region (around 1.5 keV)

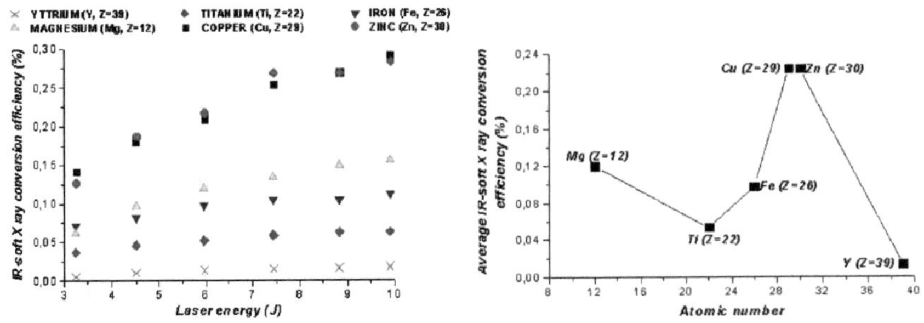

Fig. 15.18. *On the left*: in the soft X-ray spectral region around 1.5 keV, IR-soft X-ray conversion efficiency (η) versus laser energy using the Tor Vergata Nd:YAG/Glass TPLS to generate plasma on *solid targets*. *On the right*: average IR-soft X-ray conversion efficiency, $< \eta >$, in the same spectral region around 1.5 keV versus atomic number of considered element

Therefore, we have demonstrated that our experimental apparatus, based on the Tor Vergata Nd:YAG/Glass laser source, is able to produce soft X-rays from laser-induced plasma through estimation of the conversion efficiency (η) from an infrared (IR) laser radiation to soft X-rays (IR-soft X-ray conversion efficiency estimation) for six different solid targets in the soft X-ray spectral region comprised between 1.3 keV and 1.55 keV (that is 8–9.56 Å).

Through this study and using information already known from the literature, it is possible to conclude that:

1. IR–soft X-ray conversion efficiency (η) depends on the type of target used, being connected to transition probability (variable changing element) between levels involved in the emission of X radiation (see previous figure);
2. IR–soft X-ray conversion efficiency (η) increases/decreases when laser energy (and therefore laser intensity) on the target increases/decreases (see previous figure);
3. IR–soft X-ray conversion efficiency (η) depends on studied soft X-ray spectral region and h varies when considered soft X-ray spectral region changes;
4. between six solid targets taken in consideration, copper (Cu, $Z = 29$) and zinc (Zn, $Z = 30$) are those that have the best IR–soft X-ray conversion efficiency ($\eta \cong 0.22\%$ in average), while yttrium (Y, $Z = 39$) has the worst ($\eta \cong 0.013\%$ in average).

Finally, it is important to observe that remaining part of laser energy not converted into soft X-ray energy (more than 99%) is found as visible or UV radiation and, especially, as heat dissipation necessary to warm solid target from initial environment temperature (about 25 °C) to final plasma temperature (10^6 °C at least for our laser source).

15.6 Conclusions

We have demonstrated the possibility of using systems for generation of soft X radiation by dense and hot plasmas, produced through high intensity pulsed tabletop laser (such as the Tor Vergata Nd:YAG/Glass TPLS) focused with micrometric spot on solid targets, like magnesium, copper, iron, zinc, etc. Moreover, experiments of X emission spectroscopy and microscopy (in particular, concerning microradiographies of biological samples) were made in the range of wavelengths around 1 nm (soft X-rays). In addition, sensor techniques were developed for X-ray detection, based on solid state photodiodes (as PIN photodiode) or vacuum technology (as vacuum photodiode), and to record soft X emission spectra or microradiographies, based on photosensitive emulsions (as Q-plates, DEF and RAR films). Moreover, studies concerning plasma backscattering radiation in the case of solid target (as copper) and IR R soft X-ray conversion efficiency were reported.

In particular, concerning soft X-ray emission spectroscopy, we have shown emission spectrum of He-like Mg XI plasma ions in the spectral region (9.1–9.35) Å and emission spectrum of near Ne-like Cu XX plasma ions in the spectral region (7.75–8.1) Å obtained with the Tor Vergata Nd:YAG/Glass TPLS. Thanks to these spectra, it is possible to conclude that laser intensity (I) is the fundamental parameter for determination of plasma conditions and, consequently, of observed spectral lines due to line transitions. More exactly, when laser intensity increases, one observes an increase or appearance of spectral line intensity relative to higher charge states in the plasma ions; while, when laser intensity decreases, one sees a reduction or disappearance of spectral line intensity relative to higher charge states in the plasma ions.

Finally, making studies of biological samples through soft X-ray microradiographies on Saponaria officinalis leaves, we have observed that our Nd:YAG/Glass laser-plasma system is able to distinguish between biological samples with different Cd doping concentrations.

References

1. D. Strickland and G. Mourou, Opt. Commun. **56** (1985), 219.
2. P. Maine, D. Strickland, P. Bado, M. Pessot, G. Mourou, IEEE J. Quantum Electronics **24** (1988), 398.
3. P. Ciuffa, "Sistemi per la Generazione e Rivelazione di Raggi X da Plasma Indotto da Laser, per applicazioni di Microlitografia, Radiobiologia, Microscopia", PhD Thesis of Microsystem Engineering, XIII Cycle, University of Rome "Tor Vergata", Academic Year 1999–2000.
4. G. J. Tallents et al., "Film calibration for soft X-ray wavelengths", SPIE **3157**.
5. "X ray streak camera", Kentech Instruments Ltd.
6. W. L. Kruer, "Physics of Laser Plasma Interaction", Addison Wesley, Redwood City, CA (1988).
7. G. Mourou, J. Barty, M. D. Perry, "Physics Today" (Jan 1998).

8. I. C. E. Turcu, J. B. Dance, "X rays from laser plasma: Generation and Applications", Wiley, London (1998).
9. E. P. Liang, S. C. Wilks, M. Tabak, Phys. Rev. Lett. **81** (1998), 4887.
10. B. A. Remington, D. Arnett, R. P. Drake, H. Takabe, Science **284** (1998), 1488.
11. B. A. Remington, D. Arnett, R. P. Drake, H. Takabe, Phys. Plasmas **7** (2000), 1641.
12. J. D. Lindl, Phys. Plasmas **2** (1995), 3933.
13. W. Hogan et al., "Energy from Inertial Fusion", IAEA, Vienna (1994).
14. K. W. D. Ledingham, P. A. Norreys, Contemp. Phys. **40** (1999), 367.
15. T. Tajima, J. M. Dawson, Phys. Rev. Lett. **43** (1979), 267.
16. A. Pukhov, Z. M. Sheng, J. Meyer Ter-Vehn, Phys. Fluids **6** (1999), 2847.
17. M. Gavrila, Adv. At. Mol. Opt. Phys. Suppl. **1** (1992), 435.
18. P. Celliers et al., Phys Rev. Lett. **84** (2000), 5564.
19. I. Bellucci, P. Ciuffa, F. Flora, S. Martellucci, G. Petrocelli, "Spettroscopia ad alta risoluzione di un plasma di Mg", Collana di quaderni di ottica e fotonica **N. 7**, Societa italiana di ottica e fotonica – 40 Anni di laser, Firenze (28 November 2000), 159–162.
20. K. B. Fournier et al., "Observations of high-n transitions in the spectra of near-neon-like copper ions from laser-produced plasmas", J. Phys. B: At. Mol. Opt. Phys. **35** (2002), 3347–3364.
21. I. Yu. Skobelev et al., "Spectral Transitions from the Rydberg Autonionization States of a Li-Like Mg X Ion", Journal of Experimental and Theoretical Physics **95**, **N. 3** (2002), 421–428.
22. K. B. Fournier et al., "Rydberg transitions in the spectra of near-neon-like Cu and Zn ions in different laser-produced plasmas: observations and modeling", Journal of Quantitative Spectroscopy & Radiative Transfer **81** (2003), 167–182.
23. K. B. Fournier, A. Ya. Faenov, T. A. Pikuz, A. I. Magunov, I. Yu. Skobelev, V. S. Belyaev, V. I. Vinogradov, A. S. Kyrilov, A. P. Matafonov, F. Flora, S. Bollanti, P. Di Lazzaro, D. Murra, A. Reale, L. Reale, G. Tomassetti, A. Ritucci, M. Francucci, S. Martellucci, G. Petrocelli, "Identification and precise measurements of the wavelengths of high-n transitions in N-, O-, and F-like Zn ions", J. Phys. B: At. Mol. Opt. Phys. **36** (2003), 3787–3796.
24. K. B. Fournier, A. Ya. Faenov, T. A. Pikuz, A. I. Magunov, I. Yu. Skobelev, F. Flora, S. Bollanti, P. Di Lazzaro, D. Murra,, V. S. Belyaev, V. I. Vinogradov, A. S. Kyrilov, A. P. Matafonov, M. Francucci, S. Martellucci, G. Petrocelli, "Analysis of high-n dielectronic Rydberg satellites in the spectra of Na-like Zn XX and Mg-like Zn XIX", Physical Review E **70**, 016406 (2004).
25. A. Ya. Faenov et al., "X-ray Spectroscopic Observations of a Superdense Plasma in Nanoparticles Irradiated by Superintense Femtosecond Laser Radiation", JETP Letters **80**, **N. 12** (2004), 730–733.
26. L. Reale, A. Lai, A. Tucci, A. Poma, A. Faenov, T. Pikuz, F. Flora, L. Spano, T. Limongi, L. Palladino, A. Ritucci, G. Tomassetti, G. Petrocelli, M. Francucci, S. Martellucci, "Differences in X-Ray Absorption Due to Cadmium Treatment in Saponaria officinalis Leaves", Microscopy Research and Technique **64** (2004), 21–29.
27. T. Pikuz et al. "Easy spectrally tunable highly efficient X-ray backlighting schemes based on spherically bent crystals", Laser and Particle Beams **22** (2004), 289–300.

28. R. Falkone, M. Murname, AIP Conf. Proc. **147** (1986), 81.
29. M. Murname, H. C. Kapteyn, M. D. Rosen, R. Falkone, Science **251** (1991), 531.
30. A. Ya. Faenov et al., "High-performance X-ray spectroscopic devices for plasma microsources investigations", Physica Scripta **50** (1994), 333.
31. S. A. Pikuz et al., "High-luminosity monochromatic X-ray backlighting using an incoherent plasma source to study extremely dense plasmas", Rev. Sci. Instrum. **68** (1997), 740–744.
32. I. Uschmann, K. Fujita, I. Niki, R. Butzbach, H. Nishimura, J. Funakura, M. Nakai, E. Forster, K. Mima, Appl. Opt. **39** (2000), 5865.
33. J. A. Koch et al., Rev. Sci. Instrum. **74** (2003), 2130.
34. T. A. Pikuz et al., Laser and Particle Beams **19** (2001), 285.
35. A. H. Gabriel, Mon. Not. R. Astron. Soc. **160**, (1972), 99.
36. C. Biedermann, R. Radtke, K. B. Fournier, Phys. Rev. E **66** (2002), 66.
37. A. Ya. Faenov et al., Physica Scripta **T80** (1999), 536.
38. B. Rosmej, A. Ya. Faenov, Physica Scripta **T73** (1997), 106.
39. S. Ya. Khakhalin et al., JETP **78** (1994), 633.
40. S. Ya. Khakhalin et al., J. Opt. Soc. Am. B **12** (1995), 1203.
41. M. D. Rosen et al., Phys. Rev. Lett. **54** (1985), 106.

16 Induction of Permanent Structure in Transparent Materials by Ultrafast Laser and Application to Photonic Devices

Yasuhiko Shimotsuma[1], Jianrong Qiu[2], Kiyotaka Miura[1], and Kazuyuki Hirao[1]

[1] Department of Material Chemistry, Kyoto University, Kyotodaigaku-Katsura, Nishikyo-ku, Kyoto, 615-8510 Japan
[2] Department of Materials Science and Engineering, Zhejiang University, 38 Zheda road, Hangzhou 310027, China

16.1 Introduction

The ultrafast laser has been used as a powerful tool to clarify elementary processes, such as excitation-energy relaxation and both electron and proton transfer on nanosecond and picosecond time scales that occur in a micrometer-sized area. In the past decades, the ultrafast laser has been widely used in micro processing. A femtosecond laser, which is the typical ultrafast laser, has two apparent features compared with CW and long pulsed lasers: (1) elimination of the thermal effect due to extremely short energy deposition time, and (2) participation of various nonlinear processes enabled by strong localization of laser photons in both time and spatial domains. Due to the ultrashort light-matter interaction time and the high peak power density, material processing with the femtosecond laser is generally characterized by the absence of heat diffusion and consequently molten layers [1]. The photo-induced reactions are expected to occur only near the focused part of the laser beam due to multiphoton processes. In the past several years, much research has been devoted to the field of three-dimensional (3D) microscopic modifications to transparent materials by using femtosecond laser. Promising applications have been demonstrated for the formation of 3D optical memory [2, 3] and multicolor images [4], the direct writing of optical waveguides [5, 6], waveguide couplers and splitters [7, 8], waveguide optical amplifiers [9], and optical micro-gratings [10, 11]. Furthermore, in the recent applied researches, femtosecond lasers have been used as a fabrication method for a photonic crystals [12, 13]. To micromachine a transparent material in three-dimensions, a femtosecond laser beam is tightly focused into the bulk of the material. High laser intensity in the focal volume induces nonlinear absorption of laser energy by the material via multiphoton, tunneling, and avalanche ionization [14, 15]. If sufficient laser energy is deposited, permanent structural changes are produced inside a material at the location of the laser focus. Depending on laser, focusing, and material parameters, different mechanisms may play a role in producing the structural changes and lead to different morphologies due to density and refractive index modification, to color centers generation, and to void formation [16]. In this chapter, the micro- and nano-structures induced by the femtosecond laser irradiation are demonstrated. The mechanisms and possible applications of the observed phenomena are also discussed.

16.2 Various Microstructures Induced in Glass by Femtosecond Laser Irradiation

It is difficult to achieve the interaction between light and glass by a one-photon process when the wavelength of the excitation light differs from the resonant absorption wavelength of the glass. However, as shown in Fig. 16.1, various localized structural changes can be induced inside a glass sample by focusing a femtosecond laser operating at a nonresonant wavelength.

There are four examples of induced structural change: (a) coloration due to the color center formation and valence state change of active ions such as rare-earth and transition metal ions, (b) refractive index change due to local densification and atomic defect generation, (c) microvoid formation due to localized remelting and shock wave propagation, and (d) microcrack formation due to destructive breakdown or other phenomena. We know that

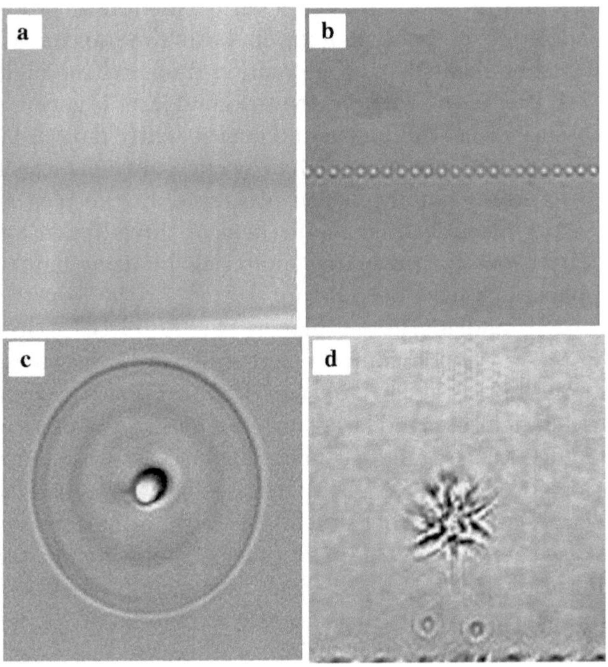

Fig. 16.1. Various femtosecond-laser-induced localized microstructures.

usually glass has no absorption at 800 nm (the wavelength of the common femtosecond laser). In other words, linear absorption of the laser radiation does not occur when the glass is irradiated by a femtosecond laser with no absorption wavelength. This is because the energy gap between the valence and conduction bands is larger than the energy of the single photon. Thus, the single photon does not contain sufficient energy to free a bound electron. However, when laser intensities are sufficiently high, an electron can simultaneously absorb the energy from the multiple photons to exceed the band gap. This nonlinear process is termed multiphoton absorption. It is a highly intensity-dependent process, with the rate $P(I) = \sigma_k I^k$, where σ_k is the multiphoton absorption coefficient for k-photon absorption. Once an electron is excited to the conduction band, it serves as a seed to a process called avalanche ionization [17]. Seeds can also arise from other process, such as electron tunneling and thermal excitation from impurity states [15].

An electron in the conduction band can absorb sufficient energy from photons, $kh\nu \geq E_g$, where k is the number of photons absorbed sequentially and E_g is the band gap energy of glass; it can then use the excess energy to ionize another electron via direct collision, also known as impact ionization. The resultant two electrons in the conduction band can continue the process of linear absorption and impact ionization to achieve an exponential growth of free electrons. Such avalanche ionization produces highly absorptive and dense plasma, facilitating the transfer of energy from the laser pulses to the glass. The resulting melting, material displacement due to plasma expansion, and possible chemical restructuring develop in various induced structures.

16.3 Valence State Manipulation of Active Ions

16.3.1 Photo-Oxidation of Transition Metal Ions

The space-selective, persistent photo-oxidation of Mn^{2+} to Mn^{3+} in a silicate glass by focusing femtosecond laser pulses through a microscope objective lens has been observed [18]. The glass composition of the Mn- and Fe-ion co-doped silicate glass sample used in this experiment was $0.05Fe_2O_3$-$0.1MnO$-$70SiO_2$-$10CaO$-$20Na_2O$ (mol %). A 4 µm spot was formed in the focused area of the laser beam in the Mn- and Fe-ion co-doped glass sample after irradiation by 120 fs laser pulses operated at a wavelength of 800 nm (repetition rate: 1 kHz, pulse energy: 0.4 mJ) on each spot for 1/63 s (i.e., 16 pulses) via a 10× objective lens with a numerical aperture of 0.30 inside the glass sample. A purple area with a diameter of about 30 µm was observed. The length of the induced structure along the laser beam propagation path was about 1.5 mm. To measure the absorption spectrum of the glass sample after the laser irradiation, we wrote a "damaged" plane of $3.0 \times 3.0\,mm^2$ inside the glass sample, which consisted of a "damaged" line at an interval of 10 µm by scanning the laser beam at a rate of 1 mm/s. The distance of the "damaged" plane from the surface of the glass sample was about 0.5 mm.

No apparent absorption was observed for the non-irradiated glass sample in the wavelength region from 400 to 1000 nm, while there was an apparent increase in the absorbance in the wavelength region from 300 to 1000 nm in the irradiated area (Fig. 16.2). There was a peak ranging from 400 to 800 nm, peaking at 520 nm, which can be assigned to the absorption of Mn^{3+} ions [19, 20]. In addition, a peak was observed at 320 nm, which can be assigned to the absorption of hole-trapped centers as observed in the X-ray irradiated silicate glass [21]. The electron spin resonance (ESR) spectrum of the non-irradiated glass sample indicates resolved hyperfine structures of six lines spread over a range of about 500 G in width centered at 3350 G (splitting coefficient $g \sim 2.0$). The spectrum showed a pattern similar to those observed for various glasses containing Mn^{2+} ions [Fig. 16.3a] [22]. The low resolution of the hyperfine lines is due to the dipolar broadening. No apparent absorption due to Mn^{3+} was observed in non-irradiated glass (Fig. 16.2). Therefore, most Mn ions are present in the divalent state in glass. Two new signals at g values of 2.010 and 2.000 were observed in the glass sample after laser irradiation. The signals can be assigned to hole-trapped centers in the glass matrix [Fig. 16.3b] [23]. From the above results, we can see that a part of Mn^{2+} was oxidized to Mn^{3+} after the femtosecond laser irradiation. The Mn- and Fe-ion co-doped silicate glass sample has no absorption in the wavelength region near 800 nm. Therefore, photo-oxidation of Mn^{2+}

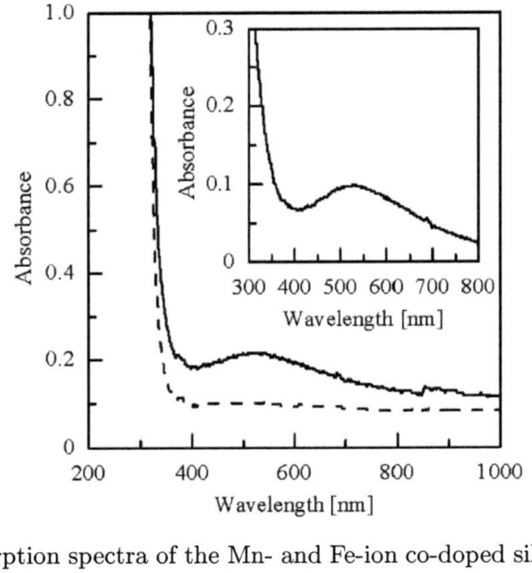

Fig. 16.2. Absorption spectra of the Mn- and Fe-ion co-doped silicate glass before (*dashed line*) and after (*solid line*) the femtosecond laser irradiation. The inset shows the difference spectrum between absorptions of the glass sample after and before the femtosecond laser irradiation.

to Mn^{3+} should be a nonlinear optical process. We suggest that multiphoton absorption is one of the mechanisms of the observed phenomenon [17]. Free electrons are generated by the multiphoton absorption of the incident photon and consequent avalanche ionization. Mn^{2+} captures a hole to form Mn^{3+}, while Fe^{3+} as well as active sites in the glass matrix may act as electron trapping centers, resulting in the formation of Mn^{3+}. Since the trap levels of defect centers may be deep, the induced Mn^{3+} ions are thermally stable. Based on the above results, since only the focal spot is colored purple, it is possible to write a three-dimensional colored image inside the transparent and colorless glass as shown in Fig. 16.3 inset.

16.3.2 Photo-Reduction of Rare-Earth Ions

The photo-induced reduction of rare-earth ions by the femtosecond laser irradiation has been also observed [24]. The composition of the Eu^{3+}-doped fluorozirconate glass sample used in this experiment was $0.1EuF_3$-$53ZrF_4$-$20BaF_2$-$3.9LaF_3$-$3AlF_3$-$20NaF$ (mol%). For comparison, we prepared a Eu^{2+}-doped fluoroaluminate glass sample with a composition of $0.1EuF_2$-$14.9YF_4$-$10MgF_2$-$20CaF_3$-$10SrF_3$-$10BaF_2$-$35AlF_3$ (mol%). A 10 µm bright spot was formed at the focal point of the laser beam in the Eu^{3+}-doped fluorozirconate glass sample after irradiation by 120 fs laser pulses operating at a wavelength of 800 nm (repetition rate: 200 kHz, pulse energy: 1 µJ) for 1 s (i.e., 2×10^5

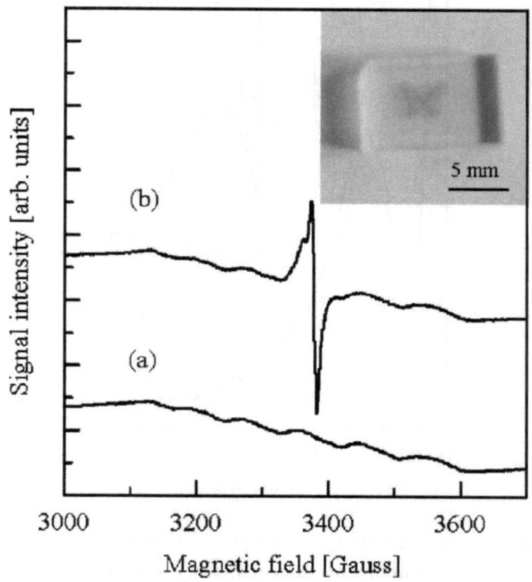

Fig. 16.3. Electron spin resonance spectra of the Mn and Fe co-doped silicate glass before (*a*) and after (*b*) the femtosecond laser irradiation. The *inset* photograph shows an image of a butterfly in purple color written inside the glass by femtosecond laser irradiation.

pulses) via a 10× objective lens with a numerical aperture of 0.30 inside the glass sample. To measure the absorption spectrum of the glass sample after the laser irradiation, we wrote a "damaged" plane of 4.0 × 4.0 mm^2 inside the Eu^{3+}-doped fluorozirconate glass sample that consisted of "damaged" lines at an interval of 10 μm by scanning the laser beam at a rate of 1 mm/s. The distance of the plane from the surface of the glass sample was about 1 mm.

A small peak at 394 nm and a strong broad peak at 250 nm can be assigned to the 4f–4f transition and the charge transfer state of Eu^{3+}, respectively (Fig. 16.4) [25]. No apparent increase in the absorbance was observed in the 400 nm to 1 μm region, whereas absorption in the 200 to 400 nm region increased after laser irradiation. From comparison of the difference absorption spectrum of Eu^{3+}-doped fluorozirconate glass before and after the femtosecond laser irradiation and the absorption spectrum of Eu^{2+}-doped fluoroaluminate glass (Fig. 16.4 inset), the broad peak at about 260 nm can be assigned mainly to the absorption that is due to the 5d–4f transition of Eu^{2+} [26].

In addition, we observed emission spectra of the Eu^{3+}-doped fluorozirconate glass sample before and after the femtosecond laser irradiation (Fig. 16.5). All emission peaks observed in the non-irradiated glass sample can be assigned to the $4f^6 \rightarrow 4f^6$ transition of Eu^{3+}, and no emission due to Eu^{2+} was detected. Eu ions are present in the trivalent state in the unirradiated

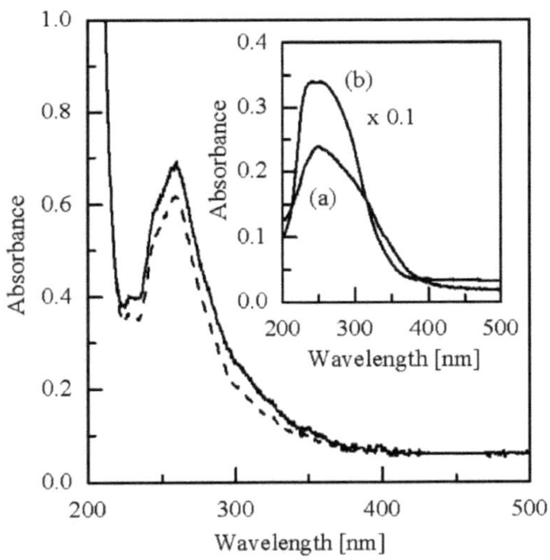

Fig. 16.4. Absorption spectra of Eu^{3+}-doped fluorozirconate glass before (*dashed line*) and after (*solid line*) the femtosecond laser irradiation. The *inset* shows (*a*) the difference absorption spectrum of Eu^{3+}-doped fluorozirconate glass before and after the femtosecond laser irradiation, together with (*b*) the absorption spectrum of a Eu^{2+}-doped fluoroaluminate glass sample fabricated in a reducing atmosphere. Note that curve (*b*) is scaled down by a factor of ×0.1.

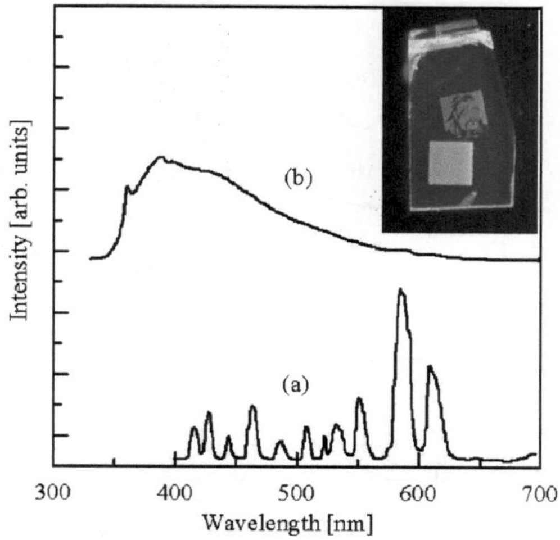

Fig. 16.5. Emission spectra of Eu^{3+}-doped fluorozirconate glass (*a*) before and (*b*) after the femtosecond laser irradiation. *Inset* image shows the photograph when the glass sample was excited by UV light at 365 nm.

glass sample. The emission peaks at 360 nm and 400 nm in the irradiated glass sample can be assigned to the ${}_6P^{7/2} \rightarrow {}_8S^{7/2}$ and $4f^6 5d^1 \rightarrow 4f^7$ transitions of Eu^{2+}, respectively. Figure 16.5 inset shows the photograph when the glass sample was excited by a UV light at 365 nm. The unirradiated area shows a red emission, while the femtosecond-laser-irradiated area shows a blue emission. From comparison of the ESR spectra of Eu^{3+}-doped fluorozirconate glass before and after the femtosecond laser irradiation and the ESR spectrum of Eu^{2+}-doped fluoroaluminate glass, no apparent signals can be observed in the spectrum of the unirradiated Eu^{3+}-doped fluorozirconate glass sample, whereas apparent signals similar to those of the Eu^{2+}-doped fluoroaluminate glass and two signals at about 3300 G (splitting coefficient $g \sim 2.0$) can be observed in the spectrum of the laser-irradiated Eu^{3+}-doped fluorozirconate glass sample (Fig. 16.6). Therefore, some of the Eu^{3+} ions were reduced to Eu^{2+} in the Eu^{3+}-doped fluorozirconate glass after the femtosecond laser irradiation. In addition, the signals at $g \sim 2.0$ can be assigned to hole-trapped V-type centers and to electrons trapped by Zr^{4+} ions, as in X-ray irradiated fluorozirconate glasses [27]. From the above results, we can see that a part of Eu^{3+} was reduced to Eu^{2+} after the femtosecond laser irradiation. The Eu^{3+}-doped fluorozirconate glass sample has no absorption in the wavelength region near 800 nm. Therefore, photo-reduction of Eu^{3+} to Eu^{2+} should be a nonlinear optical process. We suggest that the mechanism of photo-reduction is that active electrons and holes are created in the glass through a multiphoton ionization process [17]. Holes are trapped in the active sites in the glass matrix, and some electrons are trapped by Eu^{3+}, leading to the formation of

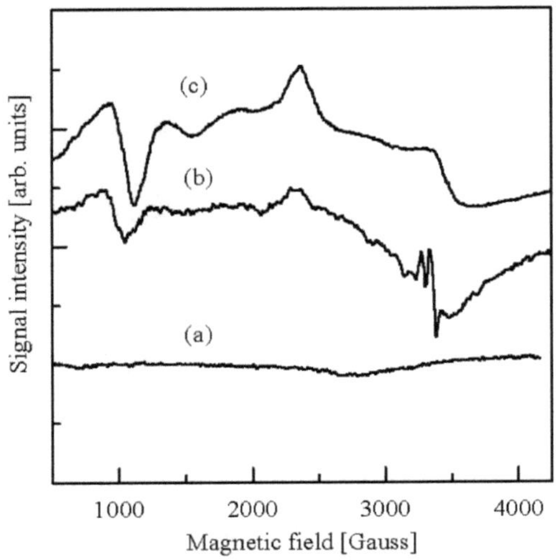

Fig. 16.6. Electron spin resonance spectra of Eu^{3+}-doped fluorozirconate glass (*a*) before and (*b*) after the femtosecond laser irradiation, together with (*c*) the spectrum of a Eu^{2+}-doped fluoroaluminate glass sample fabricated in a reducing atmosphere.

V-type hole-trapped defect centers and Eu^{2+}. In addition, the trap levels of defect centers can be deep, resulting in stable Eu^{2+} at room temperature.

16.3.3 Three-Dimensional Rewritable Optical Memory

We have also observed space-selective permanent photo-reduction of Sm^{3+} to Sm^{2+} in various glasses and crystals by femtosecond laser irradiation and demonstrated 3D optical data storage by using the femtosecond-laser-induced valence-state change of Sm ions in glasses [28–30]. Figure 16.7 shows photoluminescence spectra obtained by excitation at 488 nm for a femtosecond laser-irradiated area (a) and non-irradiated area (b) in the interior of the Sm^{3+}-doped glass sample. Comparing (a) to (b) shows that the emission in the 650–775-nm region differed appreciably. The broad bands observed around 560, 600, and 645 nm can be attributed to the ${}^4G_{5/2} \rightarrow {}^6H_{5/2,\,7/2,\,9/2}$ transitions, respectively, of the Sm^{3+} ions. On the other hand, the emissions at 680, 700, and 725 nm are attributed to the ${}^5D_0 \rightarrow {}^7F_{0,1,2}$ transitions, respectively, of the Sm^{2+} ions. This means that laser-irradiated areas (photo-reduced areas) recorded inside glass can be detected only by emissions at 680, 700, or 725 nm. By using the photo-reduction of Sm^{3+} to Sm^{2+}, alphabetical characters were recorded in the form of sub-micron size bits in a three-dimensional (layered) manner in a glass samples. Here, one recorded character consisted of 300 to 500 photo-reduction bits recorded with 5000 laser shots per bit. As a recording source we employed a regeneratively amplified 800-nm Ti: sapphire laser emitting 20 Hz or 250 kHz mode-locked pulses. We tightly focused

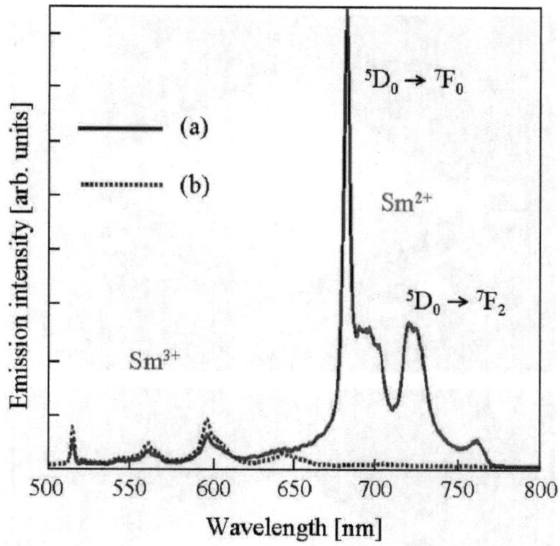

Fig. 16.7. Photoluminescence spectra obtained by excitation at 488 nm for a laser-irradiated area (*a*) and non-irradiated area (*b*) in the interior of the Sm^{3+}-doped glass.

femtosecond laser pulses inside the bulk glass to locally photo-reduce Sm^{3+} to Sm^{2+}. The spacing between alphabetical characters was 2 μm. The bits had a diameter of 400 nm, which was significantly smaller than the focal-beam size and the wavelength of the recording laser.

Figure 16.8 shows photoluminescence images of alphabetical characters recorded on different layers, which were observed by using a 40× objective lens and the 680 nm emission from Sm^{2+} with confocal detection implemented. We confirmed that the spacing of 2 μm between alphabetical character planes was sufficient to prevent cross-talk in the photoluminescence images. Although the 3D memory bits (photo-reduced areas) was recorded with a femtosecond laser, they could be read with a CW laser at 0.5 mW. The results demonstrated the possibility of selectively inducing a change of valence state of Sm^{3+} ions on the micrometer scale inside a glass sample by use of a focused nonresonant femtosecond pulsed laser. Therefore, the present technique will be useful in the fabrication of 3D optical memory devices with high storage density (approximately 10^{13} bits/cm^3). Moreover, femtosecond laser photo-reduced Sm^{3+}-doped glasses exhibited a photochemical spectral hole burning memory property. The microspot induced by the focused femtosecond laser inside a glass sample can be further used to store data information via the irradiation of laser light with different wavelengths.

As a result, the data information can be read out in the form of spectral holes. Sm^{2+}-doped glasses could become an ultimate optical memory device with an ultrahigh storage density. Optical memory using a valence state change of rare-earth ions at a spot allow data to be read out in the form of luminescence, thus providing the advantage of a high signal-to-noise

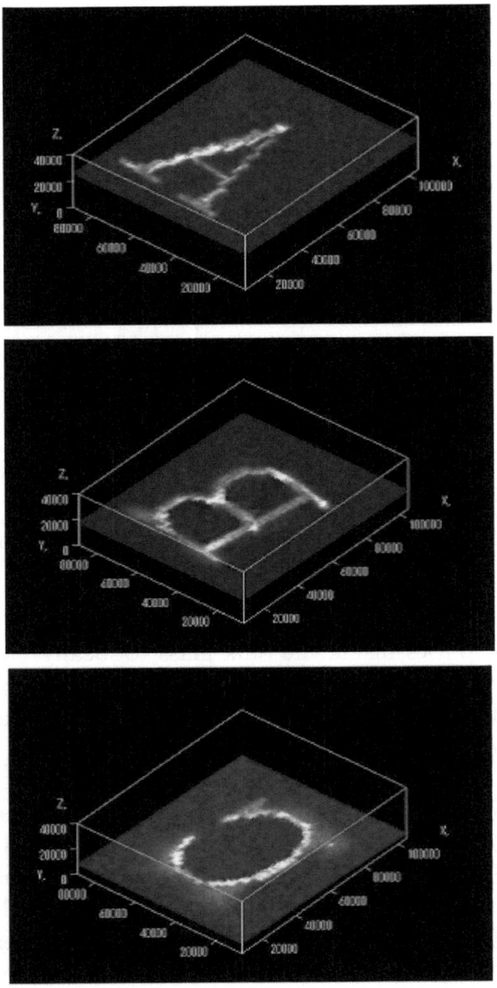

Fig. 16.8. Photoluminescence images of alphabetical characters recorded on different layers, which were observed by using a 40× objective lens and the 680 nm emission from Sm^{2+} with confocal detection implemented (excitation at 488 nm, 1 mW Ar^+ laser).

ratio. Recently, we demonstrated the 3D optical data storage by using the femtosecond laser-induced valence state change of Sm ions in glasses. We also observed that photo-reduced ($Sm^{3+} \rightarrow Sm^{2+}$) submicron bit written by using the femtosecond laser can be erased by irradiation with an Ar^+ laser at 514.5 nm (Fig. 16.9) [30]. A bit with 150 nm was formed after the femtosecond laser irradiation [Fig. 16.9a]. When an Ar^+ laser irradiated at the bit areas, the bit disappeared [Fig. 16.9b,c]. If femtosecond laser is irradiated again in the same position, the bit reverts to the previous state [Fig. 16.9d].

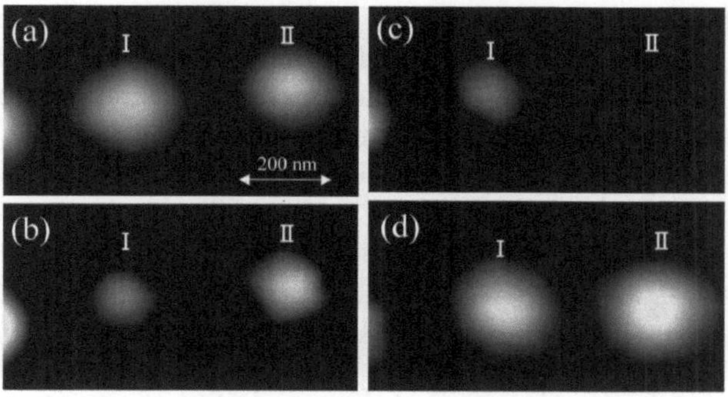

Fig. 16.9. Three-dimensional rewriteable optical memory using valence state change of Sm^{3+} by irradiation of femtosecond laser and an Ar^+ laser at 514 nm. (**a**) Photoluminescence image before the erasure. (**b**) Image after Ar^+ laser irradiation to photoreduction bit I. (**c**) Image after Ar^+ laser irradiation to bit II. (**d**) Image after femtosecond laser irradiation to areas I and II.

This result indicated the possibility of achieving a 3D optical memory with rewrite capability.

16.4 Precipitation and Control of Nanoparticles

Noble-metal-nanoparticle-doped glasses exhibit a large third-order nonlinear susceptibility and an ultrafast nonlinear response [31,32]. They are expected to be promising materials for an ultrafast all-optical switch in the THz region [33]. Stookey [34] developed photosensitive glasses in the early 1950s. These glasses contain noble metal photosensitive ions such as Ag^+ and Au^+ together with Ce^{3+}, which act as a sensitizer. After being irradiated by UV light, Ce^{3+} releases an electron to form Ce^{4+}, while Ag^+ or Au^+ captures the electron to form an Ag or Au atom. After subsequent heat treatment, crystallites, e.g., LiF and Li_2SiO_5, precipitate in the UV-irradiated area due to the nucleation by the metal cluster or colloids. Although it is possible to fabricate a three-dimensional designed structure using UV laser pulses with a resonant absorption wavelength of glass, the vertical spatial resolution is lower than in the case of femtosecond laser pulse irradiation because the UV light resonates linearly on the surface [35].

We have demonstrated three-dimensional precipitation and control of noble nanoparticles due to photo-reduction of the ion (Ag^+ or Au^+) to the atom (Ag or Au) in various transparent glasses by focused femtosecond laser pulses and successive annealing [36–41]. A typical glass composition of $77SiO_2$-$5CaO$-$18Na_2O$ (mol%) and doped with 0.01 mol% Ag_2O was used in this experiment [37]. A plane of $4.0 \times 4.0\,mm^2$ that consisted of lines at an interval of 20 μm was written inside the Ag^+-doped glass sample by scanning the laser beam at 2500 μm/s with the focused 120 fs laser pulses operated at a wavelength of 800 nm (repetition rate: 1 kHz, pulse energy: 80 mJ) via a 10× objective lens with a numerical aperture of 0.30 inside the glass sample. The position of the focal spot was about 1 mm beneath the sample surface. The irradiated sample was annealed at 500 ℃ for 10 min in air. After irradiation by the focused femtosecond laser, a gray spot was formed near the focal spot in the Ag^+-doped silicate glass sample. Absorption peaks at 430 and 620 nm can be observed and assigned to the hole trap centers (HC) at the nonbridging oxygen in the SiO_4 polyhedron with two and three nonbridging oxygen atoms, e.g., HC_1 and HC_2, respectively [Fig. 16.10b] [42]. After annealing at 500 ℃ for 10 min, the laser-irradiated area became yellow, and a peak at 408 nm due to the surface plasmon absorption of the silver nanoparticles was observed [Fig. 16.10c] [43]. The inset image in Fig. 16.10 shows the transmission electron microscopy (TEM) observation of the yellow area. Composition analysis using energy-dispersive spectroscopy (EDS) in TEM confirmed that the spherical nanoparticle is metallic Ag. Spherical silver nanoparticles with sizes ranging from 1 to 4 nm were observed in the femtosecond-laser-irradiated area after annealing at 500 ℃ for 10 min. On the basis of absorption and ESR spectra, we suggest that the Ag^+ ion is reduced to an Ag atom by capturing an electron from a nonbridging oxygen during the femtosecond laser irradiation, and silver atoms migrate and aggregate to form nanoparticles after the heat treatment [36].

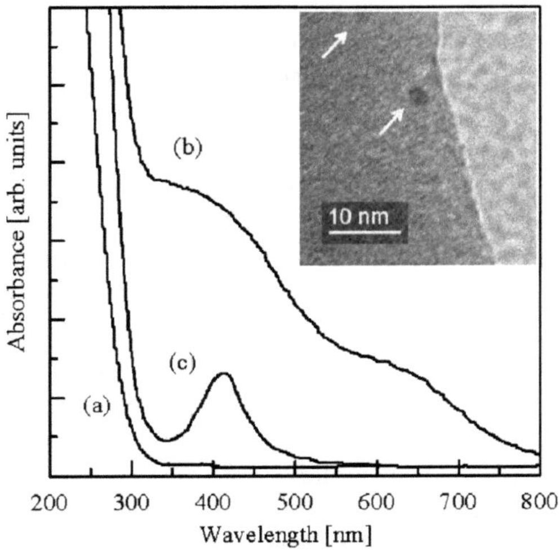

Fig. 16.10. Absorption spectra of the Ag^+-doped silicate glass sample (*a*) before and (*b*) after the femtosecond laser irradiation, together with (*c*) the spectrum of the laser-irradiated sample after annealing at 500 ℃ for 10 min. The *inset* shows a TEM image of the femtosecond-laser-irradiated area of the Ag^+-doped silicate glass after annealing at 500 ℃ for 10 min.

We also succeeded in the space-selective precipitation and control of gold nanoparticles in glass [41]. An Au^{3+}-doped silicate glass sample with the composition of $70SiO_2$-$10CaO$-$20Na_2O$ and doped with 0.01 mol% Au_2O_3 was irradiated inside the glass sample on each spot with focused 120 fs laser pulses operating at a wavelength of 800 nm (repetition rate: 1 kHz, pulse energy: 0.8 μJ) for 1/63 s (i.e., 16 pulses) via a 10× objective lens with a numerical aperture of 0.30. Gray spots of about 40 μm were observed in the focused laser-irradiated area through an optical microscope. No microcracks were observed in the glass samples. After the samples were annealed at 550 ℃ for 30 min, the gray spots became red. Absorption spectra showed that there was an apparent increase in absorption in the wavelength range from 300 to 800 nm in the irradiated area (Fig. 16.11). From the difference absorption spectrum of the Au^{3+}-doped glass sample before and after the femtosecond laser irradiation (Fig. 16.11 inset), the peaks at 245, 306, 430, and 620 nm can be assigned to E′ centers (E′ = Si), which include an electron trapped in an sp3 orbital of silicon at the site of an oxygen vacancy, a hole trapped by an oxygen vacancy that neighbors alkali-metal ions, and nonbridging oxygen hole centers, respectively.

The absorption spectra of the Au^{3+}-doped silicate glass samples, which were annealed at various temperatures for 30 min after the femtosecond laser irradiation, are also plotted in Fig. 16.11. When the annealing temperature is below 300 ℃ , the absorption (300 ∼ 800 nm) intensities induced by laser

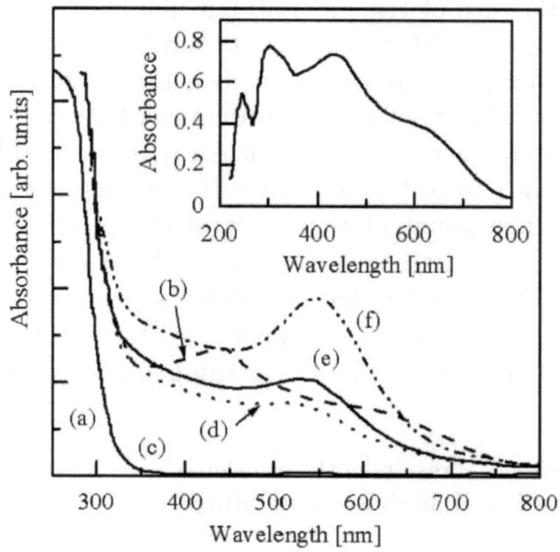

Fig. 16.11. Absorption spectra of Au^{3+}-doped silicate glass (*a: solid line*) before and (*b: dashed line*) after the femtosecond laser irradiation, together with (*c: dashed line*), (*d: dotted line*), (*e: solid line*), and (*f: double-dashed line*) the absorption spectra of the laser-irradiated sample after annealing at 300, 450, 500, and 550 ℃ for 30 min, respectively. The *inset* shows the difference spectrum between the absorption of the glass sample after and before the femtosecond laser irradiation.

irradiation decrease as the annealing temperature increases, and completely disappear when the temperatures reaches 300 ℃ . The gray spots induced by the femtosecond laser irradiation disappear at 300 ℃ and become colorless and transparent. Annealing at 450 ℃ results in the appearance of a new peak at 506 nm, and the laser-irradiated areas turn red. This peak can be assigned to the surface plasmon resonance absorption of Au nanoparticles [44].

The wavelength of the absorption peak increases from 506 to 548 nm with increasing annealing temperature at the same time that its intensity significantly increases. Based on the Mie theory, the average radii of the embedded metal nanoparticles R are proportional to the value of $\lambda_p^2/\Delta\lambda$, where λ_p^2 is the characteristic wavelength of surface plasmon resonance and $\Delta\lambda$ is the full-width at half-maximum (FWHM) of the absorption band [45, 46]. The value of $\lambda_p^2/\Delta\lambda$ increases from 1306 to 2208 nm when the annealing temperature is increased from 450 to 550 ℃ (Table 16.1). Note that no absorption peak due to the surface plasmon resonance of Au nanoparticles is observed in the femtosecond-laser-irradiated Au^{3+}-doped glass sample after annealing at 300 ℃ .

Therefore, the average size of the Au nanoparticles increases with increasing annealing temperature. We directly observed the precipitation of Au nanoparticles in the femtosecond-laser-irradiated Au^{3+}-doped glass sample after annealing at 550 ℃ for 30 min by TEM and confirmed that these spherical nanoparticles are metallic Au. The size of the Au nanoparticles ranges from 6 to 8 nm (Fig. 16.12). We studied the effect of the femtosecond laser fluence on the size of the precipitated nanoparticles. With increasing laser intensity from 6.5×10^{13} to 2.3×10^{14} to 5.0×10^{16} W/cm^2, the color of the femtosecond-laser-irradiated areas became violet, red and yellow, respectively (Fig. 16.13 inset). The absorption peak that can be assigned to the surface plasmon resonance absorption of Au nanoparticles shifts to shorter wavelengths from 568 to 534 to 422 nm with the increase of the laser intensity (Fig. 16.13). The apparent blue shift of the peak from 568 to 422 nm is due to the decrease in the average size of the Au nanoparticles. This is probably because the high irradiation intensity produces a high concentration of reduced Au atoms per unit volume, and thus a high concentration of nucleation centers. As a result, under the same annealing process, the higher

Table 16.1. Experimental results of the wavelength and FWHM of surface plasmon resonance for the various annealing temperatures.

Annealing temperature (℃)	λ_p (nm)	$\Delta\lambda$ (nm)	$\lambda_p^2/\Delta\lambda$ (nm)
300	–	–	–
450	506	196	1306
500	526	156	1774
550	548	136	2208

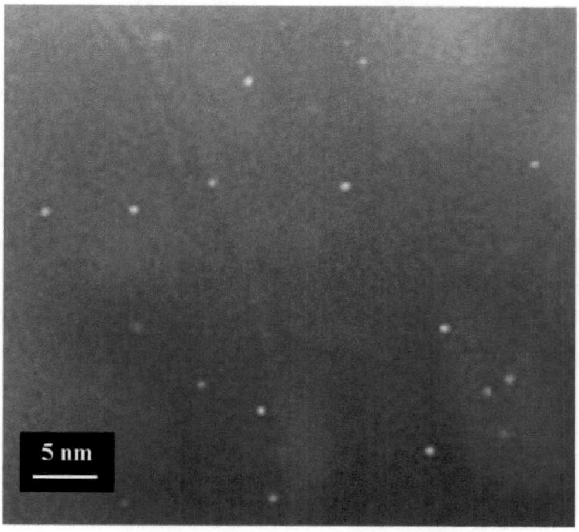

Fig. 16.12. TEM image of Au nanoparticles (*small white dots*) in the femtosecond-laser-irradiated area of the Au^{3+}-doped silicate glass after annealing at 550 °C for 30 min.

the light intensity, the smaller but denser the precipitated particles. However, the wavelength of surface plasmon resonance is not only determined by the size of nanoparticles, but also the concentration of nanoparticles. Further investigation is needed to verify the above mechanism.

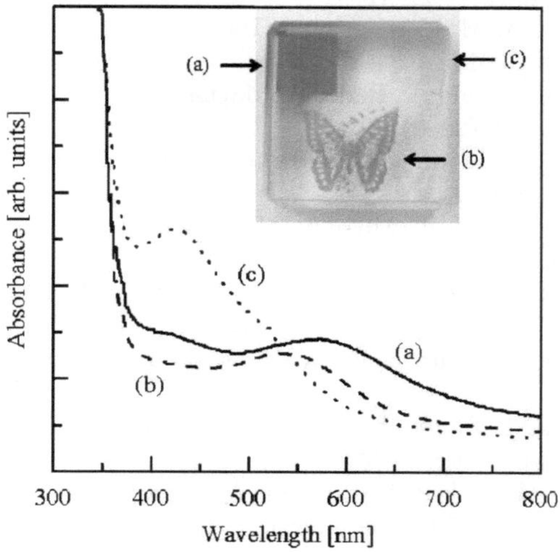

Fig. 16.13. Absorption spectra of Au^{3+}-doped silicate glass irradiated with different laser intensities: (*a: solid line*) 6.5×10^{13} W/cm^2, (*b: dashed line*) 2.3×10^{14} W/cm^2, (*c: dotted line*) 5.0×10^{16} W/cm^2. All samples were annealed at 550 °C for 1 hour. The *inset* shows the photograph of images drawn inside the Au^{3+}-doped (0.1 mol%) glass sample.

16.5 Conclusion

We have observed various femtosecond laser-induced phenomena in glasses. We have also observed the photo-oxidation of transition metal ions in a glass sample after the femtosecond laser irradiation. Furthermore, we have demonstrated the 3D optical memory with ultrahigh storage density by using photoreduction phenomenon of the rare-earth ions. We have succeeded in the space-selective precipitation and control of the nanoparticles in glasses. We have confirmed that the femtosecond laser induced microstructure will open a new possibility in the realization of novel optical functions for a transparent material.

References

1. B. N. Chickov, C. Momma, S. Nolte, F. V. Alvensleben, A. Tunnermann: Appl. Phys. A **63**, 109 (1996).
2. E. N. Glezer, M. Milosavljevic, L. Huang, R. J. Finlay, T.-H. Her, J. P. Callan, E. Mazur: Opt. Lett. **21**, 2023 (1996).
3. J. Qiu, K. Miura, K. Hirao: Jpn. J. Appl. Phys. **37**, 2263 (1998).
4. J. Qiu, K. Miura, H. Inouye, Y. Kondo, T. Mutsuyu, K. Hirao: Appl. Phys. Lett. **73**, 1963 (1998).
5. K. Hirao, K. Miura: Jpn. J. Appl. Phys. **37**, L49 (1998).
6. M. Will, S. Nolte, B. N. Chichkov, A. Tunnermann: Appl. Opt. **41**, 4360 (2002).
7. D. N. Fittinghoff, C. B. Schaffer, E. Mazur, J. A. Squier: IEEE J. Sel. Top. Quantum Electron. **7**, 559 (2001).
8. K. Minoshima, A. M. Kowalevicz, I. Hartl, E. P. Ippen, J. G Fujimoto: Opt. Lett. **26**, 1516 (2001).
9. Y. Sikorski, A. A. Said, P. Bado, R. Maynard, C. Florea, K. A. Winick: Electron. Lett. **36**, 226 (2000).
10. K. Miura, J. Qiu, T. Mitsuyu, K. Hirao: Nucl. Instr. and Meth. in Phys. Res. B **141**, 726 (1998).
11. Y. Kondo, K. Nouchi, T. Mitsuyu, M. Watanabe, P. G. Kazansky, K. Hirao: Opt. Lett. **24**, 646 (1999).
12. B. H. Cumpston, S. P. Ananthavel, S. Barlow, D. L. Dyer, J. E. Ehrlich, L. L. Erskine, A. A. Heikal, S. M. Kuebler, I.-Y. S. Lee, D. Mccord-Naughton, J. Qin, H. Rökel, M. Rumi, X.-L. Wu, S. R. Marder, J. W. Perry: Nature **398**, 51 (1999).
13. H. B. Sun, Y. Xu, S. Joudkazis, K. Sun, M. Watanabe, J. Nishii, S. Matsuo, H. Misawa: Opt. Lett. **26**, 325 (2001).
14. B. C. Stuart, M. D. Feit, S. Herman, A. M. Rubenchik, B. W. Shore, M. D. Perry: Phys. Rev. B **53**, 1749 (1996).
15. M. Lenzner, L. J. Kruger, S. Sartania, Z. Cheng, C. Spielmann, L. G. Mourou, W. Kautek, F. Krausz: Phys. Rev. Lett. **80**, 4076 (1998).
16. Y. Shimotsuma, J. Qiu, P. G. Kazansky, K. Hirao: Proc. SPIE **5662**, 173 (2004).
17. B. C. Stuart, M. D. Feit, A. M. Rubenchik, B. W. Shore, M. D. Perry: Phys. Rev. Lett. **74**, 2248 (1995).

18. J. Qiu, C. Zhu, T. Nakaya, J. Si, K. Kojima, F. Ogura, K. Hirao: Appl. Phys. Lett. **79**, 3567 (2001).
19. S. P. Singh, A. Kumar: Phys. Chem. Glasses **33**, 61 (1992).
20. J. Qiu, Y. Kondo, K. Miura, T. Mitsuyu, K. Hirao: Jpn. J. Appl. Phys. **38**, L649 (1999).
21. Y. Yokota: Phys. Rev. **95**, 1145 (1954).
22. C. Sumalatha, B. Sreedhar, M. Yamazaki, K. Kojima: Phys. Chem. Glasses **38**, 206 (1997).
23. A. Bishay: J. Non-Cryst. Solids **3**, 54 (1970).
24. J. Qiu, K. Kojima, K. Miura, T. Mitsuyu, K. Hirao: Opt. Lett. **24**, 786 (1999).
25. T. Kanou: *Handbook of Phosphors* (Ohmsya, Tokyo, 1987) p 110 [in Japanese].
26. J. Qiu, Y. Shimizugawa, Y. Iwabuchi, K. Hirao: Appl. Phys. Lett. **71**, 759 (1997).
27. D. L. Griscom, E. J. Friebele: *Fluoride Glass* (Wiley, New York, 1989) p 307.
28. J. Qiu, K. Miura, T. Suzuki, T. Mitsuyu, K. Hirao: Appl. Phys. Lett. **74**, 10 (1999).
29. J. Qiu, K. Miura, K. Nouchi, T. Suzuki, Y. Kondo, T. Mitsuyu, K. Hirao: Solid State Commun. **113**, 341 (1999).
30. K. Miura, J. Qiu, S. Fujiwara, S. Sakaguchi, K. Hirao: Appl. Phys. Lett. **80**, 2263 (2002).
31. D. Ricard, P. Roussignol, C. Flytznis: Opt. Lett. **10**, 511 (1985).
32. F. Hache, D. Ricard, C. Flytzanis: J. Opt. Soc. Am. B **3**, 1647 (1986).
33. H. Inouye, K. Tanaka, I. Tanahashi, K. Hirao: J. Phys. Soc. Jpn. **68**, 3810 (1999).
34. S. D. Stookey: Ind. Eng. Chem. **45**, 115 (1953).
35. M. Masuda, K. Sugioka, Y. Cheng, N. Aoki, M. Kawachi, K. Shihoyama, K. Toyoda, H. Helvajian, K. Midorikawa: Appl. Phys. A **76**, 857 (2003).
36. J. Qiu, M. Shirai, T. Nakaya, J. Si, X. Jiang, C. Zhu, K. Hirao: Appl. Phys. Lett. **81**, 3040 (2002).
37. H. Zeng, J. Qiu, X. Jiang, C. Zhu, F. Gan: J. Phys.: Condens. Matter **16**, 2901 (2004).
38. J. Qiu, X. Jiang, C. Zhu, H. Inouye, J. Si, K. Hirao: Opt. Lett. **29**, 370 (2004).
39. S. Qu, J. Qiu, C. Zhao, X. Jiang, H. Zeng, C. Zhu, K. Hirao: Appl. Phys. Lett. **84**, 2046 (2004).
40. S. Qu, Y. Gao, X. Jiang, H. Zeng, Y. Song, J. Qiu, C. Zhu, K. Hirao: Opt. Commun. **224**, 321 (2003).
41. J. Qiu, X. Jiang, C. Zhu, M. Shirai, J. Si, N. Jiang, K. Hirao: Angew. Chem., Int. Ed. **43**, 2230 (2004).
42. J. W. H. Schreurs: J. Chem. Phys. **47**, 818 (1967).
43. I. Tanahashi, M. Yoshida, Y. Manabe, T. Tohda: J. Mater. Res. **10**, 362 (1995).
44. H. Inouye, K. Tanaka, I. Tanahashi, K. Hirao: Phys. Rev. B **57**, 234 (1998).
45. G. Mie: Ann. Phys. **25**, 377 (1908).
46. D. Manikandan, S. Mohan, P. Magudapathy, K. G. M. Nair: Physica B **325**, 86 (2003).

Index

MIX
Papier aus verantwortungsvollen Quellen
Paper from responsible sources
FSC® C105338

If you have any concerns about our products,
you can contact us on
ProductSafety@springernature.com

In case Publisher is established outside the EU,
the EU authorized representative is:
Springer Nature Customer Service Center GmbH
Europaplatz 3, 69115 Heidelberg, Germany

Printed by Libri Plureos GmbH
in Hamburg, Germany